:LOPEDIA OF MATHEMATICS AND ITS APPLICATIONS

alo *Integral Geometry and Geometric Probability*
rews *The theory of partitions*
liece *The Theory of Information and Coding: A Mathematical Framework for Communication*
, Jr *Symmetry and separation of variables*
e *Thermodynamic Formalism: The Mathematical Structures of Classical Equilibrium Statistical*
cs
Permanents
erts *Measurement Theory with Applications to Decision making, Utility, and the Social Sciences*
denharn and J D Louck *Angular Momentum in Quantum Physics: Theory and Applications*
denharn and J D Louck *The Racah-Wigner Algebra in Quantum Theory*
llard and C N Friedman *Product Integration with Applications in Quantum Theory*
nes and W J Thron *Continued Fractions: Analytic Theory and Applications*
Martin and J W England *Mathematical Theory of Entropy*
ker, Jr and P Graves-Morris *Padé Approximants, Part I, Basic Theory*
ker, Jr and P Graves-Morris *Padé Approximants, Part II, Extensions and Applications*
ltrametti and G Cassinelli *The Logic of Quantum Mechanics*
ames and A Kerber *The Representation Theory of Symmetric Groups*
haire *Combinatorics on Words*
attorini *The Cauchy Problem*
orentz, K Jetter and S D Riemenschneider *Birkhoff Interpolation*
and H Niederreiter *Finite Fields*
Tutte *Graph Theory*
astida *Field Extensions and Galois Theory*
annon *The One-Dimensional Heat Equation*
gon *The Banach-Tarski Paradox*
omaa *Computation and Automata*
ite (ed) *Theory of Matroids*
Bingham, C M Goldie and J L Teugels *Regular Variations*
etrushev and V A Popov *Rational Approximation of Real Functions*
hite (ed) *Combinatorial Geometries*
host and H Zassenhaus *Algorithmic Algebraic Number Theory*
zel and J Dhombres *Functional Equations in Several Variables*
uczma, B Choczewski and R Ger *Iterative Functional Equations*
Ambartzumian *Factorization Calculus and Geometric Probability*
ripenberg, S-O Londen and O Staffans *Volterra Integral and Functional Equations*
asper and M Rahman *Basic Hypergeometric Series*
orgersen *Comparison of Statistical Experiments*
eumaier *Interval Methods for Systems of Equations*
orneichuk *Exact Constants in Approximation Theory*
rualdi and H Ryser *Combinatorial Matrix Theory*
White (ed) *Matroid applications*
akai *Operator Algebras in Dynamical Systems*
Hodges *Model Theory*
Stahl and V Totik *General Orthogonal Polynomials*
Schneider *Convex Bodies*
Da Prato and J Zabczyk *Stochastic Equations in Infinite Dimensions*
Björner *et al* *Oriented Matroids*
Edgar and L Sucheston *Stopping Times and Directed Processes*
Sims *Computation with Finitely Presented Groups*
W Palmer *C*-algebras I*
Borceux *Handbook of Categorical Algebra 1, Basic Category Theory*
Borceux *Handbook of Categorical Algebra 2, Categories and Structures*
Borceux *Handbook of Categorical Algebra 3, Categories of sheaves*

ENCYCLOPEDIA OF MATHEMA

Edited by G.-

Volume

Combinatorial Methods in D

ENCYC

1. L A San
2. G E An
3. R J Mc
4. W Mille
5. D Ruel
Mechani
6. H Minc
7. F S Ro
8. L C Bi
9. L C Bi
10. J D D
11. W B J
12. N F G
13. G A B
14. G A B
15. E G B
16. G D J
17. M Lo
18. H O
19. G G
20. R Lid
21. W T
22. J R B
23. J R
24. S Wa
25. A Sa
26. N W
27. N H
28. P P
29. N W
30. M P
31. J A
32. M
33. R V
34. G
35. G
36. E T
37. A
38. N
39. R
40. N
41. S
42. W
43. H
44. R
45. G
46. A
47. G
48. C
49. T
50. F
51. F
53. F

ENCYCLOPEDIA OF MATHEMATICS AND ITS APPLICATIONS

Combinatorial Methods in Discrete Mathematics

Vladimir N. Sachkov

CAMBRIDGE
UNIVERSITY PRESS

Published by the Press Syndicate of the University of Cambridge
The Pitt Building, Trumpington Street, Cambridge CB2 1RP
40 West 20th Street, New York, NY 10011-4211, USA
10 Stamford Road, Oakleigh, Melbourne 3166, Australia

First published in English 1996

Translated by V. Kolchin, Steklov Mathematical Institute

Printed in Great Britain at the University Press, Cambridge

A catalogue record of this book is available from the British Library

Library of Congress cataloguing in publication data

Sachkov, Vladimir Nikolaevich.
[Kombinatornye metody diskretnoĭ matematiki. English].
Combinatorial methods in discrete mathematics / Vladimir N. Sachkov.
p. cm. – (Enclycopedia of mathematics and its applications; 55)
Includes bibliographical references (p. –) and index.
ISBN 0 521 45513 8
1. Combinatorial analysis. I. Title. II. Series: Enclycopedia of mathematics and its applications; v. 55.
QA164.S213 1996
511′.6–dc20 94-30890 CIP

ISBN 0 521 45513 8 hardback

TAG

Contents

Preface

This book is addressed to those who are interested in combinatorial methods of discrete mathematics and their applications. A major part of the book can be used as a textbook on combinatorial analysis for students specializing in mathematics. The remaining part is suitable for use in special lectures and seminars for the advanced study of combinatorics. Those parts which are not intended for teaching include Sections 2.3, 3.6, 3.7, 5.3, 5.6, 5.8, 6.3, 6.4 and Subsection 5.1.3 where the material contains either special questions concerning applications of combinatorial methods or rather cumbersome derivations of asymptotic formulae. Of course, a course of studies in discrete mathematics can be biased towards asymptotic methods, where the selection of material can be different and where the above-mentioned sections become basic.

Some knowledge of algebra and set theory, summarized in Section 1.1, is assumed. To understand the derivations of asymptotic formulae, the reader must be familiar with those results of complex analysis usually included in standard courses for students specializing in mathematics.

For the convenience of those readers who are interested in the separate questions contained in the book, I have attempted to make the presentation of each chapter self-contained and, for the most part, independent of the other chapters.

As is usual, I acknowledge those authors whose results are presented in the book and provide the corresponding citation. The list of references is given at the end of the book.

The method of citation is unified. Citations of theorems, lemmas, corollaries, formulae, etc., include the chapter number, section number and own number within the section. For example, Theorem 1.2.3 is theorem number 3 in Section 2 of Chapter 1.

V. N. Sachkov

Preface to the English edition

A few additions, which did not change the structure and level of presentation, have been included in the preparation of this English translation. Chapter 1 is supplemented by Section 1.8. In Chapter 2 insertions have been added to Subsection 2.1.3; this chapter is supplemented by Subsection 2.2.4 and Section 2.7; Section 2.6 is supplemented by Subsections 2.6.2 and 2.6.6. In Chapter 3, Sections 3.1, 3.3 and 3.6 are supplemented, and a new section, Section 3.4, is introduced. In Chapter 5, Subsection 5.2.6, item (g) in Section 5.4 and Section 5.9 have been added. Subsection 6.1.3 has been added to Chapter 6. In the main, the option of using certain parts of the book as a textbook, as well as the independence of the chapters, is preserved.

After the Russian edition of the book had appeared, a number of significant monographs on combinatorics and closely related problems were published. These books and some papers have been included in the Bibliography.

I am grateful to Professor B. Bollobás for his kind suggestion to Cambridge University Press to publish this book in English, and to Professor V. F. Kolchin for useful discussions during the translation.

V. N. Sachkov

Introduction

With advances in cybernetics and closely related divisions of science, discrete mathematics has found increasing importance as a tool for the investigation of various models of functioning of technical devices and discretely operating systems.

A significant place in discrete mathematics is occupied by combinatorial methods whose applications can help in solving the problems of the existence and construction of arrangements of elements according to certain rules, and in the estimation of the number of such arrangements. Each arrangement determines a configuration which can be considered as a mapping of one set onto another with some restrictions posed by a particular problem. If the restrictions are complicated, then we are faced with the problem of determining conditions of existence and suggesting methods of construction of such configurations.

In Chapter 1 such questions are considered for block-designs and Latin squares. The results presented in the chapter are intended to provide an insight into the typical problems of this area of combinatorial mathematics.

Chapter 2 is devoted to transversals, usually referred to as systems of distinct representatives of a family of sets. Permanents are the main tools for calculating the number of transversals of a family of sets. In this chapter methods of calculating the values of permanents are considered. These calculations meet with more difficulties than do calculations of determinants, those objects which in respect of many other properties are close to permanents.

Much attention is given in the book to so-called enumerative problems of combinatorics. The solutions of such problems consist either of the suggestion of a method of search for combinatorial configurations of some class or in the evaluation of the number of the configurations, or both, in the search and evaluation.

In enumerative problems the main role is played by the generating function method, whose basic properties are given in Chapter 3. Depending on the problem in question the generating functions are considered either as elements of the ring of formal power series or as analytic functions of real or complex variables. If a generating function is an analytic function, we can find an asymptotic representation of the coefficients using the saddle point method, well-known in complex analysis for the estimation of contour integrals.

In Chapter 4, generating functions are used in enumerative problems of graph theory. A general method for constructing the generating functions enumerating the mappings of finite sets with some restrictions on the components is given. The enumeration of mappings with restrictions on contours and height provides a possibility for solving the corresponding enumerative problems for substitutions with restrictions on cycle lengths.

The rigorous definition of a combinatorial configuration requires formalization of the notion of the indistinguishability of elements. Currently, the approach to the introduction of the notion of indistinguishability in combinatorial analysis which uses the notion of equivalence classes generated by a group of substitutions G acting on the initial set of elements X is well known. In this approach the set X' of distinguishable elements coincides with the factor set with respect to a given equivalence relation. The problem of finding the number of distinct elements, which is usually called the enumerative problem, is reduced to the calculation of the coefficients of some polynomials depending on the residue classes of the group G. This method, known in combinatorics as Pólya's enumerative theory, began with the papers by Redfield (Redfield, 1927) and Pólya (Pólya, 1937) and was developed in de Bruijn's paper (de Bruijn, 1958). A presentation of this method and some of its applications are given in Chapter 6. It should be noted that Pólya enumerative theory is of great importance in many enumerative problems, especially in graph theory. However, in a number of important cases the application of the theory meets with considerable difficulties and proves to be inefficient. It is primarily concerned with well-known combinatorial schemes such as allocations, combinations, permutations with various additional properties, occupancy problems with restrictions on capacities of cells and different notions of distinguishability of elements, etc.

One of the central notions of this book is the so-called general combinatorial scheme introduced in Chapter 5. This scheme includes particular cases of almost all the known combinatorial models mentioned above. It allows us to formulate a unified method of construction of generating functions for special classes of enumerative combinatorial problems with some common properties. Note that some approaches suggested by Riordan in (Riordan, 1958) were used in the elaboration of this method.

The general combinatorial scheme provides a required level of mathematical rigor in formulations of various combinatorial problems and permits the unification of the methods of solution. It should be noted that in applying the general combinatorial scheme we achieve economy of presentation of material, since the necessity for separate presentation of particular combinatorial schemes no longer arises, as it usually does in monographs on combinatorial analysis.

A distinctive feature of this book is the large number of asymptotic formulae. In some cases, application of the asymptotic formulae can be considered as a tool for the simplification of the cumbersome explicit formulae obtained which cannot be used in calculations for large values of the parameters involved even if powerful computers are available. In other cases, an explicit formula cannot be obtained and the only alternative is an asymptotic formula obtained in some indirect way. In both types of case the efficiency of combinatorial methods in the absence of the use of the asymptotic formulae is problematical. At present, the use of asymptotic methods in combinatorial analysis receives wide recognition, but many interesting results are published in mathematical journals and have not yet been embodied in monographs on combinatorial analysis. It is hoped that this book will, to some extent, assist in closing this gap.

In conclusion, note that this book is not intended to present the most general results on the questions considered. It seems to me that it is more appropriate to focus our attention on typical results whose derivations well illustrate the combinatorial methods used, in order that the essence is not hidden by technical details.

1

Combinatorial configurations

The first section of this chapter is of an introductory nature and presents a summary of the main notions and results from the set theory and algebra which will be used in the book. In the sections that follow we consider various combinatorial configurations which may be introduced on the basis of the general notion of a configuration given in terms of mappings of sets. As examples of combinatorial configurations we consider Latin squares, orthogonal Latin squares, block designs and finite projective planes.

1.1 Notions of set theory and algebra

1.1.1 Boolean operations on sets

A set is a collection of elements of abstract nature, objects or notions, united by some common property. Along with the word "set" we sometimes use equivalent words such as "collection", "family", etc. A set consists of elements, and the formula $x \in X$ means that the element x belongs to the set X; otherwise we write $x \notin X$. If for each $x \in X$ the inclusion $x \in Y$ holds, then we say that X is a subset of Y and write $X \subseteq Y$. Sets X and Y are equal if $X \subseteq Y$ and $Y \subseteq X$. We say that a set X is a proper subset of Y and write $X \subset Y$ if $X \subseteq Y$ and $X \neq Y$. Any set contains, as a subset, the empty set denoted by $\varnothing$.

A set can be determined either by enumeration of its elements or by pointing out the common properties that characterize the elements. Using this second approach to the description of a set, let us define the so-called Boolean operations.

Union The union of sets X and Y is the set

$$X \cup Y = \{x : x \in X \text{ or } x \in Y\},$$

that is, $X \cup Y$ is the set of elements which belong to at least one of the sets X and Y.

Intersection The intersection or product of sets X and Y is the set

$$X \cap Y = \{x : x \in X \text{ and } x \in Y\}.$$

Difference The difference of sets X and Y is the set

$$X \setminus Y = \{x : x \in X \text{ and } x \notin Y\}.$$

Complement The complement of a set X to a set Y such that $X \subset Y$ is the set

$$\bar{X} = Y \setminus X.$$

The collection of all subsets of a set X is called the power set of X and is denoted by 2^X. Elements $X_1, \ldots, X_r$ from the power set of X constitute a partition of the set X if $X_i \neq \varnothing$, $i = 1, \ldots, r$, and

$$X = X_1 \cup \cdots \cup X_r, \quad X_i \cap X_j = \varnothing, \quad i \neq j. \tag{1.1.1}$$

The sets $X_1, \ldots, X_r$ are called the blocks of the partition.

The set of all ordered pairs (x, y) such that $x \in X$, $y \in Y$ is called the Cartesian product of the sets X and Y and is denoted by $X \times Y$, that is,

$$X \times Y = \{(x, y) : x \in X, \ y \in Y\}.$$

The Cartesian product of several sets is defined similarly:

$$X_1 \times \cdots \times X_r = \{(x_1, \ldots, x_r) : x_1 \in X_1, \ldots, x_r \in X_r\},$$

if $X_1 = \cdots = X_r = X$, then the Cartesian power is

$$X^{(r)} = X \times \cdots \times X.$$

Let X be a finite set and let $|X|$ denote the number of its elements. We give an obvious rule which is the basis of many combinatorial calculations and estimates.

The summation rule If X is a finite set and $X = X_1 \cup \cdots \cup X_r$, $X_i \in 2^X$, $i = 1, \ldots, r$, then

$$|X| \leqslant |X_1| + \cdots + |X_r|, \tag{1.1.2}$$

where the equality is attained only if $X_1, \ldots, X_r$ is a partition of the set X.

Let us now give a second simple rule which is also used in combinatorial analysis.

The multiplication rule If sets $X_1, \ldots, X_r$ are finite, then

$$|X_1 \times \cdots \times X_r| = |X_1| \cdots |X_r|. \tag{1.1.3}$$

1.1.2 Binary correspondences and binary relations

Any set R of pairs from $X \times Y$ is called a binary correspondence on the sets X and Y. If $(x, y) \in R$, then we call x and y the projections of (x, y) on X and Y, respectively, and write

$$x = \pi_1(x, y), \qquad y = \pi_2(x, y).$$

For a binary correspondence $R \subseteq X \times Y$, its projections on X and Y can be defined as

$$\pi_1(R) = \{x : x = \pi_1(x, y),\ (x, y) \in R\},$$
$$\pi_2(R) = \{y : y = \pi_2(x, y),\ (x, y) \in R\}.$$

The projections $\pi_1(R)$ and $\pi_2(R)$ are sometimes called the *domain of definition* and *range of values* of R respectively. The *image* of an element $x \in R$ in the correspondence R is the set

$$\delta_1(x; R) = \{y : y \in Y,\ (x, y) \in R\}.$$

Similarly, the set

$$\delta_2(y; R) = \{x : x \in X,\ (x, y) \in R\}$$

is the *preimage* of an element $y \in Y$ in the correspondence R.

A binary correspondence $\varphi \subseteq X \times Y$ is called a *functional* correspondence if the image of any element $x \in X$ consists of only one element. If $X = \{x_1, \ldots, x_n\}$ and $Y = \{y_1, \ldots, y_m\}$, then a binary correspondence $R \subseteq X \times Y$ can be associated with the matrix $A = \|a_{ij}\|$, $i = 1, \ldots, n$, $j = 1, \ldots, m$, where

$$a_{ij} = \begin{cases} 1, & (x_i, y_j) \in R, \\ 0, & (x_i, y_j) \notin R. \end{cases}$$

Matrices whose elements take values 0 and 1 are called (0, 1)-matrices. The (0, 1)-matrix A associated with a binary correspondence R is called the *incidence* matrix of R.

For $X = Y$ a binary correspondence $R \subseteq X \times Y$ is called a *binary relation* on the set X. The equality relation $\Delta_X = \{(x, x) : x \in X\}$, which is called the *diagonal*, is an example of a binary relation on X. Another

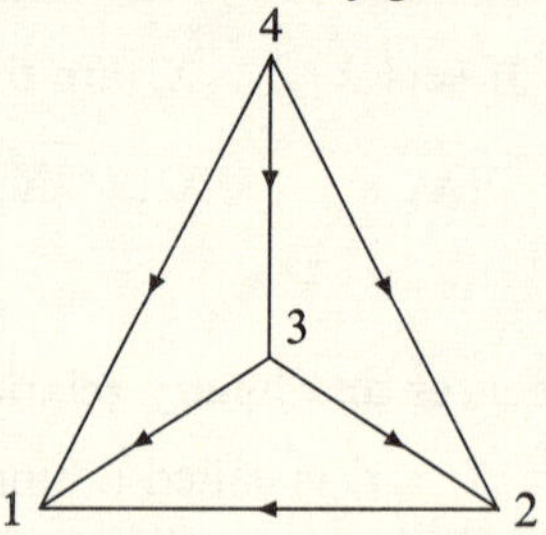

Fig. 1.1.1. The diagram $\Gamma(X,R)$

example is given by the relation of natural order R on a set X of real numbers, where $(x_1,x_2) \in R$ for $x_1,x_2 \in X$ if and only if $x_1 < x_2$.

A binary relation R on a finite set X is associated with a geometric object which is called a *directed graph* or *diagram.* Each element $x \in X$ corresponds to a point on the plane called a vertex. If $(x,x') \in R$, then the points labeled x and x' are connected by an arrow from x to x', called an *edge.* The collection of the vertices and edges formed in such a way is the diagram $\Gamma(X,R)$ of the relation R. The diagram $\Gamma(X,R)$ for $X = \{1,2,3,4\}$ and the natural order R is presented in Figure 1.1.1.

For a binary relation R on a set X we write xRx' if $(x,x') \in R$. Using this notation, we list some properties which can be satisfied by binary relations.

Reflexivity: xRx for all $x \in X$.
Antireflexivity: $R \cap \Delta_X = \varnothing$.
Symmetry: xRx' implies $x'Rx$.
Antisymmetry: xRx' and $x'Rx$ imply $x = x'$.
Transitivity: xRx' and $x'Rx''$ imply xRx''.
Dichotomy: if $x,x' \in X$, then either xRx' or $x'Rx$ holds.

A binary relation R is called an *equivalence relation* if it is reflexive, symmetric and transitive. If R is an equivalence relation and xRx', then we usually write $x \sim x'$. The set $K(x) = \{x' : x' \sim x\}$ is called the *equivalence class* of the element x, $x \in X$. The equivalence classes form a partition of the set X. Conversely, any partition of X determines an equivalence relation whose classes coincide with the blocks of the partition. The set of all equivalence classes is called the *factor set* with respect to the given equivalence relation.

A binary relation R on a set X is called a *partial order*, if R is reflexive, antisymmetric and transitive. In this case the set X is called a partially ordered set. If R is a partial order relation and xRx', then we usually write $x \preccurlyeq x'$.

A partial order relation R with the dichotomy property is called a *linear* order or, simply, order. The expression $x \leqslant x'$ means that x and x' satisfy a linear order relation R, that is, xRx'.

We can define a strict partial (linear) order relation $\prec$ ($<$) on X, setting $x \prec x'$ ($x < x'$) if $x \preccurlyeq x'$ ($x \leqslant x'$) and $x \neq x'$.

It is clear that we can always define a strict order relation on any finite set X. Each such relation is equivalent to a permutation of the elements of X and, consequently, the number of such relations is equal to $|X|!$.

We can define a partial order relation on the power set 2^X of a finite set X, setting $A \preccurlyeq A'$, $A, A' \in 2^X$, if and only if $A \subseteq A'$. In turn we can define a linear order relation on the Cartesian power $X^{(r)}$ of a set X with a strict linear order relation setting

$$(x_1, \ldots, x_r) < (x'_1, \ldots, x'_r)$$

for $(x_1, \ldots, x_r)$, $(x'_1, \ldots, x'_r) \in X^{(r)}$ if for the smallest index i such that $x_i \neq x'_i$ the relation $x < x'$ holds. This ordering is used to order entries in dictionaries and is referred to as the *lexicographical ordering*.

1.2 Mappings and composition laws

1.2.1 Mappings

A rule φ which assigns a single element $\varphi(x) \in Y$ to each $x \in X$ is called a *single-valued mapping* of the set X into the set Y or a function defined on X and taking values from Y. In such a case we write $\varphi : X \to Y$. It is clear that a single-valued mapping $\varphi : X \to Y$ is a functional binary correspondence $\varphi \subseteq X \times Y$. We define a *many-valued mapping* ψ of a set X into a set Y as a rule which assigns a set $\psi(x) \subseteq Y$ to each element $x \in X$, where the case $\psi(x) = \varnothing$ is not ruled out. It is clear that a many-valued mapping ψ is a binary correspondence $\psi \subset X \times Y$. In what follows we use single-valued mappings, unless otherwise specified. Therefore, let us consider some properties of single-valued mappings without explicitly mentioning their single-valuedness.

The most commonly used representations of a mapping are a functional form and a two-row matrix form. The *functional representation* $y = \varphi(x)$ means that under the mapping $\varphi : X \to Y$ any element $x \in X$ transfers to the element $y = \varphi(x) \in Y$. The two-row matrix representation is convenient for finite sets X and Y. Let $X = \{x_1, \ldots, x_n\}$ and $Y = \{y_1, \ldots, y_m\}$, then the two-row representation corresponding to $\varphi : X \to Y$ is of the form

$$\varphi = \begin{pmatrix} x_1 & x_2 & \ldots & x_n \\ \varphi(x_1) & \varphi(x_2) & \ldots & \varphi(x_n) \end{pmatrix}.$$

Let us consider the matrix $A = \|a_{ij}\|$, $i = 1, \ldots, n$, $j = 1, \ldots, m$, where

$$a_{ij} = \begin{cases} 1, & y_j = \varphi(x_i), \\ 0, & y_j \neq \varphi(x_i). \end{cases}$$

The matrix A is called the *incidence matrix* of the mapping φ of the set $X = \{x_1, \ldots, x_n\}$ into the set $Y = \{y_1, \ldots, y_m\}$. By virtue of the single-valuedness of the mapping φ there is exactly one unit in each row of the matrix A. Such matrices are usually called *elementary matrices.*

Two mappings $\varphi_1 : X_1 \to Y_1$ and $\varphi_2 : X_2 \to Y_2$ are considered equal if $X_1 = X_2$, $Y_1 = Y_2$ and $\varphi_1(x) = \varphi_2(x)$ for all $x \in X$. A mapping $\varphi' : X' \to Y$ is a *restriction* or *truncation* on X' of a mapping $\varphi : X \to Y$ if $X' \subseteq X$ and $\varphi'(x) = \varphi(x)$ for all $x \in X'$.

The set $\varphi^{-1}(y) = \{x : \varphi(x) = y,\ x \in X\}$ is called the *preimage* of an element y under a mapping $\varphi : X \to Y$. The preimage of an element y coincides with the preimage of this element under the functional binary correspondence $R \subseteq X \times Y$ related to the mapping φ. A mapping $\varphi : X \to Y$ is *surjective* if for any $y \in Y$ there exists $x \in Y$ such that $y = \varphi(x)$. In this case we say that φ is a mapping of X onto Y. It is clear that under a surjective mapping the preimage of any $y \in Y$ is not empty. A mapping $\varphi : X \to Y$ is *injective* if $\varphi(x) \neq \varphi(x')$ for any $x, x' \in X$ such that $x \neq x'$. The preimage of any $y \in Y$ contains no more than one element under an injective mapping. A mapping $\varphi : X \to Y$ is *bijective* if it is both surjective and injective. Such mappings are usually referred to as *one-to-one* mappings.

If $X = Y$, then a bijective mapping $\varphi : X \to X$ is called a *substitution.* A substitution of a finite set X can be associated with a square incidence matrix of size $|X|$ each row and each column of which contain exactly one unit. Such matrices are called permutation matrices.

1.2.2 Composition laws

A *composition* law or an operation (binary) on a set X is a mapping $f : X \times X \to X$. If $f(x, y) = z$, where $x, y, z \in X$, then we write $x \top y = z$.

A composition law $\top$ is *associative* if for any $x, y, z \in X$

$$(x \top y) \top z = x \top (y \top z).$$

A composition law $\top$ which satisfies the condition

$$x \top y = y \top x$$

is called *commutative.*

A composition law $\top$ is *distributive* with respect to a composition law $\perp$ if for any $x, y, z \in X$

$$x \top (y \perp z) = (x \top y) \perp (x \top z),$$
$$(y \perp z) \top x = (y \top x) \perp (z \top x).$$

We now introduce the *inverse* element for a given element, and the unit element. An element $e \in X$ is called the unit element or the *neutral* element with respect to a composition law $\top$ if for any $x \in X$

$$x \top e = e \top x = x.$$

If such an element exists then it is unique. Let a composition law $\top$ have the unit element. Then an element x^{-1} is the *inverse* or *symmetric* element for an element $x \in X$ with respect to the composition law if

$$x \top x^{-1} = x^{-1} \top x = e.$$

A set X where a composition law is defined is called a *groupoid.* A groupoid X where each of the equations $a \top x = b$ and $y \top a = b$ has a unique solution with respect to unknowns x and y for any $a, b \in X$ is called a *quasigroup.* Obviously, a quasigroup can also be defined as a set X such that each two elements from the relation $a \top b = c$, where $a, b, c \in X$, uniquely determine the third. If the composition law $\top$ is associative, then the groupoid is called a *semigroup* or *monoid.* If the operation is commutative, then the semigroup is called *Abelian.* A semigroup with unit element such that the inverse element exists for any of its elements is called a *group.* A group is *monogenic* if each of its elements can be obtained from an element other than the unit by sequential application of the composition law. A monogenic group is necessarily Abelian. A finite monogenic group is called cyclic.

For a finite group X the value $|X|$ is called the *order* of the group. Let X and Y be finite groups of arbitrary orders with operations $\top$ and $\perp$, respectively, and let $\varphi : X \to Y$ be a mapping such that for any $x, x' \in X$

$$\varphi(x \top x') = \varphi(x) \perp \varphi(x').$$

Such a mapping is called a *homomorphism* of the group X into the group Y. If the mapping φ is bijective, then the homomorphism is called an *isomorphism.*

A non-empty subset Y of a group X with a composition law $\top$ is called a *subgroup* if the following conditions are fulfilled:

(1) the inclusion $y \in Y$ implies $y^{-1} \in Y$;
(2) the inclusions $y \in Y$, $y' \in Y$ imply $y \top y' \in Y$.

If the composition of any two elements from $Y \subset X$ belongs to Y, then we say that Y is *closed* with respect to this composition law. A subgroup Y of a group X is a set closed with respect to the composition law and containing the inverse elements of all its elements.

A subgroup Y of a group X determines an equivalence relation on the group. For any $x, x' \in X$ let $x \sim x'$ if $x' \top x^{-1} \in Y$. The corresponding equivalence classes are called the *right residue classes* of X with respect to the subgroup Y. Similarly, we can put $x \sim x'$ if $x^{-1} \top x' \in Y$. The corresponding equivalence classes are called the *left residue classes* of X with respect to the subgroup Y. If x belongs to some right (left) residue class with respect to the subgroup Y, then the whole class consists of elements of the form $y \top x$ $(x \top y)$, where $y \in Y$, and is denoted by Yx (xY). The corresponding partition of X is called the *decomposition* of X into the right (left) residue classes with respect to the subgroup Y.

If X is a finite group and $x_1, \ldots, x_{k-1}$ are representatives of all residue classes, except Y itself, then the decomposition can be written as

$$X = Y \cup Yx_1 \cup \cdots \cup Yx_{k-1}.$$

For the left residue classes the decomposition has the form

$$X = Y \cup x_1 Y \cup \cdots \cup x_{k-1} Y,$$

where xY is the left residue class containing the element x. The number k is called the *index* of the subgroup Y in the group X.

A subgroup Y is an invariant subgroup in a group X if $xY = Yx$ for any $x \in X$. It is clear that the decompositions of X into the right and left residue classes with respect to an invariant subgroup coincide. For the sake of definiteness the residue classes of a group X with respect to an invariant subgroup Y are considered as the right residue classes. Let us define the composition law $*$ putting $(Yx) * (Yx') = Y(x \top x')$, where $\top$ is the composition law of the group X. The inverse element with respect to the operation $*$ is defined by the equality $(Yx)^{-1} = Yx^{-1}$; the role of the neutral element is played by Y. As a result the set of residue classes becomes a group which is called the *factor group* and is denoted by X/Y.

A set X with two operations $\perp$ and $\top$ is called a *ring* if it is an Abelian group with respect to the operation $\perp$, a semigroup with respect to the operation $\top$ and the composition law $\top$ is distributive with respect to the composition law $\perp$.

If in a power set 2^X we take the operation $\cup$ as the composition law $\perp$ and the operation $\cap$ as the composition law $\top$ and add the operation of taking complements of sets to these two operations, then we obtain the Boolean algebra.

The set of integers $\mathbf{Z}$ with ordinary addition and multiplication as the composition laws forms a ring. We say that numbers $a, b \in \mathbf{Z}$ are congruent modulo m if the difference $a - b$ is divisible by m. In this case we write $a \equiv b \pmod{m}$. The congruence is an equivalence relation and the corresponding equivalence classes are called the residue classes modulo m. Each of the classes contains exactly one of the numbers $0, 1, \ldots, m-1$ which are representatives of the classes and form the full system of the least non-negative residues modulo m. Denote by K_i the class whose representative is the number i, $0 \leqslant i \leqslant m-1$. On the set of classes $K_0, \ldots, K_{m-1}$ we can introduce operations of addition and multiplication, which for simplicity we also denote by the symbols $+$ and $\cdot$. We define these operations on the classes as follows: $K_i + K_j = K_l$ if $i + j \equiv l \pmod{m}$, $K_i \cdot K_j = K_l$ if $i \cdot j \equiv l \pmod{m}$. As a result we obtain the ring of residue classes modulo m which will be denoted by $\mathbf{Z}_m$.

In a ring the neutral elements with respect to the operations $\perp$ and $\top$ are called the *zero* and *identity* respectively.

If the set of non-zero elements with respect to the composition law $\top$ is an Abelian group, then the ring is a field. A field with a finite number of elements is called *finite*. If p is a prime number, then the set of residue classes modulo p forms a finite field with p elements.

A finite field with n elements exists if and only if $n = p^\alpha$, where p is prime and α is natural. Such a field is unique up to isomorphisms preserving both the composition laws. This field is called the *Galois field* and is denoted by $\mathrm{GF}(p^\alpha)$. If $\alpha = 1$, then $\mathrm{GF}(p)$ is isomorphic to the field of residue classes modulo p.

Let $\mathscr{P}$ be a field with operations $+$ and $\cdot$, and let X be an Abelian group with a composition law $\perp$. A mapping $\varphi : \mathscr{P} \times X \to X$ determines the so-called outer composition law $\top$, which has the following properties:

$$a \top (x \perp y) = a \top x \perp a \top y,$$
$$(a + b) \top x = a \top x \perp b \top x,$$
$$a \top (b \top x) = (a \cdot b) \top x,$$
$$e \top x = x,$$

where $a, b \in \mathscr{P}$, $x, y \in X$ and e is the identity of the field $\mathscr{P}$. An Abelian group with such operations $\perp$ and $\top$ is a *vector space* over $\mathscr{P}$, and its elements are called *vectors.* Vectors $x_1, \ldots, x_n$ are linearly independent if the equality

$$a_1 \top x_1 \perp \cdots \perp a_n \top x_n = \bar{0},$$

where $a_1, \ldots, a_n \in \mathscr{P}$ and $\bar{0}$ is the neutral element of the Abelian group X, implies that $a_1 = \cdots = a_n = 0$, where 0 is the zero of $\mathscr{P}$. A maximal set of linearly independent vectors $x_1, \ldots, x_n$ is called a basis of the space. A vector space which has at least one basis with finite number n of elements is called finite-dimensional, and the number n is its *dimension.* A vector space of dimension n is isomorphic to the vector space V_n of row-vectors of the form $v = (a_1, \ldots, a_n)$ with coordinates from the field $\mathscr{P}$, where the operation $\perp$ is the ordinary coordinate-wise addition and $\top$ is the multiplication of the coordinates by an element of $\mathscr{P}$. If $e_1, \ldots, e_n$ is a basis of the space V_n, then any vector $v \in V_n$ can be uniquely represented as a linear combination of the basis vectors in the form

$$v = \alpha_1 e_1 + \cdots + \alpha_n e_n, \quad \alpha_j \in \mathscr{P}, \quad 1 \leqslant j \leqslant n.$$

A ring X is an algebra over $\mathscr{P}$ with composition laws + and $\cdot$, if the group X with respect to the first composition law is a vector space over the field $\mathscr{P}$, and the second composition law $\top$ and multiplication $\cdot$ by elements of $\mathscr{P}$ are related by the formula

$$a \cdot (x \top y) = (a \cdot x) \top y = x \top (a \cdot y).$$

The dimension of the vector space is called the *rank* of the algebra. For example, the set of all $n \times n$ matrices with complex elements with the ordinary operations of addition and multiplication and with the operation of multiplication of a matrix by a complex number, is an algebra of rank n^2 over the field of complex numbers.

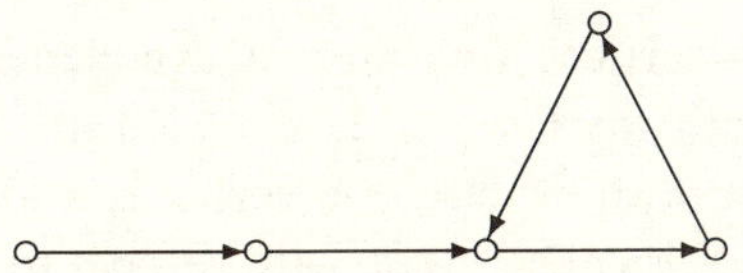

Fig. 1.2.1. The graph of a mapping

1.2.3 Transformations

A one-to-one mapping of a set X onto itself $\sigma : X \to X$ is called a *transformation* of the set. If X is a finite set, then the number $|X|$ is called the *degree* of the transformation σ. An element $x \in X$ is a fixed point of a transformation σ if $\sigma(x) = x$. The transformation e such that $e(x) = x$ for all $x \in X$ is the *identity* transformation.

On the set of all transformations of X we define the composition law $\cdot$, which is the sequential application of the transformations. More precisely, $\sigma = \sigma_1 \cdot \sigma_2$ if $\sigma(x) = \sigma_2(\sigma_1(x))$ for all $x \in X$. The family of all transformations of X with the composition law $\cdot$, which is associative, is called the *symmetric semigroup* on X. If X is a finite set, then the symmetric semigroup is denoted by $\mathfrak{S}_{|X|}$ and is called the symmetric semigroup of degree $|X|$. It is clear that the number of elements of $\mathfrak{S}_{|X|}$ is equal to $|X|^{|X|}$. Put $|X| = n$ and consider the semigroup $\mathfrak{S}_n$. Let $\sigma^k = \sigma \cdots \sigma$, where the number of multipliers is equal to k, and $\sigma^0 = e$. Define an equivalence relation on X by putting $x \equiv x'$ if there exist integers i and j such that $\sigma^i(x) = \sigma^j(x')$. The corresponding equivalence classes are called the *components* of the transformation $\sigma \in \mathfrak{S}_n$.

For describing the component structure of a transformation, it is convenient to use geometric language and the notion of the directed graph of the transformation. We assume that X is a finite set and label $|X|$ different points of the plane, called *vertices*, with the elements of the set X. Two points labeled x and x' are joined by the edge (x, x') directed from x to x' if $x' = \sigma(x)$. As a result we obtain the digraph $\Gamma(X, \sigma)$ of the transformation. We consider digraphs in more detail in Chapter 4. The main characteristic of $\Gamma(X, \sigma)$ lies in the fact that from each vertex emanates exactly one edge. The directed graph of a mapping with five vertices is presented in Figure 1.2.1.

A sequence of edges of $\Gamma(X, \sigma)$ such that the end of each preceding edge is the origin of the consequent edge is called a *path*. A path is called *elementary* if all its vertices are different. An elementary path whose first vertex coincides with the last vertex is called an *elementary contour.* An elementary contour with exactly one vertex is a *loop.*

An element $x \in X$ is called *cyclic* with respect to $\sigma \in \mathfrak{S}_{|X|}$ if there exists $p \geqslant 0$ such that $\sigma^p(x) = x$. The vertex of the graph $\Gamma(X, \sigma)$ corresponding to the cyclic element x is also called *cyclic.*

Let σ_1 be a restriction of the transformation $\sigma \in \mathfrak{S}^{|X|}$ to a component $X_1 \subseteq X$ of the transformation σ. The graph $\Gamma(X_1, \sigma_1)$ is usually referred to as a *connected component* of the graph $\Gamma(X, \sigma)$. Let us give a description of the graph $\Gamma(X_1, \sigma_1)$. All cyclic vertices of the graph are connected by an elementary contour and this contour is unique. Indeed, let x be a cyclic element and let $p > 0$ be the smallest number such that $x = \sigma^p(x)$. Then all elements of the sequence $x, \sigma(x), \ldots, \sigma^{p-1}(x)$ are different. Suppose there exists an element $x' \in X_1$ which does not belong to this sequence and $x' = \sigma^q(x')$. The elements x and x' belong to the component X_1; therefore there exist i and j such that $\sigma^i(x) = \sigma^j(x')$, and, hence, we obtain the equality $\sigma^{j(q-1)+i}(x) = \sigma^{jq}(x') = x'$, which contradicts the choice of x'.

Any vertex labeled by some element $x \in X_1$ is connected in the graph $\Gamma(X_1, \sigma_1)$ with a cyclic vertex by a unique elementary path. For a non-cyclic vertex x this elementary path has the following sequence of vertices: $x, \sigma(x), \ldots, \sigma^h(x)$, where $\sigma^h(x)$ is a cyclic vertex. The number h is called the *height* of the element x with respect to the transformation σ. The height of a cyclic element is zero by definition.

Consider a cyclic vertex labeled by an element $x \in X_1$ and denote by $\Gamma_x(X_1, \sigma_1)$ the part of the graph $\Gamma(X_1, \sigma_1)$ consisting of the non-cyclic vertices x' such that $\sigma^j(x') = x$ for some $j, l = 1, 2, \ldots$. Such a graph is a rooted tree with root x.

Thus, the graph $\Gamma(X, \sigma)$ consists of several connected components. Each component consists of an elementary contour and rooted trees whose roots are the vertices of the contour.

1.2.4 Substitutions

Recall that a bijective transformation of a set X is a substitution of this set. It is not difficult to show that the sets of surjective and injective mappings are closed with respect to the composition law, defined on the symmetric semigroup, acting on X. This implies the closure of the set of substitutions with respect to the composition law, called the multiplication of substitutions. Recall that in the action of the product

$s's$ on an element x the substitution s' acts first, that is, $s's(x) = s(s'(x))$. Each substitution s has a unique inverse substitution s^{-1} such that $s^{-1}s(x) = ss^{-1}(x) = e(x)$, where e is the identity substitution. Thus, the set of all substitutions of a set forms a group called the symmetric group acting on X. If X is a finite set and $|X| = n$, then the group is called the symmetric group of degree n and is denoted by S_n. Obviously, the order of S_n is equal to $n!$.

There exists a one-to-one correspondence between the substitutions from S_n and the permutation matrices: each substitution $s \in S_n$ can be associated with the permutation matrix $\Pi = \|\pi_{ij}\|$, $i, j = 1, 2, \ldots, n$, such that

$$\pi_{ij} = \begin{cases} 1, & s(i) = j, \\ 0, & s(i) \neq j. \end{cases}$$

Using Kronecker's symbols

$$\delta_{ij} = \begin{cases} 1, & i = j, \\ 0, & i \neq j, \end{cases}$$

we can write the matrix Π in the form

$$\Pi = \|\delta_{s(i),j}\| = \|\delta_{i,s^{-1}(j)}\|, \qquad i, j = 1, \ldots, n.$$

The family of all permutation matrices of order n whose composition law coincides with ordinary multiplication of matrices forms a group that is isomorphic to the group S_n. The restriction of a substitution on a component is called a cycle; the number of elements of this component is called the cycle length. The connected components of the graph $\Gamma(X, s)$ of a substitution s acting on X are elementary contours.

1.2.5 Linear transformations of a vector space

Consider the vector space V_n of row-vectors with coordinates from a field $\mathscr{P}$. For addition of vectors and multiplication of a vector by a scalar we use the symbols $+$ and $\cdot$ coinciding with the corresponding symbols in the field $\mathscr{P}$. A linear transformation of the space V_n is a transformation $\varphi : V_n \to V_n$ such that

$$\varphi(v_1 + v_2) = \varphi(v_1) + \varphi(v_2), \qquad \varphi(a \cdot v) = a\varphi(v),$$

where $v_1, v_2, v \in V_n$, $a \in \mathscr{P}$.

It is clear that the zero vector $\mathbf{0} = (0, \ldots, 0)$ is a fixed point under a linear transformation. A basis $e_1, \ldots, e_n$ is transformed by a linear transformation into a system of vectors $\tilde{e}_1, \ldots, \tilde{e}_n$, where $\tilde{e}_i = \varphi(e_i)$, $i = 1, \ldots, n$. Since each vector from V_n is represented as a linear combination,

$$\tilde{e}_i = \sum_{j=1}^{n} a_{ij} e_j, \qquad i = 1, \ldots, n,$$

of the basic vectors, there exists a one-to-one correspondence between the linear transformations of V_n and the square matrices $A = \|a_{ij}\|$, $i, j = 1, \ldots, n$.

Let $\mathscr{P}$ be the field of real numbers. A bijective linear transformation is called non-singular. A linear transformation is non-singular if and only if the corresponding matrix A is non-singular, that is, the determinant of the matrix A, denoted by $\det A$, is not equal to zero. The polynomial $\det(A - \lambda E)$ in λ, where E is the identity matrix, is called the characteristic polynomial, whose roots are called the characteristic roots of the matrix. A non-zero vector $v \in V_n$ is an eigenvector of the linear transformation φ if $\varphi(v) = \lambda_0 v$, where the real number λ_0 is an eigenvalue of this transformation. The real characteristic roots of the matrix A, corresponding to the transformation φ, are the eigenvalues of the transformation, and vice versa.

We can define a function $\psi : V_n \times V_n \to \mathscr{P}$, denoted by $\psi(v_1, v_2) = (v_1, v_2)$, $v_1, v_2 \in V_n$, which possesses the following properties:

$$\begin{aligned}
(v_1, v_2) &= (v_2, v_1), \\
(v_1 + v_2, v_3) &= (v_1, v_3) + (v_2, v_3), \\
(av_1, v_2) &= a(v_1, v_2), \\
(v, v) &> 0, \qquad v \neq \mathbf{0}.
\end{aligned}$$

The function ψ is called a scalar product; in this case the space V_n is called Euclidean. For $v_1 = (a_1, \ldots, a_n)$, $v_2 = (b_1, \ldots, b_n)$, the scalar product can be represented as

$$(v_1, v_2) = \sum_{i=1}^{n} a_i b_i.$$

A linear transformation φ of the Euclidean space V_n is orthogonal if it preserves a scalar product of any two vectors, that is, if $(\varphi(v_1), \varphi(v_2)) = (v_1, v_2)$. An orthogonal transformation in some basis corresponds to an orthogonal matrix Q such that $(\det Q)^2 = 1$. A quadratic form in variables $x_1, \dots, x_n$ is defined as the sum

$$f = f(x_1, \dots, x_n) = \sum_{i,j=1}^{n} a_{ij} x_i x_j, \quad a_{ij} = a_{ji}, \quad a_{ij} \in \mathscr{P};$$

the matrix $A = \|a_{ij}\|$ is called the matrix of this quadratic form. By linear transformation of the form

$$y_i = \sum_{k=1}^{n} q_{ik} x_k, \qquad i = 1, \dots, n,$$

a real quadratic form can be transformed to the canonical form

$$f = \sum_{i=1}^{n} \lambda_i y_i^2,$$

where $Q = \|q_{ik}\|$ is an orthogonal matrix and $\lambda_1, \dots, \lambda_n$ are all the characteristic roots of the matrix A counted according their multiplicities.

1.3 Combinatorial configurations

1.3.1 Configurations

In combinatorial analysis we often use such well-known constructions of elements of a finite set as combinations, arrangements, permutations, etc. A combination of size m is a subset consisting of m elements from an initial set $Y = \{y_1, \dots, y_n\}$. If the order of the chosen elements is significant, then we obtain an ordered combination or an arrangement. For the case $m = n$ an arrangement is a permutation of the elements of Y. These structures can be generalized if repetitions among the chosen elements are allowed. As a result we obtain combinations, arrangements and permutations with repetitions.

Even for these simplest combinatorial structures one feels the necessity of formalization of their definitions to avoid tedious descriptions and misunderstandings. As the complexity of structures increases, this formalization becomes more and more important. In many cases such formalization can be achieved by the introduction of the notion of a configuration (Berge, 1968).

Let $X = \{1, \dots, m\}$, $Y = \{y_1, \dots, y_n\}$, and let the linear order be defined on Y such that $y_1 < \cdots < y_n$. A mapping $\varphi: X \to Y$ satisfying some set of restrictions Λ is called a configuration. The set of restrictions Λ which the mapping φ satisfies, determines some class of configurations corresponding to the conditions on the combinatorial structures in a given problem.

Below we construct classes of configurations corresponding to the simplest combinatorial structures, such as combinations, arrangements, permutations, allocations of particles; thus we give in fact a rigorous definition of the structures that are used in the book. The notion of configuration can be used for the description of other combinatorial structures connected with choice or allocation of elements fitting some conditions.

1.3.2 Combinations, arrangements, permutations

Let $\Lambda = \varnothing$, that is, there are no restrictions on the mapping $\varphi : X \to Y$. Then each configuration φ determines a combinatorial structure, called an arrangement with unlimited repetitions of size m from n distinct elements, or an m-permutation with unlimited repetitions . It is clear that the number of such configurations is equal to

$$U(n,m) = n^m. \tag{1.3.1}$$

Let $\Lambda = I$, where I is the set of all injective mappings. In this case a configuration $\varphi : X \to Y$ is an arrangement of size m from n distinct elements or an m-permutation. The number of such configurations is equal to

$$A(n,m) = n(n-1)\cdots(n-m+1) = (n)_m, \qquad m \leqslant n, \tag{1.3.2}$$

where $(n)_m = n!/(n-m)!$ and $A(n,m) = 0$ for $m > n$. The numbers $A(n,m)$ satisfy the recurrence relation

$$A(n,m) = A(n-1,m) + mA(n-1,m-1).$$

If $n = m$, then each configuration corresponds to a permutation and the number of all permutations is equal to

$$P(n) = n!.$$

If the n elements contain q_i elements of the ith type, $q_1 + \cdots + q_k = n$, and the elements of any type are identical, then each permutation of these elements corresponds to a configuration $\varphi : X' \to Y'$, where $X' = \{1,\ldots,n\}$, $Y' = \{y_1,\ldots,y_k\}$, and y_i occurs as an image q_i times, $i = 1,\ldots,k$. The number of such configurations is equal to

$$P(q_1,\ldots,q_k) = \frac{n!}{q_1!\cdots q_k!}. \tag{1.3.3}$$

Let $\Lambda = M$, where M is the set of all strictly monotone functions $\varphi : X \to Y$ such that $\varphi(i) < \varphi(j)$ if $i < j$. In this case a configuration $\varphi : X \to Y$ is a combination of m out of n different elements or an m-combination. The number of such configurations is equal to

$$C(n,m) = \frac{(n)_m}{m!} = \binom{n}{m},$$

where the symbols $\binom{n}{m}$, usually called the binomial coefficients, possess the following obvious properties:

$$\binom{n}{m} = \binom{n}{n-m} = \frac{n!}{m!\,(n-m)!},$$

$$\binom{0}{0} = 1, \quad \binom{n}{m} = 0, \quad n < m.$$

The numbers $C(n,m)$ satisfy the recurrence relation

$$C(n,m) = C(n-1,m) + C(n-1,m-1).$$

The numbers $\binom{n}{m}$ appear in the binomial formula

$$(a+b)^n = \sum_{m=0}^{n} \binom{n}{m} a^m b^{n-m}.$$

If $|X| = n$, then the number of elements of the power set 2^X with m elements from X is equal to $\binom{n}{m}$. Therefore, putting $a = b = 1$, we obtain from the binomial formula that $|2^X| = 2^n$.

Consider the case $\Lambda = \widetilde{M}$, where $\widetilde{M}$ is the set of all monotone functions such that $\varphi(i) \leqslant \varphi(j)$ if $i < j$. In this case a configuration $\varphi : X \to Y$ is called a combination with unlimited repetitions of m out of n different elements or simply an m-combination with repetitions. The number of such configurations is equal to

$$V(n,m) = \binom{n+m-1}{m}. \tag{1.3.4}$$

A proof of this formula can easily be obtained by the introduction of a one-to-one correspondence between the configurations considered and the configurations corresponding to combinations of m out of $n+m-1$ elements.

If the additional requirement of surjectivity is imposed on these configurations, then their number is equal to

$$V^0(n,m) = \binom{n-1}{m-1}. \tag{1.3.5}$$

1.3.3 Allocations of particles to cells; urn models

Let m different particles be allocated to n different cells (urns) of unlimited capacity. The process of allocation consists in choosing the particles sequentially and allocating each of the m particles to one of the n cells, the order of particles in a cell not being taken into account. If the particles are labeled with elements of the set $X = \{1,\ldots,m\}$ and the cells are labeled with the elements of the set $Y = \{y_1\ldots,y_n\}$, then each allocation can be associated with a configuration $\varphi : X \to Y$ and we obtain a one-to-one correspondence.

If $\Lambda = \varnothing$, then the configuration obtained is an allocation with unlimited repetitions. Thus, the number of allocations of m different particles to n different cells is given by formula (1.3.1).

If we are interested in the allocations where the ith cell contains exactly m_i particles, $m_1 + \cdots + m_n = m$, then each allocation corresponds to a configuration $\varphi : X \to Y$, where $\Lambda = J$ and J is the family of mappings which use y_i as an image m_i times, $i = 1,\ldots,n$. The number of such configurations is equal to

$$U(m_1,\ldots,m_n) = \frac{m!}{m_1!\cdots m_n!}.$$

These numbers are known as the polynomial coefficients and satisfy the relation

$$(a_1 + \cdots + a_n)^m = \sum_{m_1+\cdots+m_n=m} \frac{m!}{m_1!\cdots m_n!} a_1^{m_1}\cdots a_n^{m_n},$$

where the summation is over all non-negative integer solutions of the equation $m_1 + \cdots + m_n = m$.

Let us consider a scheme which is usually called the scheme of allocating m identical particles into n different cells. Again we assume that the cells are different and have unlimited capacity, and that the particles are identical. If s_i denotes the number of particles in the ith cell, $i = 1,\ldots,n$, then each allocation of such a type can be uniquely determined by a non-negative integer solution of the equation

$$s_1 + \cdots + s_n = m.$$

On the other hand, a solution of this equation uniquely determines a configuration $\varphi : X \to Y$, where $\varphi(i) \leqslant \varphi(j)$ if $i < j$, with the additional restriction that the element y_i appears as an image exactly s_i times, $i = 1,\ldots,n$. Thus, there exists a one-to-one correspondence between the allocations of m identical particles into n different cells and the

m-combinations with unlimited repetitions from n different elements. Consequently, the number of allocations of m identical particles into n different cells is equal to $V(n,m)$ and can be calculated by formula (1.3.4).

There exists a one-to-one correspondence between the allocations of the type considered without empty cells and the combinations with repetitions possessing the surjective property. Hence it follows that the number of such allocations is equal to $V^0(n,m)$ and can be calculated by formula (1.3.5).

Some other combinatorial models interpreted in terms of allocations of particles into cells are of interest. Such models are referred to as urn models. For example, we could consider the case where the particles are distinguishable and the cells are identical, or the case where both the particles and the cells are indistinguishable. In these cases the construction of configurations and their enumeration lead to investigation of some equivalence classes. We postpone analysis of these and other more complicated models until Chapter 5.

1.4 Latin squares

1.4.1 Discordant mappings, Latin squares

Two mappings $\varphi : X \to Y$ and $\psi : X \to Y$ are called discordant on a set $X_1 \subseteq X$ if $\varphi(x) \neq \psi(x)$ for all $x \in X_1$. Mappings φ and ψ that are discordant on the whole set X are called discordant mappings; in this case we write $\varphi \uparrow \psi$. If φ and ψ are discordant substitutions, then φ^{-1} and ψ^{-1} are discordant, too.

Using the notion of discordant substitutions we can introduce a metric ρ on the symmetric group S_n. We put $\rho(s,s') = |X_1|$ for $s, s' \in S_n$ if X_1 is the maximal subset of X on which the substitutions s and s' acting on X are discordant. The function ρ satisfies all requirements of a metric:

(1) $\rho(s,s') \geqslant 0$ for any $s, s' \in S_n$ and $\rho(s,s') = 0$ if and only if $s = s'$.
(2) $\rho(s,s') = \rho(s',s)$ for any $s, s' \in S_n$.
(3) $\rho(s,s'') \leqslant \rho(s,s') + \rho(s',s'')$ for any $s, s', s'' \in S_n$ (triangle inequality).

The following property of the metric is obvious:

$$\rho(s,s') = \rho(e, s^{-1}s'),$$

where e is the identity substitution. Under this condition $\rho(e,s) = n - \gamma$, where γ is the number of unit cycles of the substitution s. Denote by $\gamma_1, \gamma_2, \gamma_3$ the number of unit cycles in substitutions $s^{-1}s'', s^{-1}s', (s')^{-1}s''$

respectively. It is clear that the inequality $s^{-1}s''(x) \neq x$ for some $x \in X$ implies at least one of the inequalities $s^{-1}s'(x) \neq x$, $(s')^{-1}s''(x) \neq x$. Hence it follows that $n - \gamma_1 \leqslant (n - \gamma_2) + (n - \gamma_3)$. This inequality is equivalent to another

$$\rho(e, s^{-1}s'') \leqslant \rho(e, s^{-1}s') + \rho(e, (s')^{-1}s''),$$

which in turn is equivalent to the triangle inequality. Thus, substitutions s and s' from S_n are discordant if and only if $\rho(s, s') = n$.

Let $s_1, \ldots, s_k$ be mutually discordant substitutions of degree n acting on the set $X = \{1, \ldots, n\}$. A family of such substitutions in a given order is called a $k \times n$ Latin rectangle and is denoted by

$$L_{kn} = [s_1, \ldots, s_k]_n.$$

We may introduce a one-to-one correspondence between Latin rectangles and $k \times n$ matrices if a Latin rectangle $L_{kn} = [s_1, \ldots, s_k]_n$ we associate with the $k \times n$ matrix

$$L_{kn} = [s_i(j)], \quad i = 1, \ldots, k, \quad j = 1, \ldots, n,$$

which will be referred to both as the matrix of the Latin rectangle and as the Latin rectangle. If $k = n$, then a Latin rectangle is called a Latin square and is denoted by $L_n = [s_1, \ldots, s_n]$. A Latin square corresponds to the square $n \times n$ matrix

$$L_n = [s_i(j)], \qquad i, j = 1, \ldots, n,$$

which is called the matrix of the Latin square and also the Latin square. These coincidences in terminology cannot lead to confusion since the right interpretation of Latin configurations is always seen from the context.

Since the substitutions $s_1, \ldots, s_n$ forming a Latin rectangle are discordant, all elements in any row and in any column of the Latin rectangle are different. The rows and columns of a Latin square are permutations of the numbers $1, \ldots, n$.

Let a triple (i, j, l) mean that the number l occupies the entry placed on the intersection of the ith row and jth column of the matrix of a Latin square. A Latin square corresponds to n^2 such triples which uniquely determine the square. This assertion follows from the fact that any two elements of a triple uniquely determine the third. It is clear that the element l is uniquely determined by i and j. Assume that two different elements j and j' correspond to the elements i and l. Then we obtain the equality $s_i(j) = s_i(j')$, $j \neq j'$, which is contradictory. If we assume that two different elements i and i' correspond to the elements j and l, then we obtain the equality $s_i(j) = s_{i'}(j)$, which is contradictory since the substitutions s_i and $s_{i'}$ are discordant for $i \neq i'$.

By the definition of a quasigroup, any two elements from the equality $a \top b = c$ for elements of a quasigroup uniquely determine the third. This means that any finite quasigroup defines some Latin square, and vice versa. Latin rectangles and Latin squares exist for any n, k $(n \leqslant k)$. Indeed, let a substitution c realize a shift of the numbers $1, 2, \ldots, n$, that is, it can be written as a cycle $c = (1, 2, \ldots, n)$. Then for any n the substitutions $c, c^2, \ldots, c^k$ determine a $k \times n$ Latin rectangle, and $c, c^2, \ldots, c^n$ determine a Latin square of size n.

It is clear that matrices of Latin rectangles and squares retain their Latin property under transpositions of their rows and columns. By transpositions of rows and columns, any Latin square can be brought to a form where the first row and the first column of its matrix are equal to the vector $(1, 2, \ldots, n)$. Such a Latin square is called normalized.

1.4.2 Orthogonal Latin squares

Latin squares

$$L_n = [s_1, \ldots, s_n], \qquad L'_n = [s'_1, \ldots, s'_n]$$

are called orthogonal if for any $(i, j) \neq (k, l)$

$$(s_i(j), s'_i(j)) \neq (s_k(l), s'_k(l)). \tag{1.4.1}$$

If Latin squares L_n and L'_n are orthogonal we write $L_n \perp L'_n$. If $L_n = [s_i(j)]$, $L'_n = [s'_i(j)]$, $i, j = 1, \ldots, n$, and $L_n \perp L'_n$, then the matrix $[(s_i(j), s'_i(j))]$, $i, j = 1, \ldots, n$, has n^2 mutually different elements.

Let us consider the properties of pairs of Latin squares given in (Mann, 1942; Mann, 1943).

Proposition 1.4.1 *Latin squares*

$$L_n = [s_1, \ldots, s_n], \qquad L'_n = [s'_1, \ldots, s'_n]$$

are orthogonal if and only if the substitutions

$$s_1^{-1} s'_1, \ldots, s_n^{-1} s'_n \tag{1.4.2}$$

form a Latin square.

Proof Suppose that $L_n \perp L'_n$ but the substitutions (1.4.2) do not form a

Latin square, that is, there exists v such that

$$s_i^{-1}s_i'(v) = s_k^{-1}s_k'(v), \qquad i \neq k. \tag{1.4.3}$$

Put $s_i^{-1}(v) = j$, $s_k^{-1}(v) = l$, then we obtain the equalities $s_i(j) = s_k(l)$ and $s_i'(j) = s_k'(l)$, which contradict the orthogonality of L_n and L_n' (see (1.4.1)).

Assume now that substitutions (1.4.2) form a Latin square, that is, equality (1.4.3) does not hold for any i, k and v. Suppose that L_n is not orthogonal to L_n', that is, for some $(i,j) \neq (k,l)$ the equality $(s_i(j), s_i'(j)) = (s_k(l), s_k'(l))$ is valid. Hence $s_i(j) = s_k(l)$, $s_i'(j) = s_k'(l)$. The first equality implies that there exists v such that $s_i^{-1}(v) = j$, $s_k^{-1}(v) = l$. Consequently, $s_i^{-1}s_i'(v) = s_k^{-1}s_k'(v)$, $i \neq k$, and we obtain a contradiction. □

We now define multiplication of a substitution of degree n by a Latin square of size n. If $L_n = [s_1, \ldots, s_n]$ is a Latin square, then

$$\varphi L_n = [\varphi s_1, \ldots, \varphi s_n], \qquad L_n\psi = [s_1\psi, \ldots, s_n\psi],$$

where φ and ψ are substitutions.

Proposition 1.4.2 *If $L_n \perp L_n'$, then $L_n\varphi \perp L_n'\psi$ for any substitutions φ and ψ of degree n.*

Proof By Proposition 1.4.1 the Latin squares $L_n\varphi$ and $L_n'\psi$ are orthogonal if and only if the substitutions

$$\varphi^{-1}s_1^{-1}s_1'\psi, \ldots, \varphi^{-1}s_n^{-1}s_n'\psi$$

form a Latin square. Suppose that there exists v such that

$$\varphi^{-1}s_i^{-1}s_i'\psi(v) = \varphi^{-1}s_k^{-1}s_k'\psi(v)$$

for some $i \neq k$. Hence it follows that

$$s_i^{-1}s_i^{-1}(\mu) = s_k^{-1}s_k'(\mu), \qquad i \neq k.$$

We obtain a contradiction, since for $L_n \perp L_n'$ substitutions (1.4.2) form a Latin square. □

A Latin square L_n is called seminormalized if it is of the form $L_n = [e, s_2, \ldots, s_n]$, where e is the identity substitution. Proposition 1.4.2 shows that any pair of orthogonal Latin squares can be brought to seminormalized forms. Indeed, if $L_n = [s_1, \ldots, s_n]$, $L_n' = [s_1', \ldots, s_n']$ and $L_n \perp L_n'$, then the Latin squares $L_n s_1^{-1}$ and $L_n'(s_1')^{-1}$ are seminormalized and orthogonal.

Let us now introduce the notion of the product of two Latin squares L_n and L'_n. If $L_n = [s_1, \ldots, s_n]$ and $L'_n = [s'_1, \ldots, s'_n]$, then, by definition, we put

$$L''_n = L_n \circ L'_n = [s_1 s'_1, \ldots, s_n s'_n].$$

Proposition 1.4.3 *Latin squares L_n and L'_n are orthogonal if and only if there exists a Latin square L''_n such that*

$$L'_n = L_n \circ L''_n. \tag{1.4.4}$$

Proof Indeed, if $L_n \perp L'_n$, then the Latin square $L''_n = [s_1^{-1}s'_1, \ldots, s_n^{-1}s'_n]$ satisfies condition (1.4.4). Conversely, if there exists a Latin square L''_n satisfying condition (1.4.4), then it is of the form given above and, consequently, $L_n \perp L'_n$. □

Proposition 1.4.3 can be generalized in the following obvious manner.

Proposition 1.4.4 *Latin squares $L_n^{(1)}, \ldots, L_n^{(r)}$ are pairwise orthogonal if and only if there exist $r(r-1)/2$ Latin squares $L_n^{(i,j)}$ such that for $i < j$*

$$L_n^{(i)} = L_n^{(j)} \circ L_n^{(i,j)}.$$

Let us now consider a class of Latin squares based on a group. A Latin square $L_n = [s_1, \ldots, s_n]$ is based on a group of substitutions G if $G = \{s_1, \ldots, s_n\}$, where the elements of the group are written without repetitions.

Proposition 1.4.5 *If Latin squares L_n and L'_n are based on a group G and are orthogonal, then the Latin square L''_n determined by the equality $L'_n = L_n \circ L''_n$ is also based on the group G.*

Proof Indeed, $L''_n = [s_1^{-1}s'_1, \ldots, s_n^{-1}s'_n]$ and $G = \{s_1^{-1}s'_1, \ldots, s_n^{-1}s'_n\}$, therefore any repetition contradicts the fact that L''_n is a Latin square. □

A maximal set of pairwise orthogonal Latin squares of order n is called complete.

Proposition 1.4.6 *A complete set of pairwise orthogonal Latin squares of order n contains no more than $n-1$ elements.*

Proof Consider a complete set of Latin squares $L_n^{(1)}, \ldots, L_n^{(r)}$. We may assume that all the Latin squares are seminormalized, since this operation preserves the orthogonality. Let us join the matrices of the Latin squares. In the matrix obtained the entry $(1, i)$ contains r elements i. In the entry $(2, i)$ all the r elements are different. Indeed, if two elements from this entry are equal to j, then the entry $(1, j)$ contains a pair of elements j, which contradicts orthogonality. None of the different elements of the entry $(2, 1)$ can coincide with the elements of the entry $(1, 1)$. Therefore, their number is no greater than $n - 1$. Thus, $r \leqslant n - 1$. □

1.4.3 The Bose–Stevens method

The Bose–Stevens method of constructing a complete set of mutually orthogonal Latin squares will be given in the proof of the following theorem.

Theorem 1.4.1 *Let $n = p^\alpha$, where p is a prime and α is a natural number. For $n \geqslant 3$ there exists a complete set of $n - 1$ mutually orthogonal Latin squares of order n.*

Proof Let $\mathrm{GF}(p^\alpha) = \{a_0 = 0, a_1 = 1, a_2, \ldots, a_n\}$. Consider $n - 1$ matrices of the form

$$L_n^{(l)} = [a_{ij}^{(l)}], \quad i, j = 0, 1, \ldots, n-1, \quad l = 1, \ldots, n-1,$$

where $a_{ij}^{(l)} = a_i a_l + a_j$. Each of the matrices determines a Latin square. Indeed, the equality $a_{ij}^{(l)} = a_{ij'}^{(l)}$ for $j \neq j'$ implies the contradiction $a_j = a_{j'}$. Similarly, the equality $a_{ij}^{(l)} = a_{i'j}^{(l)}$ for $i \neq i'$ implies that $(a_i - a_{i'})a_l = 0$. Since $a_l \neq 0$, we again obtain the equality $a_i = a_{i'}$, which is a contradiction.

Let us now prove that $L_n^{(k)}$ and $L_n^{(l)}$ are orthogonal for $k \neq l$. Suppose that

$$(a_{ij}^{(k)}, a_{ij}^{(l)}) = (a_{i'j'}^{(k)}, a_{i'j'}^{(l)})$$

for some $(i, j) \neq (i', j')$. Hence we obtain the system of equalities

$$a_i a_k + a_j = a_{i'} a_k + a_{j'},$$
$$a_i a_l + a_j = a_{i'} a_l + a_{j'}.$$

Subtracting the second equality from the first we find that

$$a_i(a_k - a_l) = a_{i'}(a_k - a_l),$$

and since $a_k - a_l \neq 0$ we obtain that $a_i = a_{i'}$, that is, $i = i'$. Now from the first equality of the system we obtain that $a_j = a_{j'}$, that is, $j = j'$. Thus we obtain a contradiction to the inequality $(i, j) \neq (i', j')$. □

For example, if $n = 3$, then, for $l = 1, 2$,

$$a_{00}^{(l)} = 0 \cdot l + 0, \quad a_{01}^{(l)} = 0 \cdot l + 1, \quad a_{02}^{(l)} = 0 \cdot l + 2,$$
$$a_{10}^{(l)} = 1 \cdot l + 0, \quad a_{11}^{(l)} = 1 \cdot l + 1, \quad a_{12}^{(l)} = 1 \cdot l + 2,$$
$$a_{20}^{(l)} = 2 \cdot l + 0, \quad a_{21}^{(l)} = 2 \cdot l + 1, \quad a_{22}^{(l)} = 2 \cdot l + 2.$$

Thus we obtain a complete set consisting of two orthogonal Latin squares.

1.5 (v, k, λ)-configurations

1.5.1 Definition and properties of (v, k, λ)-configurations

A family of subsets $X_1, \ldots, X_v$ of a set $X = \{x_1, \ldots, x_n\}$ forms a (v, k, λ)-configuration if $0 < \lambda < k < v - 1$ and

$$|X_i \cap X_j| = \begin{cases} k, & i = j, \\ \lambda, & i \neq j. \end{cases} \tag{1.5.1}$$

The subsets $X_1, \ldots, X_v$ are called blocks. The incidence matrix of a (v, k, λ)-configuration is the $(0, 1)$-matrix $A = \|a_{ij}\|$, $i, j = 1, \ldots, v$, where

$$a_{ij} = \begin{cases} 1, & x_j \in X_i, \\ 0, & x_j \notin X_i. \end{cases}$$

Denote by I the $v \times v$ identity matrix, and by J the matrix of the same size whose elements are all equal to 1. We list some properties of the incidence matrices of (v, k, λ)-configurations:

(1)
$$AJ = kJ. \tag{1.5.2}$$

This property expresses the obvious condition

$$\sum_{j=1}^{v} a_{ij} = k, \qquad i = 1, \ldots, v,$$

which is equivalent to $|X_i| = k$, $i = 1, \ldots, v$.

(2)
$$AA^T = \lambda J + (k - \lambda)I. \tag{1.5.3}$$

This property follows from equality (1.5.1), since

$$\sum_{i=1}^{v} a_{il} a_{jl} = |X_i \cap X_j| = \begin{cases} k, & i = j, \\ \lambda, & i \neq j. \end{cases} \tag{1.5.4}$$

The last equation means that

$$AA^T = \begin{bmatrix} k & \lambda & \dots & \lambda \\ \lambda & k & \dots & \lambda \\ \vdots & \vdots & \ddots & \vdots \\ \lambda & \lambda & \dots & k \end{bmatrix} = \lambda J + (k-\lambda)I. \qquad (1.5.5)$$

(3) $$\det(AA^T) = (\det A)^2 = (k + \lambda(v-1))(k-\lambda)^{v-1}. \qquad (1.5.6)$$

For a matrix of the form (1.5.5), we subtract the first column from each of the others, and add to the first row of the matrix obtained all the other rows. As a result we obtain a triangular matrix whose elements of the principle diagonal give the last relation from (1.5.6).

(4) The parameters of a (v, k, λ)-configuration are related by the equality

$$k(k-1) = \lambda(v-1). \qquad (1.5.7)$$

Multiplying both sides of equality (1.5.3) by J from the right, we find that

$$AA^T J = (k + \lambda(v-1))J. \qquad (1.5.8)$$

Since $\lambda < k$, it follows from (1.5.6) that the matrix A is non-singular and (1.5.2) implies the equality

$$A^{-1}J = \frac{1}{k}J.$$

Now from (1.5.8) we find that

$$A^T J = \frac{1}{k}(k + \lambda(v-1))J. \qquad (1.5.9)$$

Transposing both sides of this equality and multiplying by J from the right, we obtain

$$JAJ = \frac{1}{k}(k + \lambda(v-1))J.$$

On the other hand, it follows from (1.5.2) that

$$JAJ = kvJ.$$

The last two equalities imply the equality

$$\frac{v}{k}(k + \lambda(v-1)) = kv, \qquad (1.5.10)$$

which is equivalent to (1.5.7).

(5) $$JA = kJ. \tag{1.5.11}$$

This property expresses the fact that the number of subsets from the family $X_1, \ldots, X_v$ containing the element x_j is equal to k, that is,

$$\sum_{i=1}^{v} a_{ij} = k, \qquad j = 1, \ldots, v. \tag{1.5.12}$$

The validity of (1.5.11) now follows from relations (1.5.9) and (1.5.10).

(6) $$AA^T = A^T A. \tag{1.5.13}$$

The matrices which possess this property are called normal. The proof of (1.5.13) can be carried out as follows:

$$\begin{aligned} A^T A &= A^{-1} A A^T A = A^{-1}(\lambda J + (k - \lambda) I) A \\ &= \frac{\lambda}{k} JA + (k - \lambda) I = \lambda J + (k - \lambda) I = AA^T. \end{aligned}$$

Taking (1.5.3) and (1.5.4) into account, we obtain from (1.5.13) that

$$\sum_{i=1}^{v} a_{ij} a_{il} = \begin{cases} k, & j = l, \\ \lambda, & j \neq l. \end{cases} \tag{1.5.14}$$

This property means that any pair of different elements x_i and x_j belongs exactly to λ blocks from the family $X_1, \ldots, X_v$.

The properties of a (v, k, λ)-configuration considered above allow us to give an equivalent definition of this notion.

A family of subsets $X_1, \ldots, X_v$ of a set $X = \{x_1, \ldots, x_v\}$ forms a (v, k, λ)-configuration if $0 < \lambda < k < v - 1$ and each element of the set X belongs to exactly k subsets of the family and each pair of elements from X belongs to exactly λ subsets of the family.

This definition is useful for clarifying a connection between the notion of a (v, k, λ)-configuration and the more general notion of a (b, v, r, k, λ)-configuration which is considered in Section 1.7.

1.5.2 The Bruck–Chowla–Ryser theorem

Let A be the incidence matrix of a (v, k, λ)-configuration. Each block X_i can be associated with the linear form

$$F_i = F_i(u_1, \ldots, u_v) = \sum_{j=1}^{v} a_{ij} u_j, \qquad i = 1, \ldots, v.$$

Note that

$$Q_v = \sum_{i=1}^{v} F_i^2 = \sum_{j=1}^{v} u_j^2 \sum_{i=1}^{v} a_{ij} + \sum_{j\neq l} u_j u_l \sum_{i=1}^{v} a_{ij}a_{il}.$$

Hence, taking equalities (1.5.12) and (1.5.14) into account, we obtain

$$Q_v = \sum_{i=1}^{v} F_i^2 = (k-\lambda)\sum_{j=1}^{v} u_j^2 + \lambda\left(\sum_{j=1}^{v} u_j\right)^2. \tag{1.5.15}$$

Before turning to the Bruck–Chowla–Ryser theorem, which gives the necessary conditions of existence of (v,k,λ)-configurations, we present two lemmas needed for its proof.

Lemma 1.5.1 *Let c_1, c_2, c_3, c_4 be rational numbers, $c_1^2+c_2^2+c_3^2+c_4^2 \neq 0$, and let variables u_1, u_2, u_3, u_4 and w_1, w_2, w_3, w_4 satisfy the relation*

$$\begin{bmatrix} w_1 \\ w_2 \\ w_3 \\ w_4 \end{bmatrix} = \begin{bmatrix} c_1 & -c_2 & -c_3 & -c_4 \\ c_2 & c_1 & -c_4 & c_3 \\ c_3 & c_4 & c_1 & -c_2 \\ c_4 & -c_3 & c_2 & c_1 \end{bmatrix} \cdot \begin{bmatrix} u_1 \\ u_2 \\ u_3 \\ u_4 \end{bmatrix}. \tag{1.5.16}$$

Then

$$w_1^2+w_2^2+w_3^2+w_4^2 = (c_1^2+c_2^2+c_3^2+c_4^2)(u_1^2+u_2^2+u_3^2+u_4^2). \tag{1.5.17}$$

The variables u_1, u_2, u_3, u_4 are represented as a linear combination of the variables w_1, w_2, w_3, w_4 with rational coefficients.

Proof Equality (1.5.16) can be rewritten in vector form as follows:

$$\boldsymbol{w} = C\boldsymbol{u},$$

where $\boldsymbol{w}$ and $\boldsymbol{u}$ are the corresponding vectors and C is the corresponding matrix. Hence it follows that

$$\boldsymbol{w}^T\boldsymbol{w} = \boldsymbol{u}^T C^T C \boldsymbol{u}.$$

It is easy to check that

$$C^T C = \sum_{i=1}^{4} c_i^2 I,$$

where I is the identity matrix. Consequently,

$$\sum_{i=1}^{4} w_i^2 = \sum_{i=1}^{4} c_i^2 \sum_{i=1}^{4} u_i^2.$$

A direct calculation shows that

$$\det C = (c_1^2 + c_2^2 + c_3^2 + c_4^2)^2.$$

This equality means that the coefficients in the linear representation of the variables u_1, u_2, u_3, u_4 in terms of w_1, w_2, w_3, w_4 are rational, since their numerators are obtained from c_1, c_2, c_3, c_4 by additions and multiplications and the denominators are equal to $\det C$.

The lemma is proved. □

We cite without proof the following assertion from number theory due to Lagrange.

Lemma 1.5.2 *Any natural number n can be represented as a sum*

$$n = c_1^2 + c_2^2 + c_3^2 + c_4^2 \tag{1.5.18}$$

of four squares of non-negative integers.

Now we formulate and prove the Bruck–Chowla–Ryser theorem.

Theorem 1.5.1 *For the existence of a (v,k,λ)-configuration, it is necessary that*

(a) *$k - \lambda$ is a square of a natural number if v is even;*
(b) *the equation*

$$z^2 = (k-\lambda)x^2 + (-1)^{(v-1)/2}\lambda y^2 \tag{1.5.19}$$

has a solution in integers such that $x^2 + y^2 + z^2 \neq 0$ if v is odd.

Proof It follows from (1.5.6) and (1.5.7) that

$$(\det A)^2 = k^2(k-\lambda)^{v-1}.$$

Hence it follows that $(k-\lambda)^{v-1}$ is a square, and, if v is even, then $k-\lambda$ must be a square, too.

We turn to the case where v is odd. First let $v \equiv 1 \pmod 4$. Divide the variables $u_1, \ldots, u_{v-1}$ into sequential groups with four elements in each and introduce the variables $w_1, \ldots, w_{v-1}$ by the equalities

$$\boldsymbol{W}_j = C\boldsymbol{U}_j \qquad j = 1, 5, 9, \ldots, v-4, \tag{1.5.20}$$

where

$$W_j^T = (w_j, w_{j+1}, w_{j+2}, w_{j+3}), \qquad U_j^T = (u_j, u_{j+1}, u_{j+2}, u_{j+3}).$$

The matrix C is of the same form as in (1.5.16) and the numbers c_1, c_2, c_3, c_4 are defined such that $k-\lambda = c_1^2+c_2^2+c_3^2+c_4^2$. By Lemma 1.5.2 the following equalities of type (1.5.17) are valid:

$$w_j + \cdots + w_{j+3} = (k-\lambda)(u_j + \cdots + u_{j+3}), \qquad j = 1, 5, \ldots, v-4,$$

and, according to Lemma 1.5.1, the variables $u_j, u_{j+1}, u_{j+2}, u_{j+3}$ are linear combinations of $w_j, w_{j+1}, w_{j+2}, w_{j+3}$ with rational coefficients.

It follows from (1.5.15) that

$$\sum_{i=1}^{v} F_i^2 = \sum_{j=1}^{v-1} w_j^2 + (k-\lambda)w_v^2 + \lambda w^2, \tag{1.5.21}$$

where $w_v = u_v$ and $w = u_1 + \cdots + u_v$. Note that since $u_1, \ldots, u_v, w$ are linear combinations of $w_1, \ldots, w_v$ with rational coefficients, equation (1.5.21) is an identity with rational coefficients with respect to the variables $w_1, \ldots, w_v$. Let

$$F_1 = d_{11}w_1 + \cdots + d_{1v}w_v,$$

where $d_{11}, \ldots, d_{1v}$ are rational numbers. We put $F_1 = -w_1$ if $d_{11} = 1$, and $F_1 = w_1$ if $d_{11} \neq 1$. In both the cases we obtain the relation from which the variable w_1 is expressed as a linear combination of $w_2, \ldots, w_v$ with rational coefficients, and $F_i^2 = w_1^2$. Now from (1.5.21) we obtain the following rational identity with independent variables $w_2, \ldots, w_v$:

$$\sum_{i=2}^{v} F_i^2 = \sum_{j=2}^{v-1} w_j^2 + (k-\lambda)w_v^2 + \lambda w^2.$$

Continuing in a similar way and putting $F_2 = \pm w_2, \ldots, F_{v-1} = \pm w_{v-1}$ sequentially, we obtain the identity

$$F_v^2 = (k-\lambda)w_v^2 + \lambda w^2, \tag{1.5.22}$$

where F_v and w are proportional to w_v with rational coefficients, and w_v is an independent variable. Now taking a non-zero integer as w_v and multiplying both sides of (1.5.15) by the denominators of F_v and w, we obtain an equality of the form

$$z^2 = (k-\lambda)x^2 + \lambda y^2, \tag{1.5.23}$$

where $x \neq 0$.

Now consider the case where $v \equiv 3 \pmod 4$. If we introduce a new variable u_{v+1} and add $(k-\lambda)u_{v+1}^2$ to both sides of (1.5.21), then we obtain the equality

$$\sum_{i=1}^{v} F_i^2 + (k-\lambda)u_{v+1}^2 = (k-\lambda)\sum_{j=1}^{v+1} uj^2 + \lambda\left(\sum_{j=1}^{v} u_j\right)^2. \quad (1.5.24)$$

Dividing the summands in the right-hand side of (1.5.4) into groups consisting of four terms each and applying transformation (1.5.20), we obtain the following identity with rational coefficients in variables $w_1, \ldots, w_{v+1}$:

$$\sum_{i=1}^{v} F_i^2 + (k-\lambda)u_{v+1} = \sum_{j=1}^{v+1} w_j^2 + \lambda w^2,$$

where $w = u_1 + \cdots + u_v$. Reasoning as in the case where $v \equiv 1 \pmod 4$, we obtain the following identity with an independent variable w_{v+1}:

$$(k-\lambda)u_{v+1}^2 = w_{v+1}^2 + \lambda w^2, \quad (1.5.25)$$

where u_{v+1} and w are proportional to w_{v+1} with rational coefficients. Taking a non-zero integer as w_{v+1} and multiplying both sides of (1.5.25) by the denominators of u_{v+1} and w, we obtain the equality

$$z^2 = (k-\lambda)x^2 - \lambda y^2. \quad (1.5.26)$$

Uniting equalities (1.5.23) and (1.5.26), we obtain (1.5.19). The theorem is proved. □

Applying part (a) of the Bruck–Chowla–Ryser theorem, we see that there is no (v,k,λ)-configuration for $v = 44$, $k = 5$ and $\lambda = 2$, since $k - \lambda = 3$ is not the square of an integer. Similarly, there is no (v,k,λ)-configuration with $v = 43$, $k = 7$ and $\lambda = 1$ according to part (b) of the theorem, since the equation $z^2 = 6x^2 - y^2$ has no integer non-zero solution.

We now consider some particular cases of (v,k,λ)-configurations.

1.5.3 Perfect difference sets

Let $D = \{d_1, \ldots, d_k\}$ be a set of different, smallest, non-negative residues modulo v which satisfy the condition that for any $a \not\equiv 0 \pmod v$ there exist exactly λ ordered pairs (d_i, d_j), $i \neq j$, such that $d_i - d_j \equiv a \pmod v$, where $0 < \lambda < k < v - 1$. Such a set D is called a perfect difference set. Let

$$D_l = \{d_1 + l, \ldots, d_k + l\}, \qquad l = 0, 1, \ldots, v - 1,$$

where the addition is carried out modulo v.

Let us show that $D_0, D_1, \ldots, D_{v-1}$ as subsets of $D = \{0, 1, \ldots, v - 1\}$ form a (v, k, λ)-configuration. To this end it is sufficient to show that

$$|D_i \cap D_j| = \begin{cases} k, & i = j, \\ \lambda, & i \neq j. \end{cases}$$

For $i \neq j$

$$\begin{aligned} |D_i \cap D_j| &= |\{d_1 + i, \ldots, d_k + i\} \cap \{d_1 + j, \ldots, d_k + j\}| \\ &= |\{(d_s, d_l) \mid d_s + i \equiv d_l + j \pmod{v}\}| \\ &= |\{(d_s, d_l) \mid d_s - d_l \equiv a \pmod{v},\ a = j - i\}| = \lambda. \end{aligned}$$

It follows from the above-cited properties of (v, k, λ)-configurations that the parameters of any perfect difference set are related by the equality $k(k-1) = \lambda(v-1)$.

A matrix $A = \|a_{ij}\|$, $i, j = 1, \ldots, n$, is called cyclic if any of its rows is obtained from the preceding row by a cyclic shift of one step to the right, that is, $a_{i+1,j+1} = a_{ij}$, where the addition of indices is carried out modulo n. It is not difficult to see that the incidence matrix of the (v, k, λ)-configuration corresponding to a perfect difference set is a cyclic matrix.

It is easy to check that $D = \{1, 2, 4\}$ is a perfect difference set modulo 7. The corresponding (v, k, λ)-configuration with parameters $v = 7$, $k = 3$ and $\lambda = 1$ is of the form

$$\begin{array}{llll} \{1,2,4\}, & \{2,3,5\}, & \{3,4,6\}, & \\ \{4,5,0\}, & \{5,6,1\}, & \{6,0,2\}, & \{0,1,3\}. \end{array}$$

1.5.4 Hadamard matrices and Hadamard configurations

An $n \times n$ square matrix H is called a Hadamard matrix if its elements take values 1 and -1 and

$$HH^T = nI, \tag{1.5.27}$$

where I is the identity matrix and T is the symbol of transposition. It is clear that if $H_1, \ldots, H_n$ are the row-vectors of the matrix H, then, according to (1.5.27), the scalar product of the two rows is equal to

$$(H_i, H_j) = \begin{cases} n, & i = j, \\ 0, & i \neq j, \end{cases} \tag{1.5.28}$$

that is, the rows of the matrix H are orthogonal. It follows from equality (1.5.27) that the matrix H is non-singular, since

$$\det(HH^T) = (\det H)^2 = n^2.$$

The matrix H is normal; indeed,

$$H^T H = H^{-1} H H^T H = H^{-1}(nI)H = nI = HH^T.$$

A Hadamard matrix is transformed into another by transposition of the rows and columns and by multiplication of any column or row by -1. Indeed, if $H = \|h_{ij}\|$, $i, j = 1, \ldots, n$, then (1.5.28) is equivalent to the equalities

$$\sum_{l=1}^{n} h_{il} h_{jl} = \begin{cases} n, & i = j, \\ 0, & i \neq j. \end{cases}$$

It is clear that these equalities are invariant with respect to the above-mentioned operations, by which the matrix H can be transformed to the normalized form with the first row and the first column consisting of positive units.

Theorem 1.5.2 *An $n \times n$ Hadamard matrix can exist only if $n = 1, 2$ or $n \equiv 0 \pmod 4$.*

Proof For $n = 1, 2$ the Hadamard matrices are

$$[1] \quad \text{and} \quad \begin{bmatrix} 1 & 1 \\ 1 & -1 \end{bmatrix},$$

and the theorem is true.

Consider a normalized Hadamard matrix with $n \geqslant 3$. The columns of the first three rows can have one of the following forms:

$$\begin{bmatrix} 1 \\ 1 \\ 1 \end{bmatrix}, \quad \begin{bmatrix} 1 \\ 1 \\ -1 \end{bmatrix}, \quad \begin{bmatrix} 1 \\ -1 \\ 1 \end{bmatrix}, \quad \begin{bmatrix} 1 \\ -1 \\ -1 \end{bmatrix}.$$

Assume that the numbers of columns of these four types in the first three rows are equal to x, y, z, w respectively. Then taking into account that the total number of the columns is n and that the rows are orthogonal, we obtain the system of equations

$$x+y+z+w=n, \qquad x-y+z-w=0,$$
$$x+y-z-w=0, \qquad x-y-z+w=0.$$

From this system we find that

$$x=y=z=w=n/4.$$

Thus $n=4\mu$, where μ is a natural number, and the theorem is proved. □

We now define the Kronecker product of two square matrices. If $A=\|a_{ij}\|$, $i,j=1,\dots,n$, and $B=\|b_{ij}\|$, $i,j=1,\dots,m$, then the Kronecker product $A\otimes B$ of the matrices A and B is the $mn\times mn$ square matrix

$$A\otimes B=\begin{bmatrix} a_{11}B & a_{12}B & \dots & a_{1n}B \\ a_{21}B & a_{22}B & \dots & a_{2n}B \\ \vdots & \vdots & \ddots & \vdots \\ a_{n1}B & a_{n2}B & \dots & a_{nn}B \end{bmatrix}.$$

In other words,

$$C=A\otimes B=\|c_{s_i,s_j}\|, \qquad i,j=1,\dots,mn,$$

where

$$c_{s_i,s_j}=a_{i_1j_1}b_{i_2j_2}, \quad s_i=(i_1,j_1), \quad s_j=(j_1,j_2)$$

and $s_1,s_2,\dots,s_{mn}$ are pairs of the form (i,j), with $i=1,\dots,n$, $j=1,\dots,m$, arranged in lexicographical order:

$$((1,1),\dots,(1,m),(2,1),\dots,(2,m),\dots,(n,1),\dots,(n,m)).$$

The Kronecker product of matrices possesses the following properties:

(1) If T is the symbol of transposition, then

$$(A\otimes B)^T=A^T\otimes B^T.$$

(2) If A and B are $n\times n$ matrices and C is an $m\times m$ matrix, then

$$(A+B)\otimes C=A\otimes C+B\otimes C.$$

These follow directly from the definition of the Kronecker product.

(3) If A and B are $n \times n$ matrices and C and D are $m \times m$ matrices, then

$$(AB) \otimes (CD) = (A \otimes C)(B \otimes D).$$

The validity of this property follows from the equality

$$\sum_{\mu} a_{i_1\mu} b_{\mu j_1} \sum_{\nu} c_{i_2\nu} d_{\nu j_2} = \sum_{\mu,\nu} a_{i_1\mu} c_{i_2\nu} b_{\mu j_1} d_{\nu j_2}.$$

(4) The Kronecker product of two Hadamard matrices is again a Hadamard matrix. Indeed, if H and H' are $n \times n$ and $m \times m$ Hadamard matrices, respectively, and I_n is the $n \times n$ identity matrix, then

$$\begin{aligned}(H \otimes H')(H \otimes H')^T &= (H \otimes H')(H^T \otimes H'^T) \\ &= HH^T \otimes H'H'^T = nI_n \otimes mT_m \\ &= mnI_{mn}.\end{aligned}$$

Theorem 1.5.3 *For any natural number α there exists an $n \times n$ Hadamard matrix with $n = 2^\alpha$.*

Proof Indeed, the matrix

$$H_2 = \begin{bmatrix} 1 & 1 \\ 1 & -1 \end{bmatrix}$$

is a Hadamard matrix. According to Property 4 the matrix

$$H_4 = H_2 \otimes H_2 = \begin{bmatrix} 1 & 1 & 1 & 1 \\ 1 & -1 & 1 & -1 \\ 1 & 1 & -1 & -1 \\ 1 & -1 & -1 & 1 \end{bmatrix}$$

is also a Hadamard matrix. The product

$$H_8 = H_2 \otimes H_4 = H_2 \otimes H_2 \otimes H_2$$

is again a Hadamard matrix. In general, the product

$$H_{2^\alpha} = H_2 \otimes \cdots \otimes H_2$$

with α factors is a Hadamard matrix. □

Theorem 1.5.4 *A normalized $n \times n$ Hadamard matrix with $n = 4\mu$ is equivalent to a (v, k, λ)-configuration with parameters*

$$v = 4\mu, \quad k = 2\mu - 1, \quad \lambda = \mu - 1.$$

Proof Let us erase the first row and the first column from a normalized Hadamard matrix H and substitute 0 for -1. We thereby obtain a $(0,1)$-matrix A of order $v = 4\mu - 1$, each row of which contains exactly $2\mu - 1$ units, since the initial Hadamard matrix contained 2μ positive units in each row. Any two rows of the matrix A contain $\mu - 1$ units in common columns, since any two rows of the matrix H have μ positive units in the same columns. These properties show that the matrix A is the incidence matrix of a (v, k, λ)-configuration with $v = 4\mu - 1$, $k = 2\mu - 1$ and $\lambda = \mu - 1$. This configuration is called the Hadamard configuration.

Beginning with the incidence matrix of a (v, k, λ)-configuration with the parameters given above and reasoning in reverse order, we can construct a normalized $n \times n$ Hadamard matrix with $n = 4\mu$. □

1.6 Finite projective planes

1.6.1 Main properties of finite projective planes

A finite projective plane of order n is a (v, k, λ)-configuration with

$$v = n^2 + n + 1, \quad k = n + 1, \quad \lambda = 1. \tag{1.6.1}$$

As an example we give the finite projective plane

$$X_1 = \{1,2,4\}, \quad X_2 = \{2,3,5\}, \quad X_3 = \{3,4,6\},$$
$$X_4 = \{4,5,7\}, \quad X_5 = \{5,6,1\}, \quad X_6 = \{6,7,2\},$$
$$X_7 = \{7,1,3\}.$$

Geometric language is customary in the description of finite projective planes. The elements of the set X are called points, and the blocks $X_1, \ldots, X_v \subseteq X$ are lines. If $x_j \in X_i$, then we say that the point x_j lies on the line X_i or that the line X_i goes through the point x_j. The inclusion $x_j \in X_i \cap X_k$ means that the lines X_i and X_k intersect in the point x_j.

This terminology lets us reformulate the properties of (v, k, λ)-configurations for the case of finite projective planes as follows:

(1) Each line goes through $n + 1$ points.
(2) Two lines intersect at one point.
(3) Each point is an intersection of $n + 1$ lines.
(4) Only one line goes through two given points.

Properties 2 and 4 correspond to the properties of points and lines on the ordinary plane and justify the use of geometric terminology.

Let us give the necessary conditions of existence of a finite projective plane of order n, known as the Bruck–Ryser theorem.

Theorem 1.6.1 *For the existence of a finite projective plane of order n with* $n \equiv 1,\ 2 \pmod 4$ *it is necessary that there exist integers a and b which satisfy the equation*

$$n = a^2 + b^2. \tag{1.6.2}$$

Proof In fact, this theorem is a corollary of the Bruck–Chowla–Ryser theorem, where $v = n_2+n+1$, $k = n+1$, $n = k-\lambda$. The value $t = (v-1)/2$ is an integer, therefore $v = 2t + 1$ is even. According to Theorem 1.5.1 there exist integers x, y, z such that $x^2 + y^2 + z^2 \neq 0$ and

$$z^2 = nx^2 + (-1)^{n(n+1)/2}y^2. \tag{1.6.3}$$

First let $n \equiv 0,\ 3 \pmod 4$. Then $n(n+1)/2$ is even and

$$nx^2 = z^2 - y^2. \tag{1.6.4}$$

If $n \equiv 3 \pmod 4$, then the triple

$$x = 1, \quad y = (n-1)/2, \quad z = (n+1)/2$$

is an integer-valued solution of equation (1.6.4). If $n \equiv 0 \pmod 4$, then equation (1.6.4) has the integer-valued solution

$$x = 1, \quad y = (n-4)/4, \quad z = (n+4)/4.$$

Now let $n \equiv 1,\ 2 \pmod 4$. In this case $n(n+1)/2$ is odd, and equation (1.6.3) takes the form

$$nx^2 = z^2 + y^2. \tag{1.6.5}$$

It is known from number theory that for the existence of an integer solution of equation (1.6.5) it is necessary that there exist integers a and b such that $n = a^2 + b^2$. Thus the theorem is proved. □

1.6.2 Orthogonal Latin squares and finite projective planes

Consider the matrix

$$C = \|c_{ij}\|, \quad i = 1,\ldots,n^2, \quad j = 1,\ldots,t+2,$$

where c_{ij} take values $1,\ldots,n$ and

$$(c_{ik}, c_{il}) \neq (c_{jk}, c_{jl}), \quad i \neq j, \quad k \neq l. \tag{1.6.6}$$

Condition (1.6.6) means that the rows of each $n^2 \times 2$ submatrix of C represent n^2 different pairs constructed from the numbers $1,\dots,n$.

Rearrange the rows of C in such a way that the row elements of the $n^2 \times 2$ submatrix corresponding to the first two columns follow the lexicographical order

$$(1,1),\dots,(1,n),(2,1)\dots,(2,n),\dots,(n,1),\dots,(n,n).$$

These two columns of C will be called standard and the matrix obtained will be called the standard matrix. Note that transpositions of rows of the matrix C preserve the main property (1.6.6).

We now consider sets of t mutually orthogonal Latin squares of order n with substitutions acting on the set $\{1,\dots,n\}$.

The following lemma will be used later in this section.

Lemma 1.6.1 *For $n \geqslant 3$, $t \geqslant 2$ a set $\mathcal{M}_{nt}$ consisting of t pairwise orthogonal Latin squares of order n is equivalent to an $n^2 \times (t+2)$ matrix C of the form* (1.6.6), *that is, the matrix can be constructed for a given set $\mathcal{M}_{nt}$ and vice versa.*

Proof Suppose we are given a set $\mathcal{M}_{nt}$ consisting of t pairwise orthogonal Latin squares $L_n^{(1)},\dots,L_n^{(t)}$. For example, if $n=3$, $t=2$, then

$$L_3^{(1)} = \begin{bmatrix} 1 & 2 & 3 \\ 2 & 3 & 1 \\ 3 & 1 & 2 \end{bmatrix}, \qquad L_3^{(2)} = \begin{bmatrix} 1 & 2 & 3 \\ 3 & 1 & 2 \\ 2 & 3 & 1 \end{bmatrix}. \tag{1.6.7}$$

We construct the matrix $C = \|c_{ij}\|$, $i = 1,\dots,n^2$, $j = 1,\dots,t+2$, corresponding to the set $\mathcal{M}_{nt}$ in the following way. First we construct the $n^2 \times 2$ submatrix consisting of the first two standard columns, whose rows are the lexicographically ordered pairs of numbers $1,\dots,n$. Further, the lth column of the matrix is obtained by sequentially placing the rows of the Latin square $L_n^{(l-2)}$ from top to bottom, $l = 3,4,\dots,t+2$. For the

Latin squares (1.6.7) the matrix is of the form

$$C = \begin{bmatrix} 1 & 1 & 1 & 1 \\ 1 & 2 & 2 & 2 \\ 1 & 3 & 3 & 3 \\ 2 & 1 & 2 & 3 \\ 2 & 2 & 3 & 1 \\ 2 & 3 & 1 & 2 \\ 3 & 1 & 3 & 2 \\ 3 & 2 & 1 & 3 \\ 3 & 3 & 2 & 1 \end{bmatrix}. \tag{1.6.8}$$

The matrix C obtained by this procedure satisfies property (1.6.6). Indeed, the equality $(c_{i1}, c_{i2}) = (c_{j1}, c_{j2})$, $i \neq j$, contradicts the property of the rows of the standard columns. The equalities

$$(c_{i1}, c_{ik}) = (c_{j1}, c_{jk}), \quad (c_{i2}, c_{ik}) = (c_{j2}, c_{jk}), \quad i \neq j, \quad k = 3, \ldots, t+2,$$

contradict that property of Latin squares which states that there are no repetitions of elements in the rows and in the columns. Finally, the equalities $(c_{ik}, c_{il}) = (c_{jk}, c_{jl})$, $i \neq j$, $k = 3, \ldots, t+2$, contradict the orthogonal property of the initial Latin squares.

Suppose now we are given a matrix $C = \|c_{ij}\|$, $i = 1, \ldots, n^2$, $j = 1, \ldots, t+2$, with elements $1, \ldots, n$, satisfying condition (1.6.6). We bring the matrix to the standard form, preserving the main property (1.6.6). For $n = 3$, $t = 2$ the standard form of the matrix C is given by (1.6.8). Further, we associate the $(l+2)$th column of the standard form of the matrix C with the Latin square $L_n^{(l)}$ whose rows are taken as the sequential parts of length n of the column. The properties determining a Latin square are satisfied, since

$$(c_{ik}, c_{il}) \neq (c_{jk}, c_{jl}), \quad i \neq j, \quad l = 1, \ldots, t, \tag{1.6.9}$$

for $k = 1, 2$. The pairwise orthogonality of the Latin squares obtained, $L_n^{(1)}, \ldots, L_n^{(t)}$, follows from condition (1.6.9), which is valid for all $k \neq l$. The lemma is proved. □

Theorem 1.6.2 *If there exist sets $\mathcal{M}_{nt}$ and $\mathcal{M}_{mt}$ of t pairwise orthogonal Latin squares of orders n and m, respectively, then there exists a set $\mathcal{M}_{mn,t}$ of pairwise orthogonal Latin squares of order mn.*

Proof Let

$$C_1 = \|c_{ij}^{(1)}\|, \quad i = 1,\dots,n^2, \quad j = 1,\dots,t+2,$$
$$C_2 = \|c_{ij}^{(2)}\|, \quad i = 1,\dots,m^2, \quad j = 1,\dots,t+2,$$

be the standard matrices corresponding to the sets $\mathscr{M}_{nt}$ and $\mathscr{M}_{mt}$ of pairwise orthogonal Latin squares respectively. On the basis of the matrices C_1 and C_2 we construct the matrix C' with the following $(mn)^2$ rows:

$$[(c_{i1}^{(1)}, c_{j1}^{(2)}), (c_{i2}^{(1)}, c_{j2}^{(2)}), \dots, (c_{i,t+2}^{(1)}, c_{j,t+2}^{(2)})], \quad i = 1,\dots,n^2, \quad j = 1,\dots,m^2.$$

The matrix C' has $(mn)^2$ rows and $t+2$ columns, and its elements are mn ordered pairs

$$(1,1),\dots,(1,m),\dots,(n,1),\dots,(n,m). \tag{1.6.10}$$

This matrix possesses the main property (1.6.6). Indeed, let $C' = \|c'_{\alpha,\beta}\|$, where α runs the $(mn)^2$ ordered pairs of the form (i,j), $i = 1,\dots,n^2$, $j = 1,\dots,m^2$, and $\beta = 1,\dots,t+2$. Then the equality

$$(c'_{\alpha k}, c'_{\alpha l}) = (c'_{\alpha' k}, c'_{\alpha' l}), \quad \alpha \neq \alpha', \quad k \neq l,$$

for $\alpha = (i,j)$, $\alpha' = (p,q)$, means that

$$c_{ik}^{(1)} = c_{pq}^{(1)}, \qquad c_{il}^{(1)} = c_{pl}^{(1)},$$
$$c_{jk}^{(2)} = c_{qk}^{(2)}, \qquad c_{jl}^{(2)} = c_{ql}^{(2)}.$$

This contradicts property (1.6.6) of the matrices C_1 and C_2.

Now it follows from Lemma 1.6.1 that the matrix C' corresponds to a set of t pairwise orthogonal Latin squares of order mn and the substitutions corresponding to these Latin squares act on the set of ordered pairs (1.6.10). The theorem is proved. □

We presented above the Bose–Stevens method, which allows us to construct a complete set of $p^\alpha - 1$ pairwise orthogonal Latin squares of order p^α, where p is a prime number. Taking this assertion into account and applying Theorem 1.6.2 sequentially, we obtain the following assertion.

Corollary 1.6.1 *Let $n = p_1^{\alpha_1} p_2^{\alpha_2} \dots p_r^{\alpha_r}$ be the canonical expansion of n and $t = \min_{1 \leqslant i \leqslant r}(p_i^{\alpha_i} - 1)$. Then for $t \geqslant 2$ there exists a set of t pairwise orthogonal Latin squares of order n.*

Lemma 1.6.1, proved above, leads to the following result on the connection between the existence of finite projective planes of order n and the existence of complete sets of pairwise orthogonal Latin squares.

Theorem 1.6.3 *A projective plane of order $n \leqslant 3$ can be constructed if and only if there exists a complete set of $n-1$ pairwise orthogonal Latin squares of order n.*

Proof Assume we are given a projective plane of order n consisting of lines $X_1, \ldots, X_{n^2+n+1}$. Denote the points of the line X_1 by $x_1, \ldots, x_{n+1}$ and the remaining points of the plane by $y_1, \ldots, y_{n^2}$. Label by the numbers $1, 2, \ldots, n$ the lines going through the point x_1, then the lines going through x_2 and so on; finally we label the lines going through the point x_{n+1}. If c_{ij} is the label of the line going through the points y_i and x_j, then the matrix

$$C = \|c_{ij}\|, \quad i = 1, \ldots, n^2, \quad j = 1, \ldots, n+1,$$

possesses the main property given above:

$$(c_{ik}, c_{il}) \neq (c_{jk}, c_{jl}), \quad i \neq j, \quad k \neq l.$$

Indeed, suppose that $c_{ik} = c_{jk}$, $c_{il} = c_{jl}$. Then the first equality implies that the line going through the points y_i and x_k and the line going through the points y_j and x_k have the same label, and, consequently, the points y_i, y_j and x_k lie on one line. Similarly, the second equality implies that y_i, y_j and x_l lie on one line. But this means that y_i and y_j lie on the line containing x_k and x_l, that is, they lie on the line X_1, a contradiction. Thus, for a given finite projective plane of order n we can construct a matrix C possessing the main property (1.6.6) and, consequently, according to Lemma 1.6.1, we can construct a complete set of pairwise orthogonal Latin squares.

Now assume we are given a matrix $C = \|c_{ij}\|$, $i = 1, \ldots, n^2$, $j = 1, \ldots, n+1$, with elements taking values $1, \ldots, n$ and possessing the main property (1.6.6). Construct the projective plane corresponding to the matrix. Consider the line $X_1 = \{x_1, \ldots, x_{n+1}\}$ and let $y_1, \ldots, y_{n^2}$ be the points which do not lie on this line. Label with the number c_{ij} the line L_{ij} going through the points y_i and x_j. It follows from the labeling rule that each line contains exactly $n+1$ points, since the jth column of the matrix C has n identical elements. Moreover, the intersection of two lines has exactly one point. Indeed, suppose that

$$\{y_\mu, y_\nu\} \subseteq X_k \cap X_l, \quad \mu \neq \nu, \quad k \neq l.$$

Then there exist k' and l' such that $c_{\mu k'} = c_{\nu k'}$, $c_{\mu l'} = c_{\nu l'}$, $k' \neq l'$. This contradicts property (1.6.6). The two properties pointed out above determine a projective plane; therefore we have constructed the projective plane of order n corresponding to the matrix C or, what amounts to the same, to the complete set of pairwise orthogonal Latin squares. □

The finite projective plane of order 3 corresponding to the complete set of pairwise orthogonal Latin squares or matrix (1.6.8) is of the form

$$\begin{aligned} X_1 &= \{x_1, x_2, x_3, x_4\}, & X_8 &= \{x_3, y_1, y_6, y_8\}, \\ X_2 &= \{x_1, y_1, y_2, y_3\}, & X_9 &= \{x_3, y_2, y_4, y_9\}, \\ X_3 &= \{x_1, y_4, y_5, y_6\}, & X_{10} &= \{x_3, y_3, y_5, y_7\}, \\ X_4 &= \{x_1, y_7, y_8, y_9\}, & X_{11} &= \{x_4, y_1, y_5, y_9\}, \\ X_5 &= \{x_2, y_1, y_4, y_7\}, & X_{12} &= \{x_4, y_2, y_6, y_7\}, \\ X_6 &= \{x_2, y_2, y_5, y_8\}, & X_{13} &= \{x_4, y_3, y_4, y_8\}. \\ X_7 &= \{x_2, y_3, y_6, y_9\}, \end{aligned}$$

1.7 Block designs

In this section we consider a generalization of (v, k, λ)-configurations, the so-called (b, v, r, k, λ)-configurations, which, in turn, belong to the more general class of configurations which are known as block designs and which are widely used in the design of experiments.

1.7.1 (b, v, r, k, λ)-configurations

A family of subsets $X_1, \ldots, X_b$ of a set $X = \{x_1, \ldots, x_v\}$ is called a (b, v, r, k, λ)-configuration if the following conditions hold:

(1) $|X_i| = k$, $i = 1, \ldots, b$.

(2) For any $x_i \in X$ there exist exactly r subsets $X_{\mu_1}, \ldots, X_{\mu_r}$ containing x_i.

(3) For any $x_i, x_j \in X$, $i \neq j$, there exist exactly λ subsets $X_{\nu_1}, \ldots, X_{\nu_\lambda}$ containing x_i and x_j.

(4) The numbers v, k and λ satisfy the conditions

$$\lambda > 0, \qquad k < v - 1.$$

Let $A = \|a_{ij}\|$, $i = 1,\dots,b$, $j = 1,\dots,v$, be the incidence matrix of a (b,v,r,k,λ)-configuration, that is,

$$a_{ij} = \begin{cases} 1, & x_i \in X_i, \\ 0, & x_j \notin X_i, \end{cases}$$

let J_v be the $v \times v$ and J_{bv} be $b \times v$ matrices whose elements are all equal to 1, and let I_v be the $v \times v$ identity matrix.

The following properties are valid:

(1) $$AJ_v = kJ_{bv}.$$

This property is equivalent to the equalities

$$|X_i| = \sum_{j=1}^{v} a_{ij} = k, \qquad i = 1,\dots,b,$$

which follow from the definition of a (b,v,r,k,λ)-configuration.

(2) $$J_bA = rJ_{bv}.$$

This property means that

$$\sum_{i=1}^{b} a_{ij} = r, \qquad j = 1,\dots,v,$$

and these equalities are equivalent to Condition 2 in the definition of a (b,v,r,k,λ)-configuration.

(3) $$A^TA = \lambda J_v + (r-\lambda)I_v.$$

This property is equivalent to the equalities

$$\sum_{l=1}^{b} a_{li}a_{lj} = \begin{cases} r, & i = j, \\ \lambda, & i \neq j, \end{cases}$$

which express Conditions 2 and 3 in the definition of a (b,v,r,k,λ)-configuration.

(4) $$bk = vr.$$

Indeed, a fixed element $x_i \in X$ appears in exactly r subsets; therefore the number of elements in $X_1,\dots,X_b$, with regard to their repetitions, is equal to vr. On the other hand, since $|X_i| = k$, $i = 1,\dots,b$, this number is equal to bk.

(5) $$r(k-1) = \lambda(v-1).$$

We calculate the number of pairs containing a fixed element $x_i \in X$ using two methods. In each of r subsets which contain x_i, the element x_i forms $k-1$ pairs. Thus the total number of pairs is equal to $r(k-1)$. On the other hand, the element x_i in a pair with the element x_j belongs to λ subsets. Since the element x_j can be chosen in $v-1$ ways, the total number of pairs is $\lambda(v-1)$.

(6) $$\det A^T A = (r + \lambda(v-1))(r-\lambda)^{v-1}.$$

According to Property 3,

$$\det A^T A = \det(\lambda J_v + (r-\lambda)I_v).$$

Now Property 6 follows from formula (1.5.6), which expresses Property 3 of (v,k,λ)-configurations.

(7) $$b \geqslant v.$$

This property is known as the Fisher inequality. By the condition $k < v-1$ Property 5 implies the inequality $r > \lambda$; therefore it follows from Property 6 that $\det A^T A \neq 0$. Since the rank of the product of matrices is no greater than the rank of any factor, we obtain

$$v = \operatorname{rank}(A^T A) \leqslant \operatorname{rank} A \leqslant b.$$

Fisher's inequality and Property 4 imply the following property:

(8) $$br \geqslant k.$$

Note that a (v,k,λ)-configuration is a particular case of a (b,v,r,k,λ)-configuration if $b = v$ and $r = k$. Under these conditions all the properties of (b,v,r,k,λ)-configurations are transformed into the corresponding properties of (v,k,λ)-configurations, or else become trivial.

1.7.2 Steiner triples

A family of subsets of a set X, $|X| = v$, containing three elements each, is called a Steiner triple system if each subset of X consisting of two elements belongs to exactly one of the triples. A Steiner triple system for $v > 3$ is a (b,v,r,k,λ)-configuration with $k = 3$ and $\lambda = 1$. It follows from Properties 4 and 5 that $2r = v-1$, $3b = vr$ and thus $b = v(v-1)/6$, $r = (v-1)/2$. Hence it follows that $v \equiv 1,\ 3 \pmod 6$. Indeed, since $v = 2r+1$, the parameter v is odd, and, as follows from the condition

$b = (2r+1)r/3$, either $r = 3\mu$ or $2r + 1 = 3(2\nu + 1)$, where μ and ν are natural numbers. Thus, either $v = 6\mu + 1$, or $v = 6\nu + 3$.

Note that for $v = 7$, $k = 3$, we have $b = 7$, $r = 3$ and the Steiner triple system coincides with a projective plane of order $n = k - \lambda = 2$.

Theorem 1.7.1 *If there exist a Steiner triple system S_1 of order v_1 and a Steiner triple system S_2 of order v_2, then there exists a Steiner triple system S of order v_1v_2.*

Proof Let S_1 be defined on a set $A = (a_1, \dots, a_{v_1})$ and S_2 be defined on $B = (b_1, \dots, b_{v_2})$.

Consider the Cartesian product $A \times B$ and define a system S on this set as follows. A triple $((a_i, b_r), (a_j, b_s), (a_k, b_t))$ belongs to S if one of the following conditions holds:

$$(a_i, a_j, a_k) \in S_1, \qquad r = s = t,$$
$$(b_r, b_s, b_t) \in S_2, \qquad i = j = k,$$
$$(a_i, a_j, a_k) \in S_1 \quad \text{and} \quad (b_r, b_s, b_t) \in S_2.$$

It is not difficult to check that there exists exactly one triple in S containing a fixed pair of elements. The order of S is $|A \times B| = v_1v_2$. The number of triples in S is equal to

$$b = \frac{v_1v_2(v_1v_2 - 1)}{6}.$$

The number of triples containing a fixed element is equal to

$$r = \frac{v_1v_2 - 1}{2}.$$

□

1.7.3 Block designs and their applications

The configurations considered above such as Latin squares, orthogonal Latin squares, (v, k, λ)-configurations and (b, v, r, k, λ)-configurations are widely used in the design of statistical experiments. In statistics (b, v, r, k, λ)-configurations are called balanced incomplete block designs and the particular case of (v, k, λ)-configurations is called symmetric balanced incomplete block designs.

Consider the designs based on Latin squares. Suppose that we have to carry out an experiment to compare the yield of n types of wheat. To this end a square field is divided into n^2 equal square plots placed in n

rows and n columns. Each type has to be cultivated on n plots so that it appears exactly one time in each row and each column. The arrangement of the types labeled by numbers $1,\ldots,n$ into the n^2 plots corresponds to the construction of an $n \times n$ matrix of a Latin square with elements $1,\ldots,n$. It is natural to suppose that the harvest of each of the plots depends on the productivity of the type and the richness of the soil in the plot. In addition to the above, in our model we usually assume that the richness of the soil is a linear function $ax + bx + c$ of the Cartesian coordinates on the field. Under such conditions a random choice of Latin square allows us to eliminate the effect of soil richness in estimating the mean productivity of the types. The methods of estimation of mean values of random variables depending on several factors are developed in that area of mathematical statistics called the analysis of variance.

In the designs based on a Latin square we have three factors and, while estimating one of them, we eliminate the influence of the two remaining. In the example considered above we had such factors as the harvest, the productivity of the different types of wheat and the richness of the soil; and in estimating the productivity of types we eliminated the influence of richness which depended on the Cartesian coordinates of the plots. In cases where the number of factors is greater than three, designs based on orthogonal Latin squares can be used.

Let there be n types of wheat and n types of fertilizer and let both the types be labeled by numbers $1,\ldots,n$. The types of wheat and fertilizer are arranged into n^2 plots according to Latin squares $L_n^{(1)}$ and $L_n^{(2)}$ respectively. If $L_n^{(1)}$ and $L_n^{(2)}$ are orthogonal, then each of the n types of wheat meets each of the n types of fertilizer on exactly one plot. By a random choice of the orthogonal Latin squares $L_n^{(1)}$ and $L_n^{(2)}$ this design lets us estimate the influence of the fertilizers on the harvest of the types of wheat, avoiding the effect of the richness of the soil.

Now consider incomplete balanced block designs used in planning of experiments. Using agricultural terminology again we determine the design as follows. Let the field consist of b blocks and let each of the blocks be divided into k plots. There are v treatments of the soil which are used exactly r times each in the experiment. In each block any treatment is used no more than one time and each pair of treatments appears exactly λ times in all blocks.

The design considered defines, in an obvious way, a (b, v, r, k, λ)-configuration, and the random choice of such a configuration again lets us eliminate the effect of richness of the soil in estimating the efficiency of the treatments.

1.8 Sperner's theorem and completely separating families of sets

A family of different subsets $X_1, \dots, X_m$ of a set $X = \{x_1, \dots, x_n\}$ possesses the Sperner property if

$$X_i \not\subset X_j, \qquad 1 \leqslant i, j \leqslant m,$$

for any $i \neq j$. Such a family is called an antichain. The following lemma was proved independently by Yamamoto (Yamamoto, 1951), Meshalkin (Meshalkin, 1963) and Lubell (Lubell, 1966).

Lemma 1.8.1 *If $X_1, \dots, X_m$ is an antichain of a set X with n elements, then*

$$\sum_{i=1}^{m} \binom{n}{|X_i|}^{-1} \leqslant 1.$$

Proof A chain $C_0, C_1, \dots, C_n$ of subsets of a set X with n elements such that

$$\varnothing = C_0 \subset C_1 \subset \cdots \subset C_n = X$$

is called a complete chain. It is clear that $|C_i| = i$ for any $i = 0, 1, \dots, n$, and the total number of complete chains is equal to $n!$. If $|X_i| = r$, $1 \leqslant r \leqslant n$, then the chain

$$\varnothing = C_0 \subset C_1 \subset \dots, C_{r-1} \subset X_i \subset C_{r+1} \subset \cdots \subset C_n = X$$

is called a complete chain passing through the subset X_i. The number of such chains is equal to $r!(n-r)!$. Since complete chains passing through different subsets of an antichain are different, the estimate

$$\sum_{i=1}^{m} |X_i|!(n - |X_i|)! \leqslant n!$$

is true. The lemma is thus proved. □

Since

$$\binom{n}{|X_i|} \leqslant \binom{n}{[n/2]}, \qquad i = 1, \dots, m,$$

the lemma implies the following theorem due to Sperner (Sperner, 1928).

Theorem 1.8.1 *The number m of subsets of an antichain of a set with n elements satisfies the inequality*

$$m \leqslant \binom{n}{[n/2]}.$$

As an application of Sperner's theorem we obtain an estimate of the maximum cardinality of a set of divisors of n such that none of the divisors from the set is a divisor of any of the remaining, provided that $n = p_1 \cdots p_r$, where $p_1, \ldots, p_r$ are different prime numbers. If $X = \{1, \ldots, r\}$, then each divisor $d = p_{i_1} \cdots p_{i_k}$ can be associated with the subset $\{i_1, \ldots, i_k\} \subseteq X$. Since any divisor is not divisible by any other, the corresponding subsets possess the Sperner property and, consequently,

$$m \leqslant \binom{r}{[r/2]}.$$

Another example of the application of Sperner's theorem concerns completely separating families of subsets. A family of subsets $X_1, \ldots, X_r$ of a set $X = \{x_1, \ldots, x_n\}$ is a completely separating family if for any $x_i \neq x_j$ there exist X_k and X_l with $k \neq l$ such that $x_i \in X_k$, $x_j \notin X_k$ and $x_j \in X_l$, $x_i \notin X_l$. Let $A = \|a_{ij}\|$, $i = 1, \ldots, r$, $j = 1, \ldots, n$, be the incidence matrix of this family, that is,

$$a_{ij} = \begin{cases} 1, & x_j \in X_i, \\ 0, & x_j \notin X_i. \end{cases}$$

Let $A^{(1)}, \ldots, A^{(n)}$ be the columns of the matrix A and let $Y_1, \ldots, Y_n$ be a family of subsets of the set $Y = \{1, \ldots, r\}$ such that $i \in Y_j$ if and only if the ith coordinate of the column $A^{(j)}$ is equal to 1, $i = 1, \ldots, r$, $j = 1, \ldots, n$. Since $X_1, \ldots, X_r$ is a completely separating family, for any two columns $A^{(i)}$ and $A^{(j)}$ with $i \neq j$ there exist coordinates k and l such that $a_{ki} = 1$, $a_{kj} = 0$ and $a_{li} = 0$, $a_{lj} = 1$. Thus $Y_i \not\subseteq Y_j$ for any $i \neq j$, that is, the family $Y_1, \ldots, Y_n$ possesses the Sperner property. According to Sperner's theorem the number of subsets r of the completely separating family satisfies the inequality

$$\binom{r}{[r/2]} \geqslant n.$$

2
Transversals and permanents

A wide range of the so-called combinatorial problems of choice can be reduced to finding a system of distinct representatives for a given family of subsets of a set. In what follows, such a system will be called a transversal. We prefer this term because it is short and so has an advantage over the corresponding, more detailed, conventional term. The main questions considered in this chapter are related to the existence and number of transversals. The basis for the answers to the first series of questions is an existence theorem due to P. Hall, and various applications of it. To determine the number of transversals, a notion of a permanent is used which is a modification of the well-known notion of a determinant playing an important role in algebra.

The theorem of P. Hall is the basis for the proofs of the theorem of M. Hall on the existence of Latin squares and rectangles and Birkhoff's theorem on the representation of a stochastic matrix as a weighted sum of permutation matrices. Birkhoff's theorem is connected with a number of assertions about the decomposition of probabilistic automata and Markov chains with doubly stochastic matrices of transition probabilities.

2.1 Transversals

2.1.1 The main theorems

Let X be an arbitrary, generally speaking, infinite set; let $X_1, \ldots, X_n$ be a family of subsets of X containing, in general, infinite subsets. Note that the equalities $X_i = X_j$ for $i \neq j$ are allowed. We denote this family by $(X_i : i \in I)$, where $I = \{1, \ldots, n\}$. A system of elements $(x_i : i \in I)$ is called a system of distinct representatives or a transversal of the family

$(X_i : i \in I)$ if

$$x_i \in X_i, \quad i \in X_i, \quad x_i \neq x_j, \quad i \neq j.$$

Let us give a criterion for the existence of a transversal of a family of sets $(X_i : i \in I)$. To this end we formulate and prove the following, known as P. Hall's theorem (Hall, 1935).

Theorem 2.1.1 *A transversal of a family of sets* $(X_i : i \in I)$ *exists if and only if*

$$|X_{i_1} \cup \cdots \cup X_{i_k}| \geqslant k, \tag{2.1.1}$$

for all $1 \leqslant k \leqslant n$, $1 \leqslant i_1 < \cdots < i_k \leqslant n$.

Proof Conditions (2.1.1) will be referred to as P. Hall's conditions. If the equality

$$|X_{i_1} \cup \cdots \cup X_{i_k}| = k \tag{2.1.2}$$

holds for some subfamily $X_{i_1}, \ldots, X_{i_k}$ of the family $(X_i : i \in I)$ with $1 \leqslant k < n$, then this subfamily is called critical.

We now turn to the proof of P. Hall's theorem. The necessity of the conditions of the theorem is obvious, since for any $1 \leqslant k \leqslant n$, $1 \leqslant i_1 < \cdots < i_k \leqslant n$, at least the equality in (2.1.1) follows from the existence of a transversal.

The sufficiency will be proved by induction on n. For $n = 1$ the assertion is true. Suppose that it is true for any subfamily containing no more than $n-1$ subsets from the family $(X_i : i \in I)$.

First consider the case where the family $(X_i : i \in I)$ does not contain a critical subfamily. Then for all $1 \leqslant k < n$, $1 \leqslant i_1 < \cdots < i_k \leqslant n$ the inequality

$$|X_{i_1} \cup \cdots \cup X_{i_k}| \geqslant k+1$$

holds. According to P. Hall's conditions, $X_1 \neq \varnothing$; therefore we take $x_1 \in X_1$ and consider the family $(X_i \setminus x_1 : i \in I')$, where $I' = I \setminus \{1\}$. The symbol $\setminus$ means subtraction of sets; in particular here it means erasing the corresponding element from the corresponding set. The family $(X_i \setminus x_1 : i \in I')$ satisfies P. Hall's conditions, and consequently, there exists a transversal $(x_i : i \in I')$ of this family. Hence it follows that $(x_i : i \in I)$ is a transversal of the family $(X_i : i \in I)$.

Now consider the case where a critical subfamily exists. Without loss of generality, we may assume that this is the family $(X_i : 1 \leqslant i \leqslant k)$,

$1 \leqslant k < n$, that is, $|X_1 \cup \cdots \cup X_k| = k$. The family $(X_i : 1 \leqslant i \leqslant k)$ satisfies P. Hall's conditions, and by the induction hypothesis there exists a transversal $(x_i : 1 \leqslant i \leqslant k)$ of this family. Put

$$X_i' = X_i \setminus \{x_1, \dots, x_k\}, \qquad k+1 \leqslant i \leqslant n,$$

and consider the family $(X_i' : k+1 \leqslant i \leqslant n)$. This family satisfies P. Hall's conditions, since for all $1 \leqslant l \leqslant n-k$, $1 \leqslant v_1 < \cdots < v \leqslant n-k$, the relations

$$\begin{aligned} |X'_{k+v_1} \cup \cdots \cup X'_{k+v_l}| &= |X'_{k+v_1} \cup \cdots \cup X'_{k+v_l}| + |X_1 \cup \cdots \cup X_k| - k \\ &= |X_1 \cup \cdots \cup X_k \cup X'_{k+v_1} \cup \cdots \cup X'_{k+v_l}| - k \\ &\geqslant (k+l) - k = l \end{aligned}$$

hold, because the family $(X_1, \dots, X_k, X'_{k+v_1}, \dots, X'_{k+v_l})$ satisfies P. Hall's conditions. Now the induction hypothesis implies that the family $(X_i' : k+1 \leqslant i \leqslant n)$ possesses a transversal $(x_i : k+1 \leqslant i \leqslant n)$. Uniting the two transversals obtained, we construct the transversal $(x_i : i \in I)$ of the family $(X_i : i \in I)$. □

It should be noted that the finiteness of the number of sets in a family is essential in P. Hall's theorem. For example, the family $\{1, 2, \dots\}$, $\{1\}, \{2\}, \dots$ satisfies P. Hall's conditions, but the theorem is not valid. P. Hall's theorem remains true if the index set I becomes countable, but the subsets are finite.

Now we formulate and prove Rado's theorem, which gives a bound for the number of transversals of a family of finite sets (Rado, 1967).

Theorem 2.1.2 *Let a family of finite sets* $(X_i : i \in I)$ *satisfy* (2.1.1) *and* $|X_1| \leqslant \cdots \leqslant |X_n|$. *Then for the number of transversals* $R(X_1, \dots, X_n)$ *the inequality*

$$R(X_1, \dots, X_n) \geqslant \prod_{0 < v \leqslant \min(n, |X_1|)} (|X_v| - v + 1) \tag{2.1.3}$$

holds.

Proof We prove the inequality by induction on n. For $n = 1$ inequality (2.1.3) is reduced to the obvious inequality

$$R(X_1) \geqslant |X_1|.$$

As in the proof of P. Hall's theorem we consider two cases.

Suppose, first, that there is no critical subfamily. Denote by $(\bar{X}_i : 2 \leqslant$

$i \leqslant n$) the subfamily $(X_i \setminus x_1 : 2 \leqslant i \leqslant n)$, and, without loss of generality, let $|\bar{X}_2| \leqslant \cdots \leqslant |\bar{X}_n|$. Using the induction hypothesis, we obtain

$$R(X_1, \ldots, X_n) \geqslant \sum_{x_1 \in X_1} \prod_{0<v \leqslant \min(n-1, |\bar{X}_2|)} (|\bar{X}_{v+1}| - v + 1).$$

Returning to the sets of the family $(X_i : 2 \leqslant i \leqslant n)$, we see that

$$R(X_1, \ldots, X_n) \geqslant \sum_{x_1 \in X_1} \prod_{0<v \leqslant \min(n-1, |X_1|-1)} (|X_{v+1}| - v).$$

Introducing the new index $v' = v + 1$, we obtain the inequality

$$R(X_1, \ldots, X_n) \geqslant |X_1| \prod_{0<v' \leqslant \min(n, |X_1|)} (|X_{v'}| - v' + 1),$$

which coincides with (2.1.3).

Suppose now that there is a critical subfamily $(X_i : 1 \leqslant i \leqslant k)$. By the definition of a critical subfamily, $|X_1 \cup \cdots \cup X_k| = k$. Let $|X_1| \leqslant \cdots \leqslant |X_k|$. Then

$$R(X_1, \ldots, X_n) \geqslant R(X_1, \ldots, X_k).$$

By the induction hypothesis,

$$R(X_1, \ldots, X_k) \geqslant \prod_{0<v \leqslant \min(k, |X_1|)} (|X_v| - v + 1).$$

The set X_1 belongs to the critical subfamily; therefore $|X_1| \leqslant k < n$. Hence

$$\min(k, |X_1|) = \min(n, |X_1|) = |X_1|,$$

and

$$R(X_1, \ldots, X_k) \geqslant \prod_{0<v \leqslant \min(n, |X_1|)} (|X_v| - v + 1).$$

Thus inequality (2.1.3) is proved. □

From Theorem 2.1.2 we obtain the following well-known estimates due to M. Hall.

Corollary 2.1.1 *Let* $d = |X_1| \leqslant |X_2| \leqslant \cdots \leqslant |X_n|$. *Then*

$$R(X_1, \ldots, X_n) \geqslant \begin{cases} d!, & d \leqslant n, \\ (d)_n, & d \geqslant n. \end{cases} \tag{2.1.4}$$

Proof Indeed, in the first case $\min(n, |X_1|) = d$; therefore,

$$R(X_1, \ldots, X_k) \geqslant \prod_{0<v\leqslant d} (d - v + 1) = d!.$$

In the second case $\min(n, |X_1|) = n$; therefore,

$$R(X_1, \ldots, X_k) \geqslant \prod_{0<v\leqslant n} (d - v + 1) = (d)_n.$$

□

Let us give two further estimates for the number of transversals.

Theorem 2.1.3 *If* $|X_i| > 0$, $1 \leqslant i \leqslant n$, *then*

$$R(X_1 \ldots, X_n) \geqslant \prod_{i=1}^{n} |X_i|, \tag{2.1.5}$$

where equality is attained if and only if

$$X_i \cap X_j = \varnothing, \qquad 1 \leqslant i < j \leqslant n.$$

If $|X_i| \geqslant i$, $1 \leqslant i \leqslant n$, *and* $|X_1| \leqslant \cdots \leqslant |X_n|$, *then*

$$R(X_1, \ldots, X_n) \geqslant \prod_{i=1}^{n} (|X_i| - i + 1), \tag{2.1.6}$$

where equality is attained if and only if

$$X_i \subseteq X_j, \qquad 1 \leqslant i < j \leqslant n.$$

Proof Inequality (2.1.5) is obvious. Let us clarify the conditions when the equality holds. Let

$$x \in X_i \cap X_j, \qquad 1 \leqslant i < j \leqslant n,$$

and let x be a representative of X_i in some transversal. Then x cannot be a representative of X_j, and the number of representatives of X_j is less than $|X_j|$. Thus, (2.1.5) is a strict inequality, and the equality holds only if $X_i \cap X_j = \varnothing$.

Now we prove inequality (2.1.6). For $k < n$ let $(x_i : 1 \leqslant i \leqslant k)$ be a transversal of the family $(X_i : 1 \leqslant i \leqslant k)$ and let $(x_i : 1 \leqslant i \leqslant k+1)$ be a transversal of the family $(X_i : 1 \leqslant i \leqslant k+1)$. Hence it follows that x_{k+1} is chosen from the set $X_{k+1} \setminus \{x_1, \ldots, x_k\}$ and

$$R(X_1, \ldots, X_{k+1}) \geqslant (|X_{k+1}| - k) R(X_1, \ldots, X_k).$$

Using this recurrence relation the required number of times, we obtain inequality (2.1.6). Verification of the conditions when equality is attained is left to the reader. □

2.1.2 Switching circuits

Switching circuits are widely used in various areas of communication and engineering. Consider a simple switching circuit with n labeled input terminals and one output terminal. A demand for communication with the output terminal comes on an input. The set of labels of inputs with demands is called a list of demands. Along with the list of demands there exists a list of unoccupied links, that is, a list of unoccupied communication paths leading to the output terminal. Without loss of generality, we can assume that the demands appear on the first n inputs. Denote by X_i the set of labels of the commutation paths, from the list of unoccupied links, which connect the ith input with the output. If the family of sets $X_1, \ldots, X_n$ has a transversal, then the n demands can be served by the switching circuit. Otherwise, if there is no transversal, then a recombination of the existing communication paths is needed to serve the n demands. In such situations transversals also play an important role.

2.1.3 An existence theorem for Latin squares

Let a Latin rectangle $[\varphi_1, \ldots, \varphi_k]_m$ be an ordered set of k mutually discordant substitutions $\varphi_1, \ldots, \varphi_k$ of degree m. M. Hall investigated the possibility of enlarging the set of substitutions by adding a substitution φ_{k+1} so that $[\varphi_1, \ldots, \varphi_k, \varphi_{k+1}]$ would be a Latin rectangle. He proved the following theorem (Hall, 1945).

Theorem 2.1.4 *For any $k \times m$ Latin rectangle $[\varphi_1, \ldots, \varphi_k]$ with $k < m$ there exists a substitution φ_{k+1} of degree m such that $[\varphi_1, \ldots, \varphi_k, \varphi_{k+1}]$ is a Latin rectangle.*

Proof Let the substitutions $\varphi_1, \ldots, \varphi_k, \varphi_{k+1}$ act on the set $\{1, \ldots, m\}$. Put

$$Q_j = \{\varphi_i(j),\ i = 1, \ldots, k\},$$
$$\bar{Q}_j = X \setminus Q_j, \qquad 1, \ldots, m.$$

A fixed element $j \in X$ belongs to exactly k sets of the family $Q_1, \ldots, Q_m$ and, consequently, to exactly $m - k$ sets of the family $\bar{Q}_1, \ldots, \bar{Q}_m$. Thus, for any collection $1 \leqslant v_1 \cdots < v_l \leqslant m$, $1 \leqslant l \leqslant m$, the element $j \in X$ enters into $\bar{Q}_{v_1} \cup \cdots \cup \bar{Q}_{v_l}$ no more than $m-k$ times. The number of elements with repetitions in $\bar{Q}_{v_1} \cup \cdots \cup \bar{Q}_{v_l}$ is equal to $(m-k)l$. Therefore, the number of distinct elements among them is no less than l. Hence it follows that

$$|\bar{Q}_{v_1} \cup \cdots \cup \bar{Q}_{v_l}| \geqslant l,$$

that is, the family $(\bar{Q}_i : 1 \leqslant i \leqslant m)$ satisfies P. Hall's conditions. If $(r_i : 1 \leqslant i \leqslant m)$ is a transversal of the family $(\bar{Q}_i : 1 \leqslant i \leqslant m)$, then the substitution φ_{k+1} determined by the equalities $\varphi_{k+1} = r_j$, $j = 1, \ldots, m$, is discordant to the substitutions $\varphi_1, \ldots, \varphi_k$ and, consequently, $[\varphi_1, \ldots, \varphi_k, \varphi_{k+1}]$ is a Latin rectangle. □

As a corollary of Theorem 2.1.4 we obtain the following existence theorem for Latin squares.

Theorem 2.1.5 *For any Latin rectangle $L_{k,m} = [\varphi_1, \ldots, \varphi_k]_m$ there exists a Latin rectangle $L_{m-k,m} = [\varphi_{k+1}, \ldots, \varphi_m]_m$ such that $[\varphi_1, \ldots, \varphi_m]$ is a Latin square.*

The following assertion, which gives a lower bound for the number of Latin rectangles and squares, is also a corollary of Theorem 2.1.4.

Corollary 2.1.2 *If $L(k,m)$ is the number of $k \times m$ Latin rectangles and $L(m)$ is the number of Latin squares of order m, then*

$$L(k,m) \geqslant m!\,(m-1)! \cdots (m-k+1)!, \qquad k = 1, \ldots, m, \tag{2.1.7}$$

$$L(m) \geqslant m!\,(m-1)! \cdots 2!\,1!. \tag{2.1.8}$$

Proof Inequality (2.1.7) can be proved by induction on k. It is clear that $L(1,m) = m!$. Suppose that $L(k,m) \geqslant m! \cdots (m-k+1)!$. Then

$$L(k+1,m) = L(k,m) R(\bar{Q}_1, \ldots, \bar{Q}_m),$$

where $\bar{Q}_1, \ldots, \bar{Q}_m$ are the sets defined in the proof of Theorem 2.1.4. Note that $|\bar{Q}_j| = m-k \leqslant m$; therefore, by Corollary 2.1.1 to Rado's theorem,

$$R(\bar{Q}_1, \ldots, \bar{Q}_m) \geqslant (m-k)!.$$

Thus $L(k+1,m) \geqslant m!(m-1)! \cdots (m-k+1)!$. Inequality (2.1.7) for $k = m$ gives (2.1.8). □

It should be noted that inequality (2.1.7) provides a rough lower bound. For large m and comparatively not so large k, a more precise estimate is given by the following asymptotic formula obtained by Erdős and Kaplansky and Yamamoto (Erdős and Kaplansky, 1946; Yamamoto, 1951). If $k < m^{1/3-\varepsilon}$, $\varepsilon > 0$, then

$$L(k,m) = (m!)^k e^{k(k-1)/2}(1+\gamma_{km}), \tag{2.1.9}$$

where $\gamma_{km} \to 0$ as $m \to \infty$. Subsequent investigations were directed towards extending the admissible range of the parameter k. The paper (Godsil and McKay, 1984) contains the asymptotic formula

$$L(k,m) = \frac{(m!)^{k+m}}{m^{km}((m-k)!)^k} e^{k(k-1)l(k,m)},$$

where $k = o(m^{6/7})$ and

$$l(k,m) = \frac{1}{4m} + \frac{k-1}{6m^2} + \frac{k^2-k-1}{8m^3} + \frac{12k^3-13k^2-13k-6}{12m^4} + \frac{15k^4-18k^3-18k^2-28k+47}{180m^5} + O\left(\frac{k^5}{m^6}\right).$$

The problem of enumeration of the Latin squares for small values of m was developed as follows. MacMahon (MacMahon, 1915) found that

$$L(m) = m!\,(m-1)!\,u(m),$$

where $u(m)$ is the number of normalized Latin squares of order m. In 1939, Norton (Norton, 1939) first found the value $u(7)$, which was improved upon in 1951 by Sade (Sade, 1951). The number $u(8)$ was pointed out by Wells (Wells, 1967) in 1967 and was later verified in a number of papers, particularly in (Kolesova, Lam and Thiel, 1990). Finally, the number $u(9)$ was found by Bammel and Rothstein (Bammel and Rothstein, 1975).

2.1.4 Common transversals

Let two partitions of a finite set X be given:

$$X = X_1 \cup \cdots \cup X_n = Y_1 \cup \cdots \cup Y_n.$$

A set $(x_i : 1 \leqslant i \leqslant n)$ is called a common transversal or a common system of representatives if there exists a renumeration of the sets $Y_1, \dots, Y_n$ such that after this renumeration $x_i \in X_i \cap Y_i$, $i = 1, \dots, n$. By virtue of the equality

$$(X_i \cap Y_i) \cap (X_j \cap Y_j) = \varnothing, \tag{2.1.10}$$

the set $(x_i : 1 \leqslant i \leqslant n)$ satisfies the conditions $x_i \neq x_j$ for all $i \neq j$. In particular, these conditions imply that $(x_i : 1 \leqslant i \leqslant n)$ is a transversal common to the families $(X_i : 1 \leqslant i \leqslant n)$ and $(Y_i : 1 \leqslant i \leqslant n)$.

Theorem 2.1.6 *A common transversal for two partitions*

$$X_1 \cup \cdots \cup X_n = Y_1 \cup \cdots \cup Y_n \tag{2.1.11}$$

exists if and only if for any $1 \leqslant \nu_1 < \cdots < \nu_k \leqslant n$ *and* $1 \leqslant \mu_1 < \cdots < \mu_l \leqslant n$ *the inclusion*

$$X_{\nu_1} \cup \cdots \cup X_{\nu_k} \subseteq Y_{\mu_1} \cup \cdots \cup Y_{\mu_l} \tag{2.1.12}$$

is impossible for $l < k$.

Proof Let us prove the necessity. Let a common transversal exist but allow inclusion (2.1.12) to hold. Hence it follows that $X_{\nu_1}, \ldots, X_{\nu_k}$ can have common representatives only with the sets $Y_{\mu_1}, \ldots, Y_{\mu_l}$; this is impossible if $l < k$.

We now prove the sufficiency. Consider the family of sets $(S_i : 1 \leqslant i \leqslant n)$, where $S_i = \{j : X_i \cap Y_j \neq \varnothing\}$, $i = 1, \ldots, n$. This family satisfies P. Hall's conditions. Indeed, suppose that for $1 \leqslant \nu_1 < \cdots < \nu_k \leqslant n$, $1 \leqslant k \leqslant n$,

$$|S_{\nu_1} \cup \cdots \cup S_{\nu_k}| < k.$$

This inequality means that $X_{\nu_1}, \ldots, X_{\nu_k}$ have non-empty intersections with $l < k$ sets $Y_{\mu_1}, \ldots, Y_{\mu_l}$ only. Since the sets $X_{\nu_1}, \ldots, X_{\nu_k}$ have no common elements with the remaining sets of the family $Y_1, \ldots, Y_n$, we obtain the inclusion

$$X_{\nu_1} \cup \cdots \cup X_{\nu_k} \subseteq Y_{\mu_1} \cup \cdots \cup Y_{\mu_l},$$

which contradicts the hypotheses of the theorem. □

If $(j_i : 1 \leqslant i \leqslant n)$ is a transversal of the family $(S_i : 1 \leqslant i \leqslant n)$, then the transversal determines a renumeration of the sets $Y_1, \ldots, Y_n$ such that $x_i \in X_i \cap Y_{j_i}$, $i = 1, \ldots, n$, and $(x_i : 1 \leqslant i \leqslant n)$ is a common transversal for the partitions.

Corollary 2.1.3 *Let*

$$G = g_1 H + \cdots + g_n H = H\bar{g}_1 + \cdots + H\bar{g}_n$$

be a decomposition of a finite group G *into residue classes with respect to a subgroup* H. *Then there exist elements* $\tilde{g}_1, \ldots, \tilde{g}_n \in G$ *such that*

$$G = \tilde{g}_1 H + \cdots + \tilde{g}_n H = H\tilde{g}_1 + \cdots + H\tilde{g}_n.$$

Since decompositions of a group G into right and left residue classes are partitions of the group G, the assertion of the corollary can be proved by direct application of Theorem 2.1.6.

2.2 Decomposition of non-negative matrices

A matrix A with real elements is called non-negative if all its elements are non-negative.

Let $\mathfrak{A}_n$ be a collection of non-negative $n \times n$ matrices and $\mathfrak{B}_n$, a collection of matrices such that $\mathfrak{B}_n \subset \mathfrak{A}_n$. A decomposition of a matrix $A \in \mathfrak{A}_n$ is any representation of the form

$$A = \alpha_1 B_1 + \cdots + \alpha_s B_s, \tag{2.2.1}$$

where $B_i \in \mathfrak{B}_n$, $i = 1, \ldots, s$, and $\alpha_1, \ldots, \alpha_s$ are real positive numbers. In this section we study the conditions of decomposition for some collections $\mathfrak{A}_n$ and $\mathfrak{B}_n$.

2.2.1 Birkhoff's theorem

A real non-negative matrix $A = \|a_{ij}\|$, $i, j = 1, \ldots, n$, is called doubly stochastic if

$$\sum_{j=1}^{n} a_{ij} = 1, \qquad i = 1, \ldots, n,$$

$$\sum_{i=1}^{n} a_{ij} = 1, \qquad j = 1, \ldots, n.$$

Consider as the collection $\mathfrak{A}_n$ the set Ω_n^t of non-negative matrices $A = \|a_{ij}\|$, $i, j = 1, \ldots, n$, such that

$$\sum_{j=1}^{n} a_{ij} = t, \qquad i = 1, \ldots, n,$$

$$\sum_{i=1}^{n} a_{ij} = t, \qquad j = 1, \ldots, n.$$

Such matrices will be referred to as multiple doubly stochastic matrices. Doubly stochastic matrices constitute the set $\Omega_n = \Omega_n^1$. Let us investigate the conditions of representability of $A \in \Omega_n^t$ as a weighted sum of permutation matrices. Recall that a matrix $\Pi = \|\pi_{ij}\|$, $i, j = 1, \ldots, n$, is a permutation matrix if its elements take values 0 and 1 and each row and each column contains exactly one unit. It is clear that a permutation matrix is doubly stochastic. If φ is the permutation which corresponds to a permutation matrix Π, then

$$\Pi = \|\delta_{\varphi(i),j}\| = \|\delta_{i,\varphi^{-1}(j)}\|, \qquad i,j = 1,\ldots,n, \tag{2.2.2}$$

where δ_{ij} is Kronecker's symbol. The following assertion was obtained by Birkhoff.

Theorem 2.2.1 *Let $A = \|a_{ij}\|$, where*

$$a_{ij} \geqslant 0, \quad \sum_{j=1}^{n} a_{ij} = \sum_{j=1}^{n} a_{ij} = t, \quad i,j = 1,\ldots,n.$$

Then there exist an s and non-negative $\alpha_1,\ldots,\alpha_s$ such that

$$\alpha_1 + \cdots + \alpha_s = t, \qquad A = \alpha_1\Pi_1 + \cdots + \alpha_s\Pi_s, \tag{2.2.3}$$

where $\Pi_1,\ldots,\Pi_s$ are $n \times n$ permutation matrices.

Proof We prove the theorem by induction on the number ω of non-zero elements in the matrix A. If A consists of zeros only, then $t = 0$ and the representation (2.2.3) exists with zero coefficients. If the matrix A contains positive elements, then, obviously, $\omega \geqslant n$. If $\omega = n$, then the matrix A is multiple to a permutation matrix, that is, $A = t\Pi$, and representation (2.2.3) is true with $s = 1$. Suppose now that representation (2.2.3) exists for all matrices with the number of positive elements no greater than $\omega - 1$.

Consider the family of sets $(Q_i : 1 \leqslant i \leqslant n)$, where $Q_i = \{j : a_{ij} > 0\}$, $i = 1,\ldots,n$. This family satisfies P. Hall's conditions. Indeed, if for some $1 \leqslant v_1 < \cdots < v_k \leqslant n$, $1 \leqslant k \leqslant n$,

$$|Q_{v_1} \cup \cdots \cup Q_{v_k}| < k,$$

then the positive elements in the rows with numbers $v_1,\ldots,v_k$ occupy no more than $k-1$ columns. The row summation of these elements gives the value kt, but the column sum is no greater than $(k-1)t$. We thus obtain a contradiction.

Now let $(x_i : 1 \leqslant i \leqslant n)$ be a transversal of the family $(Q_i : 1 \leqslant \imath \leqslant n)$. Define the permutation matrix $\Pi_1 = \|\pi_{ij}^{(1)}\|$, $i,j = 1,\ldots,n$, putting

$$\pi_{ij}^{(1)} = \begin{cases} 1, & j = r_i, \\ 0, & j \neq r_i. \end{cases}$$

It follows from the definition of $Q_1,\ldots,Q_n$ that $a_{ir_i} > 0$, $i = 1,\ldots,n$, and, consequently,

$$\alpha_1 = \min_{1\leqslant i\leqslant n} a_{ir_i} > 0.$$

The matrix $A_1 = A - \alpha_1\Pi_1 = \|a_{ij}^{(1)}\|$, $i, j = 1,\ldots,n$, satisfies the conditions of the theorem, namely, it has non-negative elements,

$$\sum_{j=1}^{n} a_{ij}^{(1)} = t - \alpha_1, \qquad i = 1,\ldots,n,$$

$$\sum_{i=1}^{n} a_{ij}^{(1)} = t - \alpha_1, \qquad j = 1,\ldots,n,$$

and the number ω_1 of its positive elements is no greater than $\omega - 1$. By the induction hypothesis

$$A_1 = \alpha_2\Pi_2 + \cdots + \alpha_s\Pi_s,$$

where $\Pi_2,\ldots,\Pi_s$ are permutation matrices. The theorem is thus proved. □

Corollary 2.2.1 *For any doubly stochastic $n \times n$ matrix $A = \|a_{ij}\|$ there exists a representation*

$$A = \alpha_1\Pi_1 + \cdots + \alpha_s\Pi_s, \qquad \sum_{j=1}^{n} \alpha_j = 1, \tag{2.2.4}$$

where $\Pi_1,\ldots,\Pi_s$ are permutation matrices.

The corollary follows from Birkhoff's theorem if we put $t = 1$.

Elements a_{ij} and a_{kl} of a matrix A are called non-collinear if $i \neq k$ and $j \neq l$. A set of n non-collinear elements of a square $n \times n$ matrix A is called a diagonal of the matrix. A diagonal is called positive if all its elements are positive.

Corollary 2.2.2 *Any finite doubly stochastic matrix has a positive diagonal.*

2.2.2 The convex polytope of doubly stochastic matrices

Let U be a d-dimensional vector space over the field of real numbers. If $x, y \in U$ and θ is a real number, then the set $\{z : z = \theta x + (1-\theta)y \in U,\ 0 \leqslant \theta \leqslant 1\}$ is called the interval joining x and y. A subset $S \subseteq U$ is convex if for any $x, y \in S$ the interval joining x and y is contained in S. If $U = M_{nm}(R)$, where $M_{nm}(R)$ is the set of non-negative $n \times m$ matrices over the field of real numbers, then $d = nm$. The set Ω_n of all doubly stochastic matrices is convex in the space $M_{nm}(R)$.

If $u, v \in U$ and $u = (u_1, \ldots, u_d)$, $v = (v_1, \ldots, v_d)$, the quantity

$$(u, v) = \sum_{i=1}^{d} u_i v_j$$

is the scalar product of u and v. The set $\{z : (z, x) = \alpha,\ z \in U\}$, where x is a fixed element from U and α is a fixed number, is called a hyperplane.

A linear combination $\theta_1 u_1 + \cdots + \theta_s u_s$, where $u_1, \ldots, u_s \in U$, is called a convex linear combination of $u_1, \ldots, u_s$ if $\theta_1, \ldots, \theta_s$ are non-negative and $\theta_1 + \cdots + \theta_s = 1$. The set of all convex combinations of elements $u_1, \ldots, u_s$ is called the convex polytope spanned on $u_1, \ldots, u_s$ and is denoted by $H(u_1, \ldots, u_s)$. A convex polytope is a convex set. For $X \subset U$ we denote by $H(X)$ the set of all convex combinations of finite sets formed from the elements of X. The set $H(X)$ is called the convex hull of the set X.

The following properties of convex hulls are obvious:

(1) If $X \subset Y$, then $H(X) \subset H(Y)$.
(2) $H(H(X)) = H(X)$ for any X.
(3) If $x \in H(u_1, \ldots, u_s)$, then

$$H(x, u_1, \ldots, u_s) = H(u_1, \ldots, u_s).$$

Let us prove a theorem on the existence and uniqueness of the vertices of a polytope.

Theorem 2.2.2 *For any polytope $H(x_1, \ldots, x_s)$ spanned on $x_1, \ldots, x_s \in U$ there exists a unique finite set of elements $y_j \in H(x_1, \ldots, x_s)$, $j = 1, \ldots, r$, such that*

$$y_j \notin H(y_1, \ldots, y_{j-1}, y_{j+1}, \ldots, y_r), \qquad j = 1, \ldots, r,$$

and

$$H(y_1, \ldots, y_r) = H(x_1, \ldots, x_s).$$

Proof The elements $y_1, \ldots, y_r$ are called the vertices of the polytope $H(x_1, \ldots, x_s)$. We prove the existence of the vertices by induction on s. For $s = 1$ the assertion is trivial. If for $s > 1$ none of x_i is a linear combination of the remaining x_j, $j \neq i$, then $x_1, \ldots, x_s$ can be taken as the vertices of the polytope. Otherwise, without loss of generality, we suppose that $x_1 \in H(x_2,\dots,x_s)$. By the induction hypothesis there exist vertices $y_1,\dots,y_r$ of the polytope $H(x_2,\dots,x_s)$, that is, no y_j is a linear combination of y_i, $i \neq j$, and $H(y_1,\dots,y_r) = H(x_2,\dots,x_s)$. Therefore, $x_1 \in H(y_1,\dots,y_r)$ and $H(y_1,\dots,y_r) = H(x_1,\dots,x_s)$.

We prove the uniqueness of the vertices. Suppose that there exists another set of vertices $u_1,\dots,u_p$, that is, none of the elements $u_1,\dots,u_p$ is a linear combination of the remaining elements and $H(u_1,\dots,u_p) = H(x_1,\dots,x_s)$. Then

$$y_1 = \sum_{j=1}^{p} \alpha_j u_j, \quad \alpha_j \geqslant 0, \quad \sum_{j=1}^{p} \alpha_j = 1,$$

$$u_j = \sum_{t=1}^{r} \beta_{jt} y_t, \quad \beta_{jt} \geqslant 0, \quad \sum_{t=1}^{r} \beta_{jt} = 1, \quad j = 1,\dots,p.$$

Hence it follows that

$$y_1 = \sum_{t=1}^{r}\sum_{j=1}^{p} \alpha_j \beta_{jt} y_t.$$

This equality implies that

$$\sum_{j=1}^{p} \alpha_j \beta_{j1} = 1, \quad \sum_{j=1}^{p} \alpha_j \beta_{jt} = 0, \quad t = 2,\dots,r.$$

Hence it follows that

$$1 = \sum_{j=1}^{p} \alpha_j \beta_{j1} \leqslant \max_{1 \leqslant j \leqslant p} \beta_{j1} \leqslant 1,$$

and there exists k, $1 \leqslant k \leqslant p$, such that $\beta_{k1} = 1$ and, consequently, $\beta_{kt} = 0$ for $t = 2,\dots,r$. Thus, we obtain

$$y_1 = u_k.$$

Similarly, it can be proved that $\{y_2,\dots,y_r\}$ is a subset of the set $\{u_1,\dots,u_p\}$. If we change the roles of the sets $u_1,\dots,u_p$ and $y_1,\dots,y_r$ we obtain the inverse inclusion. □

It follows from Birkhoff's theorem that the set Ω_n of doubly stochastic $n \times n$ matrices is a convex polytope spanned on the permutation matrices. Since none of the permutation matrices can be represented as a convex combination of the other permutation matrices, the permutation matrices are the unique set of vertices of the polytope Ω_n.

A polytope $H(x_0, x_1,\dots,x_p)$ such that $x_1 - x_0,\dots,x_p - x_0$ are linearly independent is called a p-simplex. The vectors $x_0, \dots, x_p$ form a basis of the p-simplex. Note that the role of x_0 can be played by any other vector x_r, $1 \leqslant r \leqslant p$, since the vectors $x_j - x_r$, $j \neq r$, $j = 0, \dots, p$, are linearly independent if and only if the vectors $x_j - x_0$, $j = 1, \dots, p$, are linearly independent.

Theorem 2.2.3 *Any element of a simplex is uniquely represented as a convex combination of its basic elements.*

Proof Let us prove first that the elements of a basis $x_0, x_1, \dots, x_p$ are the vertices of the p-simplex $H(x_0, \dots, x_p)$. To this end we show that for any j, the element x_j cannot be represented as a convex combination of the remaining x_i, $i \neq j$. Suppose that

$$x_j = \sum_{i=0, i \neq j}^{p} \theta_i x_i, \quad \sum_{i=0, i \neq j}^{p} \theta_i = 1, \quad \theta_i \geqslant 0, \quad i = 0, \dots, p.$$

Then we obtain the equality

$$\sum_{i=0, i \neq j} \theta_i (x_i - x_j) = 0,$$

which contradicts the linear independence of the vectors $x_i - x_j$, $i \neq j$.

Let us now prove that the representation of any p-simplex $H(x_0, \dots, x_p)$ in the form of a convex combination of its basic elements is unique. Suppose that the representation is not unique. Then

$$\sum_{i=0}^{p} \alpha_i x_i = \sum_{i=0}^{p} \beta_i x_i, \qquad \sum_{i=0}^{p} \alpha_i = \sum_{i=0}^{p} \beta_i = 1.$$

Hence it follows that

$$\sum_{i=0}^{p} (\alpha_i - \beta_i)(x_i - x_0) = \sum_{i=1}^{p} (\alpha_i - \beta_i) x_i = 0.$$

The linear independence of the vectors $x_i - x_0$, $i = 1, \dots, p$, implies the equalities $\alpha_i = \beta_i$, $i = 1, \dots, p$, which mean that $\alpha_0 = \beta_0$, too.

The set Ω_n for $n \geqslant 3$ is not a simplex. Indeed, if $\Pi_1, \dots, \Pi_N$, where $N = n!$, are the permutation matrices of order n, then

$$\Pi_i \notin H(\Pi_1, \dots, \Pi_{i-1}, \Pi_{i+1}, \dots, \Pi_N), \qquad i = 1, \dots, N,$$

and $\Pi_1, \dots, \Pi_N$ are the vertices of the polytope $H(\Pi_1, \dots, \Pi_N)$. Let us show that $\Pi_j - \Pi_1$, $j = 2, \dots, N$, are not linearly independent. The dimension of the vector space $M_{nn}(R)$ which includes Ω_n is equal to n^2.

Therefore, Ω_n cannot contain more than n^2 linearly independent matrices. But $n! - 1 > n^2$ for $n \geqslant 4$, and, consequently, $\Pi_j - \Pi_1$, $j = 2, \ldots, N$, are linearly dependent. Consideration of the case where $n = 3$ is left as an exercise for the reader. □

We put the dimension of p-simplex equal to p, and define the dimension of a convex set as the maximum dimension of the complexes contained in the set.

Theorem 2.2.4 *The dimension of the convex polytope Ω_n of doubly stochastic $n \times n$ matrices is no greater than $(n-1)^2$.*

Proof The polytope Ω_n is the intersection of $2n$ hyperplanes

$$\sum_{j=1}^{n} x_{ij} = 1, \qquad i = 1, \ldots, n,$$

$$\sum_{i=1}^{n} x_{ij} = 1, \qquad j = 1, \ldots, n.$$

Let us show that the space of solutions of this system of $2n$ equations has dimension equal to $(n-1)^2$. Indeed, if

$$X \in \Omega_n, \quad X = \|x_{ij}\|, \quad J = \|1\|, \quad i, j = 1, \ldots, n,$$

then the system can be written in matrix form:

$$XJ = J, \qquad JX = J.$$

The corresponding homogeneous system has the form

$$XJ = O_{n,n}, \qquad JX + O_{n,n},$$

where $O_{n,n}$ is the $n \times n$ matrix with zero elements.

It is well known that if A is a real symmetric $n \times n$ matrix, then there exists a real orthogonal matrix S such that

$$S^{-1}AS = \operatorname{diag}(r_1, \ldots, r_n),$$

where r_j, $j = 1, \ldots, n$, are the eigenvalues of the matrix A.

The matrix J is symmetric and has the unique non-zero eigenvalue equal to n. Therefore, there exists an orthogonal matrix S such that $J = nSE_{1,1}S^{-1}$, where $E_{1,1}$ is the matrix with unit in the entry $(1, 1)$ and with zeros in all the other entries. Now the homogeneous system takes the form

$$YE_{1,1} = O_{n,n}, \qquad E_{1,1}Y = O_{n,n},$$

where $Y = S^{-1}XS$. Any matrix Y whose first row and first column consist of zeros is a solution of this system. Thus, a solution of the homogeneous system considered as a vector has $2n-1$ non-zero coordinates and, consequently, the dimension of the solution space is equal to $n^2-(2n-1)=(n-1)^2$. Further, if Ω_n contains a p-simplex with vertices $Q_0,\dots,Q_p$, then Q_j-Q_0, $j=1,\dots,p$, are linearly independent and satisfy the system

$$(Q_j-Q_0)J = O_{n,n}, \quad J(Q_j-Q_0) = O_{n,n}, \quad J = 1,\dots,p.$$

Hence it follows that $p \leqslant (n-1)^2$. □

Note that this theorem can be strengthened, since, in fact, the dimension of Ω_n is equal to $(n-1)^2$. To this end it is sufficient to construct a polytope which is an $(n-1)^2$-simplex. This will be done in Subsection 2.2.3.

Corollary 2.2.3 *For any matrix $A \in \Omega_n$ there exists a representation in the form of a convex combination of permutation matrices $\Pi_1,\dots,\Pi_s$*

$$A = \alpha_1\Pi_1 + \cdots + \alpha_s\Pi_s$$

such that

$$s \leqslant (n-1)^2+1.$$

Proof Any $n\times n$ matrix A is an element of some simplex $H(\Pi_1,\dots,\Pi_s)$ from Ω_n. According to Theorem 2.2.4 the dimension of this simplex, equal to $s-1$, satisfies the inequality $s-1 \leqslant (n-1)^2$. □

2.2.3 Linearly independent permutation matrices

Consider the problem of constructing a set consisting of $(n-1)^2+1$ linearly independent permutation matrices. Determine a set of matrices $\{A_1,\dots,A_{m^2}\}$ as follows. We order the entries of an $m\times m$ matrix, assigning the number $i+(j-i)m \pmod{m^2}$ to the entry (i,j). This enumeration uses distinct numbers, since the condition

$$i_1+(j_1-i_1)m \equiv i_2+(j_2-i_2)m \pmod{m^2}$$

implies that $i_1 \equiv i_2 \pmod m$, that is, $i_1 = i_2$, and $j_1 m \equiv j_2 m \pmod{m^2}$. It follows from the last congruence that $j_1 = j_2$.

For the $m\times m$ matrix A_i we place units in the entries with numbers $i, i+1, \dots, i+m-2 \pmod{m^2}$ and zeros in all remaining entries. It is easy to see that each matrix A_i can be obtained from a permutation matrix by the substitution of zero for one of its units.

For $m = 4$ the numbers of entries of the matrix can be written in the following matrix form:

$$\begin{array}{rrrr} 1 & 5 & 9 & 13 \\ 14 & 2 & 6 & 10 \\ 11 & 15 & 3 & 7 \\ 8 & 12 & 0 & 4 \end{array}$$

According to this ordering of the entries the matrices $A_1, \ldots, A_{16}$ take the form

$$A_1 = \begin{bmatrix} 1 & 0 & 0 & 0 \\ 0 & 1 & 0 & 0 \\ 0 & 0 & 1 & 0 \\ 0 & 0 & 0 & 0 \end{bmatrix},$$

$$A_2 = \begin{bmatrix} 0 & 0 & 0 & 0 \\ 0 & 1 & 0 & 0 \\ 0 & 0 & 1 & 0 \\ 0 & 0 & 0 & 1 \end{bmatrix}, \ldots, A_{16} = \begin{bmatrix} 1 & 0 & 0 & 0 \\ 0 & 1 & 0 & 0 \\ 0 & 0 & 0 & 0 \\ 0 & 0 & 1 & 0 \end{bmatrix}. \qquad (2.2.5)$$

Lemma 2.2.1 *The matrices $A_1, \ldots, A_{m^2}$ are linearly independent.*

Proof Suppose that

$$\sum_{i=1}^{m^2} \alpha_i A_i = [0].$$

Taking into account that the addition in the entries is carried out modulo m^2, we calculate the elements with indices $i+m-2$ and $i+m-1$ of the matrix from the left-hand side of this equality. We obtain

$$\alpha_1 + \alpha_{i+1} + \cdots + \alpha_{i+m-2} = 0,$$

$$\alpha_{i+1} + \cdots + \alpha_{i+m-2} + \alpha_{i+m-1} = 0. \qquad (2.2.6)$$

Hence it follows that $\alpha_i = \alpha_{i+m-1}$. This equality holds for any i, therefore $\alpha_i = \alpha_{i+j(m-1)}$ for any i and j. Since $m-1$ and m^2 are relatively prime, we obtain $\alpha_1 = \alpha_2 = \cdots = \alpha_{m^2}$. It follows from (2.2.6) that they are all equal to zero. The lemma is thus proved. □

Now consider the set of $n \times n$ permutation matrices $\{I, \Pi_1, \ldots, \Pi_{(n-1)^2}\}$, where I is the identity matrix, and Π_i is the permutation matrix such that deleting the first row and the first column of the matrix we obtain the $(n-1) \times (n-1)$ matrix A_i, $i = 1, \dots, (n-1)^2$.

For example, if $n = 5$, then, according to (2.2.5),

$$\Pi_1 = \begin{bmatrix} 0 & 0 & 0 & 0 & 1 \\ 0 & 1 & 0 & 0 & 0 \\ 0 & 0 & 1 & 0 & 0 \\ 0 & 0 & 0 & 1 & 0 \\ 1 & 0 & 0 & 0 & 0 \end{bmatrix},$$

$$\Pi_2 = \begin{bmatrix} 0 & 1 & 0 & 0 & 0 \\ 1 & 0 & 0 & 0 & 0 \\ 0 & 0 & 1 & 0 & 0 \\ 0 & 0 & 0 & 1 & 0 \\ 0 & 0 & 0 & 0 & 1 \end{bmatrix}, \dots, \Pi_{16} = \begin{bmatrix} 0 & 0 & 0 & 0 & 1 \\ 0 & 1 & 0 & 0 & 0 \\ 0 & 0 & 1 & 0 & 0 \\ 1 & 0 & 0 & 0 & 0 \\ 0 & 0 & 0 & 1 & 0 \end{bmatrix}.$$

Theorem 2.2.5 (Blakley, Coppage and Dixon (1967)) *The matrices $I, \Pi_1, \dots, \Pi_{(n-1)^2}$ are linearly independent.*

Proof The element with indices $(1, 1)$ is equal to zero in each of the matrices $\Pi_1, \dots, \Pi_{(n-1)^2}$, since in constructing the matrix Π_i from A_i we add only two units corresponding to one zero row and one zero column in each A_i. Therefore, in any relation of the form

$$\alpha_0 I + \alpha_1 \Pi_1 + \cdots + \alpha_{(n-1)^2} \Pi_{(n-1)^2} = [0], \tag{2.2.7}$$

where not all the coefficients are equal to zero, the coefficient α_0 is zero. From (2.2.7) we obtain the equality

$$\alpha_1 A_1 + \cdots + \alpha_{(n-1)^2} A_{(n-1)^2} = [0],$$

where not all the coefficients are zeros, which contradicts the linear independence of $A_1, \dots, A_{(n-1)^2}$. The theorem is thus proved. □

2.2.4 König's theorems

The following theorem due to König is a corollary to Birkhoff's theorem.

Theorem 2.2.6 *Any non-negative integer matrix A with row and column sums equal to $t > 0$ can be represented as a sum of t permutation matrices $\Pi_1, \dots, \Pi_t$:*

$$A = \Pi_1 + \cdots + \Pi_t. \tag{2.2.8}$$

Note that Theorem 2.2.6 can also be proved directly (Ryser, 1963) by induction on t.

We introduce the notion of a semipermutation matrix. A $(0,1)$-matrix is called a semipermutation matrix if each row and each column of the matrix contains no more than one unit. The following generalization of Theorem 2.2.6 was also proved by König.

Theorem 2.2.7 *Any non-negative integer $m \times n$ matrix A with row and column sums no greater than t can be represented as a sum of t semipermutation matrices $P_1, \ldots, P_t$:*

$$A = P_1 + \cdots + P_t. \tag{2.2.9}$$

Proof The shortest proof of Theorem 2.2.7 is given in (Brualdi and Csima, 1992). Let the row and column sums of A be equal to $r_1, \ldots, r_m$ and $s_1, \ldots, s_n$ respectively. Consider the $(m+n) \times (m+n)$ matrix

$$A' = \begin{bmatrix} \begin{matrix} t-r_1 & & & \\ & t-r_2 & & \\ & & \ddots & \\ & & & t-r_m \end{matrix} & A \\ A^T & \begin{matrix} t-s_1 & & & \\ & t-s_2 & & \\ & & \ddots & \\ & & & t-s_m \end{matrix} \end{bmatrix},$$

where A^T is the transposed matrix A. The row and column sums of A' are equal to t. Therefore, according to Theorem 2.2.6,

$$A' = \Pi_1 + \cdots + \Pi_t,$$

where $\Pi_1, \ldots, \Pi_t$ are $(m+n) \times (n+m)$ permutation matrices.

Now let P_i be the submatrix of Π_i formed by the first m rows and the last n columns. It is clear that $P_1, \ldots, P_t$ are semipermutation matrices and (2.2.9) holds. □

2.3 Decomposition of probabilistic automata

2.3.1 Probabilistic automata

A probabilistic automaton is a tuple

$$\mathfrak{A} = \langle X, Y, S, \{P(y \mid x)\} \rangle,$$

where $X = \{x_1, \ldots, x_l\}$, $Y = \{y_1, \ldots, y_m\}$ are the input and output alphabets, $S = \{s_1, \ldots, s_n\}$ is the set of states, $\{P(y \mid x)\}$ is a family of $n \times n$ matrices

$$P(y \mid x) = \|p_{ij}(y \mid x)\|, \qquad i, j = 1, \ldots, n,$$

and $p_{ij}(y \mid x)$ is the conditional probability of transition from the state s_i to the state s_j, provided that $x \in X$ is at the input and $y \in Y$ is at the output of the automaton. It is clear that the matrix

$$P(x) = \sum_{y \in Y} P(y \mid x) = \|p_{ij}(x)\|$$

is such that

$$\sum_{i=1}^{n} p_{ij}(x) = 1, \qquad j = 1, \ldots, n,$$

that is, $P(x)$ is a stochastic matrix.

A probabilistic automaton acts in the following way. If $u = x_{i_1} x_{i_2} \ldots x_{i_t}$ is a word over the input alphabet, $v = y_{j_1} y_{j_2} \ldots y_{j_t}$ is a word over the output alphabet and $p_{ij}(v \mid u)$ is the probability of transition of the probabilistic automaton from the state s_i to the state s_j provided the word u is accepted at the input and the word v appears at the output, then

$$p_{ij}(v \mid u) = \sum_{r=1}^{n} p_{ir}(v_1 \mid u_1) p_{rj}(v_2 \mid u_2),$$

where $u = u_1 u_2$ and $v = v_1 v_2$, that is, u and v are composed from two subwords and the lengths of u_i and v_i coincide; $i = 1, 2$.

2.3.2 Decomposition

Consider a particular case of probabilistic automaton, the so-called probabilistic automaton without output:

$$\mathfrak{A} = \langle X, S, \{P(x)\} \rangle, \tag{2.3.1}$$

such that the transition from state s_i to state s_j, provided a word x is accepted at the input of the automaton, is realized with probability $p_{ij}(x)$ and

$$P(x) = \|p_{ij}(x)\|, \qquad i, j = 1, \ldots, n.$$

We now give some definitions and prove a lemma which will be needed later.

Recall that an $m \times n$ matrix Θ is called elementary if each row of the matrix contains exactly one unit and all other elements are zeros. Let us extend the notion of a stochastic matrix to rectangular matrices. An $m \times n$ non-negative matrix $A = \|a_{ij}\|$ is called stochastic if

$$\sum_{j=1}^{n} a_{ij} = 1, \qquad i = 1,\dots,m.$$

Lemma 2.3.1 *Any $m \times n$ stochastic matrix A can be represented in the form*

$$A = \sum_{j=1}^{r} \alpha_j \Theta_j, \quad \sum_{j=1}^{r} \alpha_j = 1, \quad \alpha_j \geqslant 0, \quad j = 1,\dots,r, \tag{2.3.2}$$

where $\Theta_1,\dots,\Theta_r$ are elementary matrices and

$$r \leqslant m(n-1) + 1. \tag{2.3.3}$$

Proof For an $m \times n$ stochastic matrix $A = \|a_{ij}\|$ put

$$\alpha_1 = \min_i \max_j a_{ij}$$

and consider the $m \times n$ elementary matrix $\Theta_1 = \|\theta_{ij}^{(1)}\|$, where $\theta_{ij}^{(1)}$ is equal to one if a_{ij} is the first maximal element in the ith row, and zero otherwise. It is clear that $A - \alpha_1 \Theta_1 \geqslant 0$, the matrix

$$A_1 = \frac{1}{1-\alpha_1}(A - \alpha_1 \Theta_1)$$

is stochastic and the number of positive elements in A_1 is less than the corresponding number in A. By induction we obtain

$$A_1 = \sum_{j=2}^{r} \alpha'_j \Theta_j, \quad \sum_{j=2}^{r} \alpha'_j = 1, \quad \alpha'_j \geqslant 0, \quad j = 2,\dots,r.$$

This representation implies (2.3.2), since $A = \alpha_1 \Theta_1 + (1-\alpha_1) A_1$. The number of steps in this procedure does not exceed $m(n-1)$; this yields inequality (2.3.3). □

Now consider a decomposition of a probabilistic automaton of the form of (2.3.1). Denote by P the stochastic $nl \times n$ matrix obtained as the union of $l = |X|$ stochastic matrices $P(x)$, $x \in X$. By Lemma 2.3.1 the matrix P can be represented as

$$P = \sum_{\nu=1}^{r} \alpha_\nu \Theta_\nu, \quad \sum_{\nu=1}^{r} \alpha_\nu = 1, \quad \alpha_\nu \geqslant 0, \quad \nu = 1,\dots,r, \tag{2.3.4}$$

where $\Theta_1, \dots, \Theta_r$ are $nl \times n$ elementary matrices.

Now the functioning of the automaton can be described as follows. Consider a sequence of independent trials with outcomes $w_1, \dots, w_r$. Let the probabilities of the outcomes $w_1, \dots, w_r$ be equal to $\alpha_1, \dots, \alpha_r$ respectively. An outcome w_v and a pair (x_t, s_i), which corresponds to the kth row of the matrix P, uniquely determine the next state of the automaton as the state s_j, where j is the number of the column which contains the unique unit of the kth row of the elementary matrix Θ_v. Then the corresponding transition probability is equal to

$$p'_{ij}(x_t) = \alpha_{v_1} + \cdots + \alpha_{v_m},$$

where $v_1, \dots, v_m$ are determined by the condition that the matrices $\Theta_{v_1}, \dots, \Theta_{v_m}$, and only these matrices, have a unit in the entry (k, j).

On the other hand, from (2.3.4) it follows that

$$p_{ij}(x_t) = \alpha_{v_1} + \cdots + \alpha_{v_m}.$$

Thus $p'_{ij}(x_t) = p_{ij}(x_t)$ and the suggested interpretation is equivalent to an automaton of form (2.3.1).

2.3.3 Probabilistic transformers

A finite, abstract, determinate automaton without input is a tuple

$$\mathfrak{A} = \langle X, S, S_0, f \rangle,$$

where $X = \{x_1, \dots, x_m\}$ is the input alphabet; $S = \{s_1, \dots, s_n\}$ is the set of states of the automaton; S_0 is the set of initial states; and $f : X \times S \to S$ is the transition function which determines the state s_t of the automaton at an instant t on the basis of the input symbol x_{t-1}, and the state s_{t-1} at the instant $t-1$ as $s_t = f(x_{t-1}, s_{t-1})$. An initial state $s_0 \in S_0$ determines the state of the automaton at the instant $t = 0$.

An automaton $\mathfrak{A}$ can be associated with a directed multigraph $\Gamma(\mathfrak{A})$, called the automaton graph. The vertices of the graph are the elements of the set S. A directed edge with a label $x \in X$ connects a vertex $s \in S$ and a vertex s' if when accepting the symbol x at the input the automaton $\mathfrak{A}$ goes from state s to state s'.

Consider a sequence of independent trials with outcomes $x_1, \dots, x_m$ which have probabilities $q_1, \dots, q_m$, $q_1 + \cdots + q_m = 1$, respectively. If the symbols obtained as a result of such trials go to the input of the automaton, then the states of the automaton form a random sequence which is a simple Markov chain with stationary transition probabilities.

Indeed, let the automaton graph $\Gamma(\mathfrak{A})$ have l multiple edges from s_i to s_j with labels $x_{i_1},\ldots,x_{i_l}$. Then the automaton $\mathfrak{A}$ transits into state s_j at an instant t, provided it is in state s_i at instant $t-1$, with probability

$$p_{ij} = q_{i_1} + \cdots + q_{i_l}, \tag{2.3.5}$$

since only one symbol appears at the input of the automaton $\mathfrak{A}$. These probabilities depend only on the state of the automaton at the preceding instance of time and define the matrix of transition probabilities

$$P = \|p_{ij}\|, \quad i,j = 1,\ldots,n, \quad \sum_{j=1}^{n} = 1,$$

of the corresponding Markov chain. A determinate automaton whose input accepts a sequence of symbols obtained by independent trials will be called a probabilistic transformer.

Suppose now that $S = \{1,\ldots,n\}$ and the matrix P is doubly stochastic. Then, as follows from Birkhoff's theorem,

$$P = \alpha_1\Pi_1 + \cdots + \alpha_r\Pi_r, \qquad \alpha_1 + \cdots + \alpha_r = 1, \tag{2.3.6}$$

where $\Pi_k = \|\delta_{\varphi_k(i),j}\|$, $i,j = 1,\ldots,n$, $k = 1,\ldots,r$, are permutation matrices and $\varphi_1,\ldots,\varphi_r$ are the corresponding permutations. Put

$$\{v_1,\ldots,v_d\} = \{v_l : \varphi_{v_l}(i) = j\}.$$

It is clear that

$$p_{ij} = \alpha_{v_1} + \cdots + \alpha_{v_d}. \tag{2.3.7}$$

Now consider an automaton $\mathfrak{A}'$. The set of states of $\mathfrak{A}'$ is S, the input alphabet is $X = \{\delta_1,\ldots,\delta_r\}$, where $\delta_k = (0,\ldots,0,1,0,\ldots,0)$ with 1 in the kth place, $k = 1,\ldots,r$. The transition function is determined by the substitutions $\varphi_1,\ldots,\varphi_r$:

$$f(\delta_k, s) = \varphi_k(s), \qquad s \in S.$$

Suppose we are given a device realizing independent trials with outcomes $1,\ldots,r$ and corresponding probabilities $\alpha_1,\ldots,\alpha_r$, $\alpha_1 + \cdots + \alpha_r = 1$. If an outcome i is realized, then the input of the automaton $\mathfrak{A}'$ accepts δ_i and the automaton transits from a state s to the state $\varphi_i(s)$.

If the input of the automaton $\mathfrak{A}'$ accepts a sequence of independent symbols from the set $1,\ldots,r$, then the sequence of its states is a simple Markov chain with the matrix of stationary transition probabilities = $\|p_{ij}\|$, $i,j = 1,\ldots,n$.

Indeed, the transition from a state i to state j occurs if and only if one of the outcomes $v_1,\ldots,v_d$ such that $\varphi_{v_k}(i) = j$, $k = 1,\ldots,d$, is realized. Since the outcomes are mutually exclusive, the corresponding probability is equal to $\alpha_{v_1} + \cdots + \alpha_{v_d}$. From (2.3.7) it follows that this probability is equal to p_{ij}, that is, it coincides with the transition probability of the Markov chain corresponding to the automaton $\mathfrak{A}$.

Thus the Markov chain corresponding to the finite automaton $\mathfrak{A}$ can be obtained by use of the automaton $\mathfrak{A}'$.

2.4 Permanents

2.4.1 Definition and properties of a permanent

Let $A = \|a_{ij}\|$, $i = 1,\ldots,n$, $j = 1\ldots,m$, $n \leqslant m$, be a matrix with elements from a ring F. The permanent of A is the function taking values from F determined by the equality

$$\operatorname{per} A = \sum_{j_1,\ldots,j_n} a_{1j_1} \cdots a_{nj_n}, \tag{2.4.1}$$

where the summation is over all different arrangements of size n from m distinct elements. If $m = n$, then the summation is carried out over all permutations of the elements $1,\ldots,n$, and the permanent of a matrix can be obtained from the determinant of the matrix if all summands are taken to be positive.

Consider some obvious properties of permanents similar to the properties of determinants:

(1) If a row of an $n \times m$ matrix A with $n \leqslant m$ consists of zeros, then $\operatorname{per} A = 0$. If $m = n$, then the same assertion is also valid for columns.

(2) The multiplication of one of the n rows of A with $n \leqslant m$ by an element $c \in F$ changes $\operatorname{per} A$ to $c \operatorname{per} A$. If $m = n$, then the same assertion is also valid for columns.

(3) Transposition of rows and columns of a matrix A does not change its permanent; in other words, if Π_1 and Π_2 are permutation matrices of orders n and m, respectively, then

$$\operatorname{per}(\Pi_1 A \Pi_2) = \operatorname{per} A. \tag{2.4.2}$$

(4) If A_{ij} is obtained by deleting of the ith row and jth column from a matrix A, and A'_{ij} is obtained from A by the substitution of 0 for a_{ij}, then

$$\operatorname{per} A = \operatorname{per} A'_{ij} + a_{ij} \operatorname{per} A_{ij}. \tag{2.4.3}$$

This relation can be proved by the partition of all elements of the permanent of A into two classes such that one of the classes contains all terms with element a_{ij} and the other class consists of all remaining terms. The sum of the terms containing a_{ij} is equal to $a_{ij}\operatorname{per}A_{ij}$ and the remaining terms are summed to $\operatorname{per}A'_{ij}$.

Equality (2.4.3) will be referred to as the decomposition of the permanent with respect to the element a_{ij}.

(5) If $a_{ij_1},\ldots,a_{ij_k}$ are the non-zero elements of the ith row, then

$$\operatorname{per}A=\sum_{\nu=1}^{k}a_{ij_\nu}\operatorname{per}A_{ij_\nu}. \tag{2.4.4}$$

This equality is called the decomposition of the permanent with respect to the ith row. It is obtained by the sequential decomposition of the permanent with respect to the non-zero elements of the ith row.

(6) Let $A[i_1,\ldots,i_r \mid j_1,\ldots,j_s]$ be the submatrix of a matrix A placed on the intersection of the rows numbered $i_1,\ldots,i_r$ and the columns numbered $j_1,\ldots,j_s$, and let $A(i_1,\ldots,i_s \mid j_1,\ldots,j_s)$ be the submatrix obtained by deleting those rows and columns with numbers $i_1,\ldots,i_r$ and $j_1,\ldots,j_s$ respectively. Using this notation, for $r\leqslant s$, $n-r\leqslant m-s$,

$$\operatorname{per}A=\sum_{i_1,\ldots,i_s}\operatorname{per}A[i_1,\ldots,i_r \mid j_i,\ldots,j_s]\operatorname{per}A(i_1,\ldots,i_r \mid j_1,\ldots,j_s), \tag{2.4.5}$$

where the summation is over all combinations of size s from the set $1,\ldots,m$. Equation (2.4.5) is called the decomposition of the permanent by the Laplace formula.

Let A be a square matrix consisting of square matrices $A_1,\ldots,A_k$ placed along the main diagonal and such that the elements of A which are outside these matrices are equal to zero. In such a case we write

$$A=\operatorname{diag}(A_1,\ldots,A_k).$$

As a corollary to equality (2.4.5) we obtain

$$\operatorname{per}A=\operatorname{per}A_1\cdots\operatorname{per}A_k. \tag{2.4.6}$$

Many of the properties of a permanent are similar to the corresponding ones of the determinant of a square matrix, but in general the relation

$$\det AB=\det A\det B$$

is not true for permanents. In contrast to the determinant, the permanent of a matrix with linearly dependent rows or columns is not necessarily zero.

2.4.2 The number of transversals

The following theorem demonstrates one of the most important applications of permanents.

Theorem 2.4.1 *Let $A = \|a_{ij}\|$, $i = 1,\dots,n$, $j = 1,\dots,m$, $n \leqslant m$, be the incidence matrix of sets $A_1,\dots,A_n$ which are subsets of a set $X = \{x_1,\dots,x_m\}$, that is,*

$$a_{ij} = \begin{cases} 1, & x_j \in X_i, \\ 0, & x_j \notin X_i. \end{cases}$$

Then

$$R(X_1,\dots,X_n) = \operatorname{per} A,$$

where $R(X_1,\dots,X_n)$ is the number of transversals of the family $(X_i : 1 \leqslant i \leqslant n)$.

Proof Indeed, a set $(x_{j_1},\dots,x_{j_n})$ is a system of distinct representatives if and only if $a_{1j_1}\cdots a_{nj_n} = 1$ and the elements $a_{1j_1},\dots,a_{nj_n}$ are non-collinear. Summing the numbers $a_{1j_1}\cdots a_{nj_n}$ over all arrangements $(j_1,\dots,j_n)$ we obtain, on the one hand, the number of all transversals and, on the other, the value of $\operatorname{per} A$. □

2.4.3 The ménage problem

Consider the problem of finding the number of transversals of the family $(X_i : 1 \leqslant i \leqslant n)$, where $X_i = X \setminus \{i\}$ and $X = \{1,\dots,n\}$. This problem has various different formulations and is usually referred as the ménage problem. In this case the incidence matrix is equal to $A_n = \|1 - \delta_{ij}\|$, $i, j = 1,\dots,n$, where δ_{ij} is Kronecker's symbol.

Put $h_n = \operatorname{per}\|1 - \delta_{ij}\|$, and decompose the permanent with respect to the first row. We then find that

$$h_n = (n-1)\operatorname{per} A'_{n-1},$$

where the matrix A'_{n-1} is obtained from the matrix A by substituting 1 for

0 in the entry (1, 1). Decomposing per A'_{n-1} with respect to the element in the entry (1, 1), we find that

$$\operatorname{per} A'_{n-1} = h_{n-1} + h_{n-2}.$$

Thus, putting $h_0 = 1$, $h_1 = 0$, we obtain the recurrence relation

$$h_n = (n-1)(h_{n-1} + h_{n-2}), \qquad n = 2, 3, \ldots,$$

which can be rewritten in the form

$$\begin{aligned} h_n - nh_{n-1} &= -(h_{n-1} - (n-1)h_{n-2}) = \cdots \\ &= (-1)^k (h_{n-k} - (n-k)h_{n-k-1} = \cdots = (-1)^n. \end{aligned}$$

Putting $h_n = n!\, u_n$, we see that

$$u_n - u_{n-1} = \frac{(-1)^n}{n!}.$$

Hence it follows that

$$u_n = \sum_{k=0}^{n} \frac{(-1)^k}{k!},$$

and the final result is

$$h_n = n! \sum_{k=0}^{n} \frac{(-1)^k}{k!}.$$

2.4.4 The Fibonacci numbers

Let us find the number of transversals of the family $(X_i : 1 \leqslant i \leqslant n)$, where

$$X_1 = \{1, 2\}, \qquad X_2 = \{1, 2, 3\},$$
$$X_3 = \{2, 3, 4\}, \ldots, X_{n-1} = \{n-2, n-1, n\}, \qquad X_n = \{n-1, n\}$$

are subsets of the set $X = \{1, \ldots, n\}$. This number is equal to

$$C_n = \operatorname{per} \begin{bmatrix} 1 & 1 & & & & 0 \\ 1 & 1 & 1 & & & \\ & 1 & 1 & 1 & & \\ & & & \ddots & & \\ & & & 1 & 1 & 1 \\ 0 & & & & 1 & 1 \end{bmatrix}.$$

Decomposing this permanent with respect to the first row, we obtain the recurrence relation

$$C_n = C_{n-1} + C_{n-2}, \quad C_1 = 1, \quad C_2 = 2.$$

The numbers C_n are called the Fibonacci numbers.

2.4.5 The assignment problem

Consider a set X of tasks of m kinds and n workers numbered $1,\dots,n$ which can fulfil the tasks from sets $X_1,\dots,X_n$ respectively. Denote by d_{nm} the number of ways of assigning the n workers to fulfil n from the m tasks, $n \leqslant m$, provided each worker fulfils only one task. If A is the incidence matrix of $X_1,\dots,X_n$ and X, then it is clear that

$$d_{nm} = \operatorname{per} A.$$

Now let r_{ij} be the cost of the fulfilment of the jth task by the ith worker and $R = \|r_{ij}\|$, $i = 1,\dots,n$, $j = 1,\dots,m$. Suppose that all possible assignments are taken at random with equal probabilities, and let ξ_{nm} be the total cost of the tasks under the random assignment. Then, obviously, the mean value of the cost is equal to

$$\mathbf{E}\xi_{nm} = \frac{\operatorname{per} R}{\operatorname{per} A}.$$

2.4.6 The dimer problem

The dimer problem arises in investigations of the absorption of diatomic molecules on some surface. As is well known, absorption is the taking up of some substance from a gas or solution by the surface of another fluid or solid substance. This phenomenon is common in nature, particularly in soil and living organisms. It is used in various areas of technology, in the sugar and oil industries, for water purification, air conditioning, etc.

The dimer problem consists of finding the number of ways of combining the atoms into diatomic molecules, called dimers, so that a plane lattice whose span equals the length of the dimer, may be completely covered by the dimers, provided that each dimer covers two adjacent points of the lattice. It is clear that the number of points to be covered must be even.

Consider a regular $2M \times 2N$ rectangular lattice which obviously has $4MN$ points. A dimer connects two adjacent points on horizontal or

Fig. 2.4.1. Lattice in the case of $M = 2$, $N = 3$

vertical lines so that the $2MN$ dimers completely cover the lattice. Let $D(M, N)$ be the number of such coverings. The dimer problem consists of finding this number.

For the lattice point we introduce a coordinate system with zero at the left upper vertex of the rectangle. Denote by B and C the sets of points with even and odd sums of coordinates respectively. It is clear that a dimer covers a pair of points of the lattice such that one part of it lies in B and the other lies in C. Thus each covering corresponds to a bijection of B onto C. Such a bijection is associated with a substitution φ of degree $2MN$, acting on the set $X = \{1,\ldots,2MN\}$.

We now define a $2MN \times 2MN$ square matrix A such that each collection of $2MN$ non-collinear units of the matrix corresponds to some covering of the lattice by dimers. For the sake of clarity we consider first an example where $M = 2$ and $N = 3$. The lattice in this case is of the form given in Figure 2.4.1. We label the elements of B with circled numbers and the elements of C otherwise. In this case the set $X = \{1,\ldots,12\}$. Let us now construct the 12×12 square $(0, 1)$-matrix A. A dimer, together with the point ①, can cover points 1 and 4; therefore the first row of the matrix A contains units in those positions with numbers 1 and 4 and zeros in all the others. Point ② can be covered by a dimer together with one of the points 1, 2 and 5. Therefore, in the second row of the matrix A units are in positions 1, 2 and 5, etc. Finally, we obtain

$$
A = \begin{array}{c|cccccccccccc|}
 & 1 & 2 & 3 & 4 & 5 & 6 & 7 & 8 & 9 & 10 & 11 & 12 \\
\hline
1 & 1 & & & 1 & & & & & & & & \\
2 & 1 & 1 & & & 1 & & & & & & & \\
3 & & 1 & 1 & & & 1 & & & & & & \\
4 & 1 & & & 1 & 1 & & 1 & & & & & \\
5 & & 1 & & & 1 & 1 & & 1 & & & & \\
6 & & & 1 & & & 1 & & & 1 & & & \\
7 & & & & 1 & & & 1 & & & 1 & & \\
8 & & & & & 1 & & 1 & 1 & & & 1 & \\
9 & & & & & & 1 & & 1 & 1 & & & 1 \\
10 & & & & & & & 1 & & & 1 & 1 & \\
11 & & & & & & & & 1 & & & 1 & 1 \\
12 & & & & & & & & & 1 & & & 1 \\
\hline
\end{array},
$$

where zeros are situated in the empty entries.

Introduce the $N \times N$ matrices

$$
I_N = \begin{bmatrix} 1 & & & 0 \\ & 1 & & \\ & & \ddots & \\ 0 & & & 1 \end{bmatrix}, \qquad E_N = \begin{bmatrix} 1 & & & & 0 \\ 1 & 1 & & & \\ & 1 & 1 & & \\ & & & \ddots & \\ 0 & & & 1 & 1 \end{bmatrix}.
$$

In the case $M = 2$ and $N = 3$ the matrix A can now be written as follows:

$$
A = \begin{bmatrix} E_3 & I_3 & & 0 \\ I_3 & E_3^T & I_3 & \\ & I_3 & E_3 & I_3 \\ 0 & & I_3 & E_3^T \end{bmatrix}.
$$

Similar, but more cumbersome reasoning, shows that, in the general case,

$$
A = \begin{bmatrix} E_N & I_N & & & & 0 \\ I_N & E_N^T & I_N & & & \\ & I_N & E_N & I_N & & \\ & & & \ddots & & \\ & & & I_N & E_N' & I_N \\ 0 & & & & I_N & E_N'' \end{bmatrix},
$$

where $E_N' = E_N^T$ and $E_N'' = E_N$ if the number of matrices on the main diagonal is odd, and $E'_N = E_N$, $E''_N = E_N^T$ if that number is even. The solution of the dimer problem can now be given by the equality

$$D(M,N) = \operatorname{per} A,$$

where the matrix A is given above (Percus, 1969).

2.4.7 König's theorem

Consider a (0,1)-matrix $A = \|a_{ij}\|$, $i = 1,\ldots,n$, $j = 1,\ldots,m$, $n \leqslant m$. As usual, a row or a column of a matrix will be referred to as a line. We say that a line covers an element of a matrix if the element belongs to the line.

The minimum number of lines covering all positive elements of the (0,1)-matrix A is called the line rank of the matrix, denoted by $\rho = \rho(A)$. The maximum number of non-collinear positive elements of the (0,1)-matrix A is called the boundary rank, denoted by $r = r(A)$. It is easily seen that both notions may be directly extended to arbitrary non-negative matrices.

The permanent rank $P(A)$ of the (0,1)-matrix A is the maximum order of its square submatrix with non-zero permanent. It follows directly from the definition that

$$P(A) = r(A).$$

In particular, $\operatorname{per} A > 0$ if and only if $r(A) = n$. This implies the following assertion.

Lemma 2.4.1 *If A is the incidence matrix of a family $(X_i : 1 \leqslant i \leqslant n)$, where $X_i \subset X$, $\|X\| = m$, then this family has a transversal if and only if $r(A) = n$.*

The following theorem, proved by König using graph-theoretic methods (König, 1950), is of considerable importance in various applications.

Theorem 2.4.2 *The boundary rank of a (0,1)-matrix A coincides with its line rank, that is,*

$$r(A) = \rho(A).$$

Proof The lines determining the line rank include all the positive elements that determine the boundary rank; therefore

$$r(A) \leqslant \rho(A).$$

We prove that $r(A) \geqslant \rho(A)$. Suppose that the minimal covering determining $\rho(A)$ consists of k rows and l columns, that is, $\rho(A) = k+l$. Since $r(A)$ and $\rho(A)$ preserve their values under permutations of rows and columns, we can assume that the matrix A has the form

$$A = \begin{bmatrix} A_1 & A_2 \\ A_3 & 0 \end{bmatrix},$$

where A_1 is a $k \times l$ matrix. Let us show that $r(A_2) = k$ and $r(A_3) = l$. Consider A_2 as the incidence matrix of the subsets $X_1, \ldots, X_k$ of the set $X = \{l+1, \ldots, m\}$, where

$$X_i = \{j : a_{ij} > 0,\ j = l+1, \ldots, m\}, \qquad i = 1, \ldots, k.$$

The family $(X_i : 1 \leqslant i \leqslant k\}$ satisfies P. Hall's conditions. Indeed, the inequality

$$|X_{v_1} \cup \cdots \cup X_{v_s}| < s,$$

for some $1 \leqslant v_1 < \cdots < v_s \leqslant k$, $1 \leqslant s \leqslant k$, means that the positive elements of the rows with numbers $v_1, \ldots, v_s$ can be covered by fewer than s columns. This is impossible since the covering determining the rank is minimal. Thus a transversal of the family $(X_i : 1 \leqslant i \leqslant k)$ exists and $r(A_2) = k$ by Lemma 2.4.1.

By similar reasoning for the transpose A_3^T we obtain $r(A_3^T) = r(A_3) = l$. Hence it follows that

$$r(A) \geqslant k + l = \rho(A).$$

The theorem is thus proved. □

It should be noted that König's theorem remains true if instead of (0, 1)-matrices we consider non-negative matrices, that is, matrices with non-negative elements. Let us now clarify the conditions on the location of the positive elements of a non-negative matrix A under which $\operatorname{per} A = 0$.

Theorem 2.4.3 *Let $A = \|a_{ij}\|$, $i = 1, \ldots, n$, $j = 1, \ldots, m$, $n \leqslant n$, be a non-negative matrix. The equality* $\operatorname{per} A = 0$ *holds if and only if the matrix A has a zero $r \times s$ submatrix such that $r + s > m$.*

Proof If $\operatorname{per} A = 0$, then $r(a) = \rho(A) < n$ by Theorem 2.4.2. For a minimum covering with k rows and l columns, $\rho(A) = k + l$ and the zero submatrix which is placed out of the covering has $r = n - k$ rows and $s = m - l$ columns. Thus,

$$r + d = m + n - (k + l) > m.$$

Now let the matrix A have a zero $r \times s$ submatrix such that $r + s > m$. Without loss of generality, we may assume that the submatrix is placed at the bottom right-hand corner of the matrix. It is clear that a minimum covering with k rows and l columns satisfies the condition $k + l \leqslant (n-r)+(m-s)$. Hence it follows that $\rho(A) = k+l \leqslant m+n-(r+s) < n$, that is, $\operatorname{per} A = 0$. □

In conclusion we formulate Frobenius' theorem (Frobenius, 1912).

Theorem 2.4.4 *All terms of the sum which represents the determinant of an $n \times n$ matrix A are equal to zero if and only if the matrix A contains a zero $r \times (n - r + 1)$ submatrix for some r, $1 \leqslant r \leqslant n$.*

It is clear that this theorem is a corollary to Theorem 2.4.2.

2.5 Calculation of permanents

We want to examine methods of calculation of permanents of square matrices which possess some symmetry. This symmetry lets us reduce the problem either to the calculation of permanents of the simplest matrices or to recurrence relations by which the values of permanents can be obtained sequentially for matrices of increasing orders.

2.5.1 The permanent of a linear combination of two permutation matrices

Let φ_1 and φ_2 be substitutions of degree n and let Π_1 and Π_2 be the permutation matrices corresponding to φ_1 and φ_2 respectively. The matrix $\alpha\Pi_1 + \beta\Pi_2$, where α, β are real numbers, is a linear combination of the matrices Π_1 and Π_2. Let us show that

$$\operatorname{per}(\alpha\Pi_1 + \beta\Pi_2) = \prod_{i=1}^{k}(\alpha^{l_j} + \beta^{l_j}), \tag{2.5.1}$$

where $l_1, \ldots, l_k$ are the cycle lengths of the substitution $\varphi_1^{-1}\varphi_2$.

Suppose that the substitution $\varphi_1^{-1}\varphi_2$ is of the form

$$\varphi_1^{-1}\varphi_2 = (a_1, \ldots, a_{l_1})(b_1, \ldots, b_{l_2}) \cdots (c_1, \ldots, c_{l_k}),$$

and consider the substitution

$$\psi = \begin{pmatrix} a_1 & \ldots & a_{l_1} & b_1 & \ldots & b_{l_2} & \ldots & c_1 & \ldots & c_{l_k} \\ 1 & \ldots & l_1 & l_1+1 & \ldots & l_1+l_2 & \ldots & n-l_k+1 & \ldots & n \end{pmatrix}.$$

It is easy to check that

$$\psi^{-1}\varphi_1^{-1}\varphi_2\psi = (1,\ldots,l_1)(l_1+1,\ldots,l_1+l_2)\cdots(n-l_k+1,\ldots,n).$$

Denote by $\tilde{\Pi}$ the permutation matrix corresponding to the substitution ψ. Let E be the identity matrix and $\bar{\Pi} = \tilde{\Pi}^{-1}\Pi_1^{-1}\Pi_2\tilde{\Pi}$; then

$$\operatorname{per}(\alpha\Pi_1 + \beta\Pi_2) = \operatorname{per}(\alpha E + \beta\bar{\Pi}),$$

where $\bar{\Pi} = \operatorname{diag}(\bar{\Pi}_1,\ldots,\bar{\Pi}_k)$, and $\bar{\Pi}_i$ is the $l_i \times l_i$ permutation matrix corresponding to the cycle $(1,\ldots,l_i)$, $i = 1,\ldots,k$. Thus

$$\operatorname{per}(\alpha\Pi_1 + \beta\Pi_2) = \operatorname{per}\operatorname{diag}(A_1,\ldots,A_k),$$

where the $l_i \times l_i$ matrix A_i is of the form

$$A_i = \begin{bmatrix} \alpha & \beta & & & & 0 \\ & \alpha & \beta & & & \\ & & \alpha & \beta & & \\ & & & \ddots & & \\ 0 & & & & \alpha & \beta \\ \beta & & & & & \alpha \end{bmatrix}, \qquad i = 1,\ldots,k. \tag{2.5.2}$$

Decomposing $\operatorname{per} A_i$ with respect to the element from the lower left corner, we obtain

$$\operatorname{per} A_i = \alpha^{l_i} + \beta^{l_i}, \qquad i = 1,\ldots,k. \tag{2.5.3}$$

Thus,

$$\operatorname{per}(\alpha\Pi_1 + \beta\Pi_2) = \prod_{i=1}^{k}(\alpha^{l_i} + \beta^{l_i}).$$

Formula (2.5.1) is proved.

For $\alpha = \beta = 1$ it follows from (2.5.1) that

$$\operatorname{per}(\Pi_1 + \Pi_2) = 2^k, \tag{2.5.4}$$

where k is the number of cycles in the substitution $\varphi_1^{-1}\varphi_2$.

2.5.2 The permanent of a linear combination of three permutation matrices

Denote by C the permutation matrix corresponding to the unicyclic substitution C of degree n equal to $C = (1,\ldots,n)$. Consider a linear combination

$$Q_n = \alpha E + \beta C + \gamma C^2,$$

where α, β, γ are real numbers and E is the identity matrix. Let us show that, for $n > 3$,

$$\operatorname{per}(\alpha E + \beta C + \gamma C^2) = r_1^n + r_2^n + \alpha^n + \gamma^n, \tag{2.5.5}$$

where r_1, r_2 are the roots of the equation

$$x^2 - \beta x - \alpha\gamma = 0.$$

Put

$$D_n = \operatorname{per}\begin{bmatrix} \beta & \gamma & & & & 0 \\ \alpha & \beta & \gamma & & & \\ & \alpha & \beta & \gamma & & \\ & & & \ddots & & \\ & & & \alpha & \beta & \gamma \\ 0 & & & & \alpha & \beta \end{bmatrix}. \tag{2.5.6}$$

Decomposing this permanent with respect to the first row, we obtain the recurrence relation

$$D_n = \beta D_{n-1} + \alpha\gamma D_{n-2}. \tag{2.5.7}$$

The characteristic equation corresponding to this finite difference equation with constant coefficients is of the form

$$x^2 - \beta x - \alpha\gamma = 0. \tag{2.5.8}$$

Therefore,

$$D_n = C_1 r_1^n + C_2 r_2^n,$$

where r_1 and r_2 are the roots of the characteristic equation, is the general solution of (2.5.7). The constants C_1 and C_2 are determined by the initial conditions

$$D_0 = 1, \qquad D_1 = \beta,$$

and are equal to

$$C_1 = \frac{r_1}{\mu}, \qquad C_2 = \frac{r_2}{\mu}$$

if $\mu = \sqrt{\beta^2 + 4\alpha\gamma} \neq 0$. Hence, we finally obtain

$$D_n = \frac{1}{\mu}(r_1^{n+1} - r_2^{n+1}), \tag{2.5.9}$$

where

$$r_1 = \frac{\beta + \mu}{2}, \qquad r_2 = \frac{\beta - \mu}{2}.$$

If $\mu = 0$, then

$$D_n = (n+1)(\beta/2)^n.$$

Decomposing $\mathrm{per}(\alpha E + \beta C + \gamma C^2)$ with respect to three non-zero elements from the bottom left-hand corner, we obtain the recurrence relation

$$\mathrm{per}(\alpha E + \beta C + \gamma C^2) = \beta D_{n-1} + 2\alpha\gamma D_{n-2} + \alpha^n + \gamma^n. \qquad (2.5.10)$$

Substituting the expression of D_n from (2.5.9) into (2.5.10), and carrying out some simplifications, we obtain equality (2.5.5).

2.5.3 Matrix factorization

We may want to calculate $\mathrm{per}\, A$ by a matrix factorization which reduces the problem to multiplication of matrices of essentially greater orders than the initial matrix A. This approach was suggested in (Frobenius, 1912). Let a matrix A with complex elements be of the form

$$A = \begin{bmatrix} a_1 & a_2 & \dots & a_n \\ b_1 & b_2 & \dots & b_n \\ \vdots & \vdots & \ddots & \vdots \\ c_1 & c_2 & \dots & c_n \end{bmatrix}.$$

Consider an algebra over the field of complex numbers containing a set $E_1, \dots, E_n$ of abstract elements satisfying the conditions

$$E_i E_j = E_j E_i, \quad i \neq j, \quad E_i^2 = 0, \quad i = 1, \dots, n, \quad E_1 \cdots E_n \neq 0. \qquad (2.5.11)$$

Taking into account equalities (2.5.11) valid in this algebra, we obtain the formula

$$\mathrm{per}\, A E_1 \cdots E_n = (a_1 E_1 + \cdots + a_n E_n)(b_1 E_1 + \cdots + b_n E_n) \cdots (c_1 E_1 + \cdots + c_n E_n), \qquad (2.5.12)$$

which gives a representation of the permanent of the matrix A. If we take matrices of a special form as the elements $E_1, \dots, E_n$, the calculation of permanents by formula (2.5.12) is reduced to multiplication of matrices. Put

$$I = \begin{bmatrix} 1 & 0 \\ 0 & 1 \end{bmatrix}, \qquad L = \begin{bmatrix} 0 & 0 \\ 1 & 0 \end{bmatrix}$$

and introduce the n-multiple product

$$E_i = I \otimes \cdots \otimes I \otimes L \otimes I \otimes \cdots \otimes I,$$

where L is in the ith place and the symbol $\otimes$ means the Kronecker product of matrices defined by the rule

$$A \otimes B = \begin{bmatrix} a_{11}B & \dots & a_{1n}B \\ \vdots & \ddots & \vdots \\ a_{n1}B & \dots & a_{nn}B \end{bmatrix}.$$

Each of the matrices $E_1, \dots, E_n$ has order 2^n. The validity of properties (2.5.11) follows from the formula

$$(A_1 \otimes \cdots \otimes A_n)(B_1 \otimes \cdots \otimes B_n) = (A_1B_1) \otimes \cdots \otimes (A_nB_n),$$

which is fulfilled for the Kronecker product. It is also obvious that

$$E_1 \dots E_n = L \otimes \cdots \otimes L$$

is the $2^n \times 2^n$ matrix with unit in the entry $(2^n, 1)$ and zeros in all the other entries.

Now consider, say, the first factor in (2.5.12):

$$P(a_1, \dots, a_n) = a_1E_1 + \cdots + a_nE_n.$$

It follows from the properties of the Kronecker product that

$$P(a_1, \dots, a_n) = a_1L \otimes I \otimes \cdots \otimes I + I \otimes P(a_2, \dots, a_n).$$

This equality can be written in matrix form:

$$P(a_1, \dots, a_n) = \begin{bmatrix} P(a_2, \dots, a_n) & 0 \\ a_1I & P(a_2, \dots, a_n) \end{bmatrix}, \qquad (2.5.13)$$

where I is the identity matrix and 0 is the zero matrix of order 2^{n-1}. By (2.5.13) we can calculate $P(a_1, \dots, a_n)$ using the initial condition

$$P(a_1) = \begin{bmatrix} 0 & 0 \\ a_1 & 0 \end{bmatrix}. \qquad (2.5.14)$$

As a result we obtain a $2^n \times 2^n$ matrix. It is clear that the similarly defined $P(b_1, \dots, b_n), \dots, P(c_1, \dots, c_n)$ also satisfy relations of the form (2.5.13), which let us calculate the corresponding $2^n \times 2^n$ matrices.

Thus the calculation of a permanent of order n is reduced to multiplication of n matrices of order 2^n, which are calculated as the Kronecker products of 2×2 matrices such that the ith of these matrices depends on the ith row of the initial matrix A.

For $n = 2$,

$$A = \begin{bmatrix} a_1 & a_2 \\ b_1 & b_2 \end{bmatrix},$$

and

$$P(a_1, a_2) = \begin{bmatrix} 0 & 0 & 0 & 0 \\ a_2 & 0 & 0 & 0 \\ a_1 & 0 & 0 & 0 \\ 0 & a_1 & a_2 & 0 \end{bmatrix}, \qquad P(a_1, a_2) = \begin{bmatrix} 0 & 0 & 0 & 0 \\ b_2 & 0 & 0 & 0 \\ b_1 & 0 & 0 & 0 \\ 0 & b_1 & b_2 & 0 \end{bmatrix}.$$

The product of these matrices is a matrix with zeros in all entries, except maybe the element in the bottom left-hand corner. This element is equal to $\operatorname{per} A = a_1 b_2 + a_2 b_1$. This result coincides with equality (2.5.12) for $n = 2$.

2.6 The inclusion–exclusion method

2.6.1 Derivation of the formulae

Consider N elements $a_1, \ldots, a_N$ with weights $\omega(a_1), \ldots, \omega(a_N)$ from a ring K, and n properties $A_1, \ldots, A_n$. Each of the elements possesses some of the properties $A_1, \ldots, A_n$. Denote by $M(r)$ the total weight of the elements which possess exactly r of the properties. We show that

$$M(r) = \sum_{k=r}^{n} (-1)^{k-r} \binom{k}{r} S_k, \qquad r = 0, 1, \ldots, n,$$

$$S_k = \sum_{1 \leqslant j_1 < \cdots < j_k \leqslant n} M(A_{j_1} \ldots A_{j_k}), \tag{2.6.1}$$

where $M(A_{j_1} \ldots A_{j_k})$ is the total weight of the elements, each of which possesses all the properties $A_{j_1}, \ldots, A_{j_k}$. To prove (2.6.1), usually referred to as the inclusion–exclusion formulae, we show that the weights of the elements with exactly r properties, and only these elements, are included in the right-hand side of (2.6.1).

Indeed, the weights of the elements with exactly r properties are present once only in the sum S_r and do not enter into the remaining sums S_k with $k > r$. The weight of an element which possesses $\mu > r$ properties appears in the sum S_k, $k > r$, exactly $\binom{\mu}{k}$ times. Therefore, the number of appearances of the weight of such an element in the total sum is equal to

$$\sum_{k=r}^{n}(-1)^{k-r}\binom{k}{r}\binom{\mu}{k}=\binom{\mu}{r}\sum_{j=0}^{\mu-r}(-1)^{j}\binom{\mu-r}{j}=0, \qquad \mu>r.$$

It is clear that the weights of the elements which possess fewer than r properties do not appear at all on the right-hand side of (2.6.1). Thus formulae (2.6.1) are proved.

2.6.2 A metric in the symmetric group

Let S_n be the symmetric group of degree n. We define a function ρ setting, for any pair of substitutions $s, s' \in S_n$:

$$\rho(s,s')=|\{i: s(i)\neq s'(i),\ 1\leqslant i\leqslant n,\ s,s'\in S_n\}|.$$

The function ρ possesses all the properties of a metric:

(1) $\rho(s,s')=0$ if and only if $s=s'$.
(2) $\rho(s,s')=\rho(s',s)$ for any s, s'.
(3) $\rho(s,s'')\leqslant\rho(s,s')+\rho(s',s'')$, that is, for any s, s', s'' the triangle inequality is valid.

Indeed, if

$$A=\{i: s(i)=s'(i)\},\quad B=\{i: s'(i)=s''(i)\},\quad C=\{i: s(i)=s''(i)\},$$

then obviously $A\cap B\subseteq C$. Hence, for the complements of the sets, the inclusion $\bar{A}\cup\bar{B}\supseteq\bar{C}$ holds and therefore $|\bar{C}|\leqslant|\bar{A}|+|\bar{B}|$, which is equivalent to the triangle inequality.

We now take a substitution $s_0\in S_n$ and for $s\in S_n$ define the properties

$$A_i=\{s(i)=s_0(i)\},\qquad i=1,\dots,n.$$

The number of substitutions which possess the properties $A_{j_1},\dots,A_{j_k}$, $1\leqslant j_1<\cdots<j_k\leqslant n$, is equal to $M(A_{j_1}\dots A_{j_k})=(n-k)!$. Hence the quantity

$$h_{n,r}=|\{s: \rho(s,s_0)=n-r,\ s\in S_n\}|$$

does not depend on s_0, and by the inclusion–exclusion formula,

$$h_{n,r}=\sum_{k=r}^{n}(-1)^{k-r}\binom{k}{r}\binom{n}{k}(n-k)!,\qquad r=0,1,\dots,n.$$

After some simplification we finally have

$$h_{n,r} = \frac{n!}{r!}\sum_{k=0}^{n-r}\frac{(-1)^k}{k!}, \qquad r = 0, 1, \dots, n.$$

2.6.3 The Euler function

As is well known, the Euler totient function $\varphi(m)$ is equal to the number of integers among $0, 1, \dots, m-1$ which are mutually prime to m. If

$$m = p_1^{\alpha_1} p_2^{\alpha_2} \cdots p_r^{\alpha_r}$$

is the canonical decomposition of m into prime factors, then

$$\varphi(m) = m\left(1 - \frac{1}{p_1}\right)\left(1 - \frac{1}{p_2}\right)\cdots\left(1 - \frac{1}{p_r}\right). \tag{2.6.2}$$

Let us derive this formula by the inclusion–exclusion method. As elements we take the integers $0, 1, \dots, m-1$. An element possesses a property A_i if it is divisible by p_i, $i = 1, \dots, r$. It is not difficult to see that the number of elements possessing properties $A_{i_1}, \dots, A_{i_k}$ is equal to

$$M(A_{i_1} \dots A_{i_k}) = \frac{m}{p_{i_1} \cdots p_{i_k}}.$$

It is clear that $\varphi(m)$ is the number of elements which possess none of the properties $A_1, \dots, A_r$. Applying formulae (2.6.1), we obtain

$$\frac{\varphi(m)}{m} = 1 + \sum_{k=1}^{r}(-1)^k \sum_{1 \leqslant i_1 < \cdots < i_k \leqslant r} \frac{1}{p_{i_1} \cdots p_{i_k}}.$$

It can easily be seen that this formula is equivalent to (2.6.2).

Equality (2.6.2) implies the multiplicative property of the function $\varphi(m)$. If m and n are mutually prime, that is, $(m, n) = 1$, then

$$\varphi(mn) = \varphi(m)\varphi(n).$$

We leave the proof of this assertion to the reader.

2.6.4 The Ryser formula

Consider a matrix $A = \|a_{ij}\|$, $i = 1, \dots, n$, $j = 1, \dots, m$, $n \leqslant m$, with elements from a ring K. For the calculation of the permanent of this matrix the following formula due to Ryser (Ryser, 1963) can be used:

$$\operatorname{per} A = \sum_{k=m-n}^{m} (-1)^{k-m+n} \binom{k}{m-n} S_k, \tag{2.6.3}$$

where

$$S_k = \sum_{1 \leqslant j_1 < \cdots < j_k \leqslant m} \prod_{i=1}^{n} (a_{i1} + \cdots + a_{im} - a_{ij_1} - \cdots - a_{ij_k}).$$

We use the inclusion–exclusion method to deduce (2.6.3). As elements we take m^n mappings of the set $\{1, \ldots, n\}$ into the set $\{1, \ldots, m\}$. We assign a weight $a_{1l_1} \ldots a_{nl_n}$ to the mapping

$$\begin{pmatrix} 1 & 2 & \ldots & n \\ l_1 & l_2 & \ldots & l_n \end{pmatrix}.$$

Denote by A_j the property of the mapping that there is no j among the numbers $l_1, \ldots, l_n$, $j = 1, \ldots, m$. By definition the permanent of A is equal to the total weight of the mappings which possess exactly $m - n$ properties from the set $A_1, \ldots, A_n$. If $M(A_{j_1} \ldots A_{j_k})$ is the total weight of the mappings, each of which possesses the properties $A_{j_1}, \ldots, A_{j_k}$, then applying the inclusion–exclusion method we obtain

$$\operatorname{per} A = \sum_{k=m-n}^{m} (-1)^{k-m+n} \binom{k}{m-n} \sum_{1 \leqslant j_1 < \cdots < j_k \leqslant m} M(A_{j_1} \ldots A_{j_k}). \quad (2.6.4)$$

The total weight of all mappings is equal to

$$\prod_{i=1}^{n} (a_{i1} + \cdots + a_{im}).$$

To obtain $M(A_{j_1} \ldots A_{j_k})$ from this product, it is sufficient to substitute zeros for the elements $a_{ij_1}, \ldots, a_{ij_k}$ for $i = 1, \ldots, n$. Thus,

$$M(A_{j_1} \ldots A_{j_k}) = \prod_{i=1}^{n} (a_{i1} + \cdots + a_{im} - a_{ij_1} - \cdots - a_{ij_k}). \quad (2.6.5)$$

Inserting expression (2.6.5) into (2.6.4), we obtain (2.6.3).

For $m = n$, formula (2.6.3) takes the form

$$\operatorname{per} A = \sum_{k=0}^{n} (-1)^k \sum_{1 \leqslant j_1 < \cdots < j_k \leqslant n} \prod_{i=1}^{n} (a_{i1} + \cdots + a_{in} - a_{ij_1} - \cdots - a_{ij_k}). \quad (2.6.6)$$

2.6.5 The Bonferroni inequalities

In addition to formula (2.6.1) of the inclusion–exclusion method we derive the formula for the total number M_r of elements possessing no less than r properties from the set $A_1, \ldots, A_n$. We show that

$$M_r = \sum_{k=r}^{n} (-1)^{k-r} \binom{k-1}{k-r} S_k, \qquad r = 0, 1, \ldots, n, \tag{2.6.7}$$

where

$$S_k = \sum_{1 \leqslant j_2 < \cdots < j_k \leqslant n} M(A_{j_1} \ldots A_{j_k}), \qquad k = 0, 1, \ldots, n.$$

It is clear that

$$M_r = \sum_{l=r}^{n} M(l) = \sum_{k=r}^{n} S_k \sum_{l=r}^{k} (-1)^{k-l} \binom{k}{l}. \tag{2.6.8}$$

Note that

$$\sum_{l=r}^{k} (-1)^{k-l} \binom{k}{l} = \sum_{l=0}^{k-r} (-1)^{l} \binom{k}{l}$$
$$= \operatorname{coef}_{x^{k-r}} \frac{(1-x)^k}{1-x} = (-1)^{k-r} \binom{k-1}{k-r},$$

where $\operatorname{coef}_{x^n} f(x)$ means the coefficient of x^n in the expansion of $f(x)$ in powers of x. Substituting this expression for the inner sum in (2.6.8), we obtain (2.6.7).

From (2.6.1) it follows that

$$S_k = \sum_{j=k}^{n} \binom{j}{k} M(j), \qquad k = 0, 1, \ldots, n. \tag{2.6.9}$$

Indeed,

$$\sum_{j=k}^{n} \binom{l}{k} M(j) = \sum_{l=k}^{n} S_l \sum_{j=k}^{l} (-1)^{l-j} \binom{j}{k} \binom{l}{j}. \tag{2.6.10}$$

By simple transformations we obtain

$$\sum_{j=k}^{l} (-1)^{l-j} \binom{j}{k} \binom{l}{j} = \begin{cases} 1, & l = k, \\ 0, & l > k. \end{cases}$$

Taking into account equalities (2.6.10), we get formula (2.6.9).

Suppose now that the weights of all elements are non-negative. Under this condition we deduce the Bonferroni inequalities

$$\sum_{k=r}^{r+2v-1} (-1)^{k-r} \binom{k}{r} S_k \leqslant M(r) \leqslant \sum_{k=r}^{r+2v} (-1)^{k-r} \binom{k}{r} S_k, \tag{2.6.11}$$

where v is non-negative and takes any natural value such that $r+2v \leqslant n$. It follows from (2.6.1) that

$$M(r) - \sum_{k=r}^{d-1}(-1)^{k-r}\binom{k}{r}S_k = (-1)^{d-r}U(d,r), \qquad (2.6.12)$$

where

$$U(d,r) = \sum_{k=d}^{n}(-1)^{k-d}\binom{k}{r}S_k.$$

Let us show that $U(d,r) \geqslant 0$. It is clear that

$$U(d,r) = \sum_{j=d}^{n} M(j)\sum_{k=d}^{j}(-1)^{k-d}\binom{k}{r}\binom{j}{k}.$$

Note that

$$\sum_{k=d}^{j}(-1)^{k-d}\binom{k}{r}\binom{j}{k} = \binom{j}{r}\binom{j-r-1}{d-r-1};$$

thus

$$U(d,r) = \sum_{j=d}^{n}\binom{j}{r}\binom{j-r-1}{d-r-1}M(j) \geqslant 0.$$

Putting d equal to $r+2v$ and $r+2v+1$, we see that for $d \leqslant n$ the right-hand side of (2.6.12) becomes positive and negative respectively. Hence inequalities (2.6.11) follow.

Inequalities (2.6.11) can be supplemented by the following useful information. First, if in the sum for $M(r)$ we consider only the beginning terms, then we find that the sign of the error of approximation is the same as that of the first rejected term. This fact follows from equality (2.6.12). Moreover, the modulus of the error does not exceed the modulus of the first among the rejected terms. Indeed, taking into account that

$$U(d+1,r) \leqslant \binom{d}{r}S_d,$$

from the obvious equality

$$U(d,r) = \binom{d}{r}S_d - U(d+1,r),$$

we find that

$$U(d,r) \leqslant \binom{d}{r}S_d,$$

where the right-hand side of the inequality is just the modulus of the first of the rejected terms.

2.6.6 Boolean functions

A Boolean function $f(x_1,\dots,x_n)$ of n variables is a mapping $f : \{0,1\} \times \cdots \times \{0,1\} \to \{0,1\}$, where the Cartesian product consists of n factors. Thus the total number of such Boolean functions is $N_n = 2^{2^n}$. Let us find the number of Boolean functions which depend essentially on all n variables. Let A_j be the property that a Boolean function does not depend on x_j, $j = 1,\dots,n$. Then the number of Boolean functions which do not depend on the variables $x_{j_1},\dots,x_{j_k}$, $1 \leqslant j_1 < \cdots < j_k \leqslant n$, is equal to

$$M(A_{j_1}\dots A_{j_k}) = 2^{2^{n-k}}.$$

By the inclusion–exclusion formula the number $M(r)$ of Boolean functions which do not depend essentially on exactly r variables is equal to (Gavrilov and Sapozhenko, 1989)

$$M_n(r) = \sum_{k=r}^{n}(-1)^{k-r}\binom{k}{r}\binom{n}{k}2^{2^{n-k}}, \qquad r = 0,1,\dots,n.$$

In particular, the number of Boolean functions which depend essentially on all n variables is equal to

$$M_n(0) = \sum_{k=0}^{n}(-1)^{k}\binom{n}{k}2^{2^{n-k}}.$$

Using the Bonferroni inequalities we obtain

$$2^{2^n} - n2^{2^{n-1}} \leqslant M_n(0) \leqslant 2^{2^n}.$$

Hence it follows that

$$\lim_{n\to\infty}\frac{M_n(0)}{2^{2^n}} = 1.$$

2.7 Inequalities for permanents

2.7.1 Bounds for the permanents of non-negative matrices

From the Ryser formula and the Bonferroni inequalities it follows that for a non-negative $n \times n$ matrix $A = \|a_{ij}\|$,

$$\operatorname{per} A \leqslant \prod_{i=1}^{n}\sum_{j=1}^{n} a_{ij}. \tag{2.7.1}$$

This obvious bound is rather rough and in a number of cases it can be improved. Recall that a non-negative $n \times n$ matrix $A = \|a_{ij}\|$ is irreducible if there is no permutation matrix Π such that

$$\Pi^{-1} A \Pi = \begin{bmatrix} A_1 & A_2 \\ 0 & A_3 \end{bmatrix}, \tag{2.7.2}$$

where A_1, A_3 are square matrices and 0 stands for the zero matrix of an appropriate size. According to Frobenius' well-known theorem (Gantmakher, 1959) the non-negative irreducible matrix A has a real positive eigenvalue r which is a simple root of the characteristic equation. In some basis a unique, up to a factor, eigenvector $X = (x_1, \ldots, x_n)$ with positive coordinates corresponds to this eigenvalue. This means that

$$AX^T = rX^T, \tag{2.7.3}$$

where the symbol T stands for transposition. It follows from (2.7.3) that

$$\sum_{j=1}^{n} a_{ij} x_j = r x_i, \qquad i = 1, \ldots, n. \tag{2.7.4}$$

Taking the products of the left-hand and right-hand parts of equalities (2.7.4), we obtain

$$\prod_{i=1}^{n} \sum_{j=1}^{n} a_{ij} x_j = r^n \prod_{i=1}^{n} x_i. \tag{2.7.5}$$

Note that for any positive $x_1, \ldots, x_n$

$$\prod_{i=1}^{n} \sum_{j=1}^{n} a_{ij} x_j \geqslant \operatorname{per} A \prod_{i=1}^{n} x_i. \tag{2.7.6}$$

Putting $x_1 = \cdots = x_n = 1$ in (2.7.5) and (2.7.6), we obtain inequality (2.7.1) and the bound

$$\operatorname{per} A \leqslant r^n. \tag{2.7.7}$$

If an irreducible matrix A is doubly stochastic, then

$$0 < \operatorname{per} A \leqslant 1, \tag{2.7.8}$$

and the upper bound is attained for a permutation matrix A. The validity of the lower bound follows from Corollary 2.2.2 to Birkhoff's theorem. Obviously, inequalities (2.7.8) remain true for an arbitrary doubly stochastic matrix A.

2.7.2 Bounds for the permanents of (0, 1)-matrices

Consider upper and lower bounds for the permanents of (0, 1)-matrices with given row and column sums.

Consider an $n \times n$ (0, 1)-matrix A such that

$$\sum_{j=1}^{n} a_{ij} = r_i \geqslant 0, \qquad i = 1, \ldots, n,$$

and $\{z\} = \max\{0, z\}$. Then (Minc, 1978)

$$\operatorname{per} A \geqslant \prod_{i=1}^{n} \{r_i - n + i\}. \tag{2.7.9}$$

For $r_n = 0$ the estimate is obvious. For $r_n > 0$, without loss of generality we may assume that $a_{nj} = 1$, $j = 1, \ldots, r_n$. Decomposing the permanent of A with respect to the last row and using induction on n, we obtain

$$\operatorname{per} A = \sum_{j=1}^{r_n} \operatorname{per} A(n \mid j) \geqslant \sum_{j=1}^{r_n} \prod_{i=1}^{n-1} \{r_i - a_{ij} + i - n + 1\}$$
$$\geqslant \sum_{j=1}^{r_n} \prod_{i=1}^{n-1} \{r_i + i - n\} = \prod_{i=1}^{n} \{r_i - n + i\},$$

where $A(n \mid j)$ is the submatrix of A obtained by deletion of the nth row and jth column.

It is clear that before applying inequality (2.7.9) it is useful to rearrange the rows of A in such a way that the inequalities $r_1 \geqslant \cdots \geqslant r_n$ hold. Then bound (2.7.9) is informative if $r_i > n - i$ for all $i = 1, \ldots, n$.

For matrices of this form the bound

$$\operatorname{per} A \leqslant \prod_{i=1}^{n} (r_i!)^{1/r_i}, \tag{2.7.10}$$

which we will refer to as the Minc–Bregman inequality (Minc, 1978; Bregman, 1973; Schrijver, 1978), is valid. To prove this estimate, we use the following two lemmas.

Lemma 2.7.1 *For any non-negative real numbers* $v_1, \ldots, v_q$,

$$(v_1 + \cdots + v_q)^{v_1 + \cdots + v_q} \leqslant v_1^{v_1} \cdots v_q^{v_q} q^{v_1 + \cdots + v_q}, \tag{2.7.11}$$

where we put $0^0 = 1$.

Indeed, inequality (2.7.11) follows from the inequality

$$\frac{v_1+\cdots+v_q}{q}\ln\frac{v_1+\cdots+v_q}{q}\leqslant\frac{v_1\ln v_1+\cdots+v_q\ln v_q}{q},$$

which in turn is a consequence of convexity of the function $x\ln x$.

Lemma 2.7.2 *If $A=\|a_{ij}\|$ is an $n\times n$ (0,1)-matrix and S is the set of all substitutions s of degree n such that*

$$\prod_{i=1}^{n}a_{i,s(i)}=1,$$

then

$$\prod_{s\in S}r_i=r_i^{\operatorname{per}A},\qquad i=1,\dots,n, \tag{2.7.12}$$

and

$$\prod_{s\in S}\operatorname{per}A(i\mid s(i))=\prod_{j:\,a_{ij}=1}(\operatorname{per}A(i\mid j))^{\operatorname{per}A(i|j)},\qquad i=1,\dots,n, \tag{2.7.13}$$

where $A(i\mid j)$ is the submatrix of A obtained by deletion of the ith row and jth column.

Proof Equality (2.7.12) follows directly from the definition of a permanent. For any i and j,

$$|\{s:\,s(i)=j,\ a_{ij}=1,\ s\in S\}|=\operatorname{per}A(i\mid j);$$

therefore for any $i=1,\dots,n$

$$\begin{aligned}\prod_{s\in S}\operatorname{per}A(i\mid s(i))&=\prod_{j:\,a_{ij}=1}\ \prod_{s:\,s(i)=j}\operatorname{per}A(i\mid j)\\&=\prod_{j:\,a_{ij}=1}(\operatorname{per}A(i\mid j))^{\operatorname{per}A(i|j)}.\end{aligned}$$

□

The proof of estimate (2.7.10) is carried out by induction on n. For $n=1$ the estimate is obvious. Put $v_k=a_{ik}\operatorname{per}A(i\mid k)$ for all k such that $a_{ik}=1$. Applying Lemma 2.7.1 with $q=r_i$ to the right-hand side of the equality

$$(\operatorname{per}A)^{\operatorname{per}A}=\left(\sum_{j:\,a_{ij}=1}\operatorname{per}A(i\mid j)\right)^{\sum_{j:\,a_{ij}=1}\operatorname{per}A(i|j)},$$

we obtain

$$(\operatorname{per} A)^{\operatorname{per} A} \leqslant r_i^{\operatorname{per} A} \prod_{j:\, a_{ij}=1} (\operatorname{per} A(i \mid j))^{\operatorname{per} A(i|j)}.$$

Applying Lemma 2.7.2 now, we obtain the inequality

$$(\operatorname{per} A)^{\operatorname{per} A} \leqslant \prod_{s \in S} r_i \operatorname{per} A(i \mid s(i)). \tag{2.7.14}$$

To estimate $A(i \mid s(i))$, we use the induction hypothesis. If r'_k is the sum of elements of the kth row of the matrix $A(i \mid s(i))$, then

$$r'_k = \begin{cases} r_k - 1, & a_{k,s(i)} = 1, \\ r_k, & a_{k,s(i)} = 0. \end{cases}$$

Now, by the induction hypothesis,

$$\operatorname{per} A(i \mid s(i)) \leqslant \prod_{k \neq i,\, a_{k,s(i)}=1} ((r_k - 1)!)^{1/(r_k - 1)} \prod_{k \neq i,\, a_{k,s(i)}=0} (r_k!)^{1/r_k}.$$

Take the products of both sides of this inequality over $i = 1, \ldots, n$ and change the order of multiplication over i and k. Then we obtain

$$\prod_{i=1}^{n} \operatorname{per} A(i \mid s(i)) \leqslant \prod_{k=1}^{n} \prod_{i \neq k,\, a_{k,s(i)}=1} ((r_k - 1)!)^{1/(r_k - 1)} \prod_{i \neq k,\, a_{k,s(i)}=0} (r_k!)^{1/r_k}.$$

For any k and $s \in S$,

$$|\{i : a_{k,s(i)} = 0\}| = n - r_k,$$
$$|\{i : a_{k,s(i)} = 1\}| = r_k - 1, \qquad i \neq k.$$

This and the preceding inequalities imply that

$$\prod_{i=1}^{n} \operatorname{per} A(i \mid s(i)) \leqslant \prod_{k=1}^{n} (r_k!)^{(n - r_k)/r_k} (r_k - 1)!. \tag{2.7.15}$$

Take the products of both sides of (2.7.14) over $i = 1, \ldots, n$. Taking into account (2.7.15), we obtain

$$(\operatorname{per} A)^{n \operatorname{per} A} \leqslant \prod_{s \in S} \prod_{k=1}^{n} (r_k!)^{n/r_k}. \tag{2.7.16}$$

Since $|S| = \operatorname{per} A$, (2.7.16) yields (2.7.10).

In view of the inequality $(r_i!)^{1/r_i} \leqslant (r_i + 1)/2$, which is equivalent to the well-known relation between the geometric and arithmetic means

$$(r_i!)^{1/r_i} \leqslant \frac{1}{r_i} \sum_{k=1}^{r_i} k,$$

it follows (Minc, 1978) from (2.7.10) that

$$\operatorname{per} A \leqslant \prod_{i=1}^{n} \frac{r_i + 1}{2}. \tag{2.7.17}$$

Finally, inequality (2.7.10) implies that if k divides n and Λ_{kn} is the family of $n \times n$ (0, 1)-matrices with row and column sums equal to k, then

$$\max_{A \in \Lambda_{kn}} \operatorname{per} A = (k!)^{n/k}. \tag{2.7.18}$$

The validity of this equality was first conjectured by Ryser (Ryser, 1960).

2.7.3 The van der Waerden conjecture

In 1926 van der Waerden (van der Waerden, 1926) stated the following conjecture: for any doubly stochastic $n \times n$ matrix A

$$\operatorname{per} A \geqslant \frac{n!}{n^n} \tag{2.7.19}$$

with equality holding if and only if $A = J_n = \|1/n\|$. A series of papers devoted to this problem has appeared over several decades. Marcus and Newman (Marcus and Newman, 1959) obtained essential results for proving the conjecture in particular cases. We give a summary of these results now.

A non-negative $n \times n$ matrix is called partially reducible if it contains a $k \times (n-k)$ zero submatrix, $1 \leqslant k \leqslant n-1$. A matrix which is not partially reducible is called completely irreducible.

Let Ω_n be the set of all doubly stochastic $n \times n$ matrices. A matrix $M \in \Omega_n$ is called a minimizing matrix if

$$\operatorname{per} M = \min_{a \in \Omega_n} \operatorname{per} A. \tag{2.7.20}$$

In (Marcus and Newman, 1959) the following properties of a minimizing matrix were proved:

(1) A minimizing matrix is completely irreducible.

(2) If $M = \|m_{ij}\|$ is a minimizing matrix and $a_{ij} > 0$, then

$$\operatorname{per} M(i \mid j) = \operatorname{per} M.$$

(3) If an $n \times n$ minimizing matrix M is positive, then $\operatorname{per} M = \operatorname{per} J_n = n!/n^n$, and the minimizing matrix is unique, that is, $M = J_n$. This conjecture has been confirmed under other additional conditions (Marcus and Minc, 1964):

(4) Inequality (2.7.19) is valid for symmetric positive semidefinite matrices $A \in \Omega_n$, with equality holding if and only if $A = J_n$. Some other properties of a minimizing matrix were investigated and the conjecture was confirmed under different restrictions on the set of matrices $A \in \Omega_n$. Among these results we cite the paper by London (London, 1971) where Property 2 was improved:

(5) If M is a minimizing matrix, then, for any i and j,

$$\operatorname{per} M(i \mid j) \geqslant \operatorname{per} M. \tag{2.7.21}$$

The van der Waerden conjecture was confirmed in the general case by Egorychev in 1980 (Egorychev, 1980) using the results of an early paper by Aleksandrov (Aleksandrov, 1938) and the above-mentioned result from (London, 1971). It should be noted that a proof of inequality (2.7.19) from the van der Waerden conjecture, without the proof of the uniqueness of the minimizing matrix J_n, was found independently by Falikman (Falikman, 1981). Egorychev's proof in a modified form, using the Lorentz space, was given by van Lint (van Lint, 1981).

An $n \times n$ matrix $A = \|a_{ij}\|$ can be written in the form $A = (a_1, \ldots, a_n)$, where $a_j = (a_{1j}, \ldots, a_{nj})^T$, $j = 1, \ldots, n$. Then the particular case of Aleksandrov's theorem used in the proof of the validity of the van der Waerden conjecture can be formulated as follows (van Lint, 1981).

Theorem 2.7.1 *Let V_n be an n-dimensional vector space over the field of real numbers, and consider positive vectors $a_1, \ldots, a_{n-1} \in V_n$ and $b \in V_n$. Then*

$$(\operatorname{per}(a_1, \ldots, a_{n-1}, b))^2 \geqslant \operatorname{per}(a_1, \ldots, a_{n-1}, a_{n-1}) \operatorname{per}(a_1, \ldots, a_{n-2}, b, b), \tag{2.7.22}$$

with equality holding if and only if $b = \lambda a_{n-1}$, where λ is a constant.

Note that Theorem 2.7.1 is also true in the case where the coordinates of the vectors $a_1, \ldots, a_{n-1}$ are non-negative, but then the conditions concerning the equality are excluded from the assertion.

A space V_n with a standard basis $e_1, \ldots, e_n$ is called a Lorentz space if a symmetric scalar product $(x, y) = x^T Q y$, $x, y \in V_n$, is defined in such a way that the matrix Q has one positive and $n-1$ negative eigenvalues. A vector $x \in V_n$ is positive, negative or null if $(x, x) > 0$, $(x, x) < 0$ or $(x, x) = 0$ respectively.

Lemma 2.7.3 (van Lint (1981)) *If a is a positive vector and b is an arbitrary vector from a Lorentz space, then*

$$(a,b)^2 \geqslant (a,a)(b,b),$$

with equality holding if and only if $b = \lambda a$ for some constant λ.

For positive vectors $a_1,\ldots,a_{n-2} \in V_n$ we define a scalar product by the relation

$$(x,y) = \operatorname{per}(a_1,\ldots,a_{n-2},x,y), \qquad x,y \in V_n. \tag{2.7.23}$$

Hence it follows that $(x,y) = x^T Q y$, where $Q = \|q_{ij}\|$, $i = 1,\ldots,n$, and

$$q_{ij} = \operatorname{per}(a_1,\ldots,a_{n-1},e_i,e_j) = \operatorname{per} A(i,j \mid n-1,n),$$

where $A = (a_1,\ldots,a_n)$ and $A(i,j \mid n-1,n)$ is the submatrix obtained by deletion of the rows with numbers i and j and the columns with numbers $n-1$ and n.

Lemma 2.7.4 (van Lint (1981)) *The space V_n with the scalar product defined by* (2.7.23) *is a Lorentz space.*

Theorem 2.7.1 follows from Lemmas 2.7.3 and 2.7.4.

To obtain the final result, the following two lemmas are needed.

Lemma 2.7.5 *If $A \in \Omega_n$ is a minimizing matrix, then* $\operatorname{per} A(i \mid j) = \operatorname{per} A$ *for all i and j, $i,j = 1,\ldots,n$.*

This lemma can be proved with the help of Property 5 and Theorem 2.7.1.

Lemma 2.7.6 *If $A = (a_1,\ldots,a_n) \in \Omega_n$ is a minimizing matrix and A' is obtained from A by the substitution of $(a_i + a_j)/2$ for a_i and a_j, then A' is also a minimizing matrix in Ω_n and, consequently, Property 5 is valid for A'.*

The final part of the proof of the assertion of the van der Waerden conjecture is as follows (van Lint, 1981). Let $A \in \Omega_n$ be a minimizing matrix. Take an arbitrary column of A, say a_n. From Property 1 it follows that each row of A contains a positive element in one of the remaining columns. Therefore applying Lemma 2.7.6 repeatedly, we can obtain a minimizing matrix $A' = (a'_1,\ldots,a'_{n-1},a_n)$, where $a'_1,\ldots,a'_{n-1}$ are columns with positive coordinates. Applying Theorem 2.7.1 to the permanent of the matrix A', we obtain

$$(\operatorname{per}(a'_1,\ldots,a'_{n-1},a_n))^2 \geqslant \operatorname{per}(a'_1,\ldots,a'_{n-1},a'_{n-1})\operatorname{per}(a'_1,\ldots,a'_{n-2},a_n,a_n). \tag{2.7.24}$$

Decomposing the permanents in both sides of inequality (2.7.24) and applying Lemma 2.7.5, we see that in fact (2.7.24) is an equality. Hence it follows that $a_n = \lambda_{n-1}a'_{n-1}$, where λ_{n-1} is a constant. Similarly, we can prove that $a_n = \lambda_i a'_i$ for $i = 1,\ldots,n-1$. Since

$$\sum_{k=1}^{n} a'_{ik} = \sum_{k=1}^{n} a_{ik} = 1, \qquad i = 1,\ldots,n,$$

we obtain $\lambda_i = 1$ for all $i = 1,\ldots,n-1$. Therefore, $a_n = (1/n,\ldots,1/n)^T$. By virtue of the arbitrariness of a_n the conjecture is confirmed completely, that is, the following theorem is proved.

Theorem 2.7.2 *The minimum value of the permanents of a doubly stochastic $n \times n$ matrix A is equal to $n!/n^n$, and $\operatorname{per} A = n!/n^n$ if and only if $A = J_n = \|1/n\|$.*

Using Theorem 2.7.2 it is not difficult to obtain the following lower bound for the number of Latin rectangles of a given type:

$$L(k,m) \geqslant (m!)^k \left(\frac{(m)_k}{m^k}\right)^m, \qquad k = 1,\ldots,m. \tag{2.7.25}$$

The corresponding estimate for the number of Latin squares has the form

$$L(m,m) \geqslant \frac{(m!)^{2m}}{m^{m^2}}. \tag{2.7.26}$$

Bounds (2.7.25) and (2.7.26) are better than the corresponding estimates (2.1.7) and (2.1.8).

3
Generating functions

The generating functions considered in this chapter are important instruments for solving the so-called enumerative problems in combinatorial analysis. Enumerative problems arise if we need to be explicit about the number of ways of choosing particular elements from a finite set. The application of generating functions in this situation consists of establishing a correspondence between the elements of the set and the terms of the products of some series; the solution of enumerative problems is reduced, in fact, to finding a suitable method for the multiplication of these series. Under these conditions, the convergence of the series is not necessary, and it is natural to use a formal power series, assuming that the operations on them are properly defined. The formal power series, generally speaking, of several variables, are called the generating functions.

Note that the application of generating functions to the solution of enumerative problems connected with establishing a correspondence between the elements of a set and the terms of formal series is an intermediate problem of combinatorial analysis. To solve the main problem (consisting of the derivation of expressions for the number of elements in a set depending on the parameters determining this set) it is appropriate to consider the corresponding power series as convergent in some domain of variation of a real or complex variable. Inside such domains the power series determine analytic functions whose properties are well known in classical analysis. We do not introduce a new terminology for the analytic functions applied to the solution of enumerative problems, but, rather, we call them the generating functions also. By the way, such a notion of a generating function is used in number theory and probability theory.

The two interpretations of the generating functions mentioned above are not distinguished by any special notation. The use of a generating function as analytic in some domain is usually connected with a specification of this domain, and this is deemed to be sufficient indication.

The method of generating functions is used in this chapter in order to obtain a number of relations and formulae, and also to introduce special numbers and polynomials and to investigate their properties. Results of the kind given here will be used in subsequent chapters. The effectiveness of the method of generating functions is illustrated by solving some combinatorial problems. In conclusion, we demonstrate the potential of the method for obtaining asymptotic formulae. We deduce the asymptotic formulae for Stirling numbers of the first and second kind by the saddle point method of estimation of contour integrals of functions of a complex variable.

3.1 Generating functions

3.1.1 Convergent series

Let $\mathscr{P}$ be a number field, that is, the field of real or complex numbers, and let $\mathbf{N}_0 = [0, 1, 2, \ldots]$ be the series of natural numbers with zero. A mapping $\varphi : \mathbf{N}_0 \to \mathscr{P}$ determines a sequence $[a_n] = [a_0, a_1, \ldots]$. In a similar way we can determine a sequence of functions $[\varphi_n(x)] = [\varphi_0(x), \varphi_1(x), \ldots]$ for some set of functions Φ, $\varphi_n(x) \in \Phi$, $n = 0, 1, \ldots$. We assume that the sequence $[\varphi_n(x)]$ is linearly independent, that is, the sequences $[\varphi_0(x), \varphi_1(x), \ldots, \varphi_k(x)]$ are linearly independent for all $k = 0, 1, \ldots$. The series which converges for all $x \in X$,

$$f(x) = \sum_{n=0}^{\infty} a_n \varphi_n(x), \tag{3.1.1}$$

where $X \neq \varnothing$, is called the generating function of the sequence $[a_n]$.

The most frequently used type of sequence $[\varphi_n(x)]$ is $[x^n]$, and then the generating function

$$f(x) = \sum_{n=0}^{\infty} a_n x^n \tag{3.1.2}$$

of the sequence $[a_n]$ becomes a power series converging in some domain $-x_0 < x < x_0$, $x_0 > 0$. Such generating functions were introduced by Laplace in 1812. The function

$$f(x) = \sum_{n=0}^{\infty} a_n \frac{x^n}{n!} \tag{3.1.3}$$

is sometimes placed in correspondence with the sequence $[a_n]$; such generating functions are called exponential. If $\mathcal{P} = \mathbf{C}$, the field of complex numbers, then it is known that if the power series $\sum_{n=0}^{\infty} a_n z^n$, $z \in \mathbf{C}$, has a non-zero radius of convergence, it converges absolutely and uniformly inside its circle of convergence. A unique single-valued analytic function $f(z)$ corresponds to this series and is its sum inside this circle. The function $f(z)$ is infinitely differentiable and

$$a_n = \frac{1}{n!} f^{(n)}(0), \tag{3.1.4}$$

where $f^{(n)}(0)$ is the value of the nth derivative of $f(z)$ at $z = 0$, that is, the expansion of $f(z)$ into the power series coincides with its expansion into the Taylor series in a neighborhood of the point $z = 0$.

The coefficients of the expansion can be obtained by the Cauchy integral formula

$$a_n = \frac{1}{2\pi i} \oint_C f(z) \frac{dz}{z^{n+1}}, \tag{3.1.5}$$

where C is a closed contour in the complex plane containing the origin of coordinates and lying inside the circle where $f(z)$ is analytic. Since inside the domains of convergence the power series can be added, multiplied, differentiated and integrated term by term, substituted into one another, etc., the corresponding operations can be performed on the generating functions. The domains of existence of the generating functions obtained as a result of these operations are determined by the corresponding assertions from classical analysis.

We often consider those generating functions where the terms of the sequence $[a_n]$ themselves depend on some independent variables so that $a_n = a_n(t_1, t_2, \ldots, t_k)$, where $(t_1, t_2, \ldots, t_k) \in T$. In this case the generating function

$$f(x; t_1, t_2, \ldots, t_k) = \sum_{n=0}^{\infty} a_n(t_1, t_2, \ldots, t_k) x^n$$

is considered as a function of $k + 1$ variables; it becomes an ordinary generating function when $t_1, t_2, \ldots, t_k$ are fixed.

If we consider a continuous function $g(u)$ instead of a sequence, then the part of the generating function is played by the Laplace integral transformation

$$f(x) = \int_0^{\infty} e^{-xu} g(u)\, du.$$

Generating functions and Laplace integral transformations can be replaced by the more general notion of the Laplace–Stieltjes integral transformation

$$f(x) = \int_0^\infty e^{-xu}\, dF(u),$$

where $F(u)$ is a function of bounded variation and the right-hand side is the Stieltjes integral.

3.1.2 The ring of formal power series

When solving combinatorial problems, it may be useful to consider that power series whose radius of convergence is equal to zero. To deal with such a series, we construct a ring whose elements are formal power series.

The formal power series $a(x)$ of a sequence $[a_n]$ is the expression

$$a(x) = \sum_{n=0}^{\infty} a_n x^n, \tag{3.1.6}$$

where there are no restrictions on the domain of variation of x, and the series is not required to be convergent. For the sake of convenience we use the same symbols in the notation for the elements of a sequence and the corresponding formal power series. We say that $a(x) = b(x)$ if $a_n = b_n$, $n = 0, 1, \ldots$. We define two composition laws, addition and multiplication, on the set of formal power series for sequences with elements from the number field $\mathscr{P}$.

Addition We set $a(x) + b(x) = c(x)$ if $a_n + b_n = c_n$, $n = 0, 1, \ldots$.
Multiplication We set $a(x) \cdot b(x) = c(x)$ if

$$c_n = \sum_{k=0}^{n} a_k b_{n-k}, \qquad n = 0, 1, \ldots .$$

With respect to addition, the set of formal power series $K[x]$ is a commutative group called the additive group, where the part of the neutral element is played by the series $\sum_{k=0}^{\infty} 0 \cdot x^n$, and the inverse to a series $a(x)$ is the series $-a(x) = \sum_{n=0}^{\infty} -a_n x^n$.

With respect to multiplication, $K[x]$ is a commutative semigroup with the neutral element of the form $1 + \sum_{n=1}^{\infty} 0 \cdot x^n$. In addition, it can be easily seen that multiplication is distributive with respect to addition. Thus, $K[x]$ is a ring, which is usually called the formal power series ring. A necessary and sufficient condition for the existence of an element $a^{-1}(x) \in K[x]$, the inverse with respect to multiplication to an element

$a(x)$, in other words, an element $a^{-1}(x)$ such that $a^{-1}(x) \cdot a(x) = 1$, is the condition $a_0 \neq 0$. This immediately follows from the fact that the elements of the sequence $[a_n^{(-1)}]$ corresponding to $a^{-1}(x)$ satisfy the system of linear equations

$$a_0^{(-1)} a_0 = 1, \quad \sum_{j=0}^{n} a_j^{(-1)} a_{n-j} = 0, \quad n = 1, 2, \ldots . \tag{3.1.7}$$

The element $c(x) \in K[x]$ defined by the equality $c(x) = a^{-1}(x) \cdot b(x)$ is the quotient obtained when $b(x)$ is divided by $a(x)$.

The element inverse to $a(x)$ satisfies the following relation for $a_0 \neq 0$:

$$a^{-1}(x) = \sum_{n=0}^{\infty} a_n^{(-1)} x^n = \frac{1}{a_0} \sum_{k=0}^{\infty} \left(1 - \frac{a(x)}{a_0}\right)^k . \tag{3.1.8}$$

For $a_0 \neq 0$ the formal power series results from substituting a series into another is defined as follows:

$$c(x) = \sum_{n=0}^{\infty} c_n x^n = b(a(x)) = \sum_{k=0}^{\infty} b_k a^k(x),$$

and the formulae

$$c_0 = b_0, \quad c_n = \sum_{k=1}^{n} b_k \sum_{\substack{n_1 + n_2 + \cdots + n_k = n, \\ n_1, \ldots, n_k \geqslant 1}} a_{n_1} \cdots a_{n_k}, \quad n = 1, 2, \ldots, \tag{3.1.9}$$

hold. Considering expression (3.1.8) as the result of substituting the series $1 - a(x)/a_0$ into the series $\frac{1}{a_0} \sum_{n=0}^{\infty} x^n$ and using formulae (3.1.9), we obtain

$$a_0^{(-1)} = a_0^{-1}, \quad a_n^{(-1)} = \sum_{k=1}^{n} \frac{(-1)^k}{a_0^{k+1}} \sum_{\substack{n_1 + \cdots + n_k = n, \\ n_1, \ldots, n_k \geqslant 1}} a_{n_1} \ldots a_{n_k}, \quad n = 1, 2, \ldots . \tag{3.1.10}$$

Formulae (3.1.10) can be considered as a solution of the system of linear equations (3.1.7).

The ring $K[x]$ contains no zero divisors, that is, the condition $a(x) \cdot b(x) = 0$ implies that either $a(x) = 0$ or $b(x) = 0$. Such a ring is called an integral domain. In the same way that we have constructed the ring $K[x]$ over the field $\mathscr{P}$, by using a new variable x_1 we can construct the ring $K[x, x_1]$, which is also integral domain. This process can be continued up to any finite number of variables. It can be extended to the case of an infinite number of variables under the condition that the formal power series obtained at every step of the "extension" depends only on a finite number of variables.

We can define the operation of differentiation on the ring $K[x]$. The derivative of $a(x) \in K[x]$ is defined by the equality

$$a'(x) = \sum_{n=0}^{\infty}(n+1)a_{n+1}x^n.$$

Differentiation in $K[x]$ satisfies the well-known formulae of differentiating sums, products and fractions:

$$(a(x) + b(x))' = a'(x) + b'(x),$$
$$(a(x) \cdot b(x))' = a'(x) \cdot b(x) + a(x) \cdot b'(x),$$
$$(a(x) \cdot b^{-1}(x))' = a'(x) \cdot b^{-1}(x) - a(x) \cdot b'(x) \cdot b^{-2}(x).$$

The process of differentiation may be considered as an operator acting on the ring $K[x]$ and realizing a homomorphic mapping of the additive group of this ring into itself.

The integration of an element $a(x) \in K[x]$ is defined by the formula

$$\int_0^x a(t)\,dt = \sum_{n=1}^{\infty} \frac{a_{n-1}}{n} x^n.$$

It is not difficult to see that the operations of differentiation and integration are mutually inverse. Hence we obtain the well-known results

$$\int_0^x (a(t) + b(t))\,dt = \int_0^x a(t)\,dt + \int_0^x b(t)\,dt,$$
$$\int_0^x a(t) \cdot b'(t)\,dt = a(x) \cdot b(x) - \int_0^x a'(t) \cdot b(t)\,dt,$$

which coincide with the properties of Riemann's integrals in classical analysis.

The elements of the ring $K[x]$ along with the operations of differentiation and integration are called the generating series or the generating functions of the corresponding sequences over the field $\mathscr{P}$. Here the use of the term "generating function" is, of course, a matter of convention, since the formal power series $a(x)$ is not a function of x in the ordinary sense. The use of this term in the case of both convergent and formal power series is convenient, since, depending on the nature of the problem, one can either consider a given series as an element of $K[x]$ or establish its convergence in some neighborhood of zero, if it holds.

In conclusion we note that we can construct the ring $\bar{K}[x]$ of the exponential generating functions in a similar manner. To this end, we take the sequence $[a_n/n!]$ everywhere instead of $[a_n]$. The corresponding formal power series will be denoted by capital letters. It is obvious that $A(x) \cdot B(x) = C(x)$ means

$$c_n = \sum_{k=0}^{n} \binom{n}{k} a_k b_{n-k}, \qquad n = 0, 1, \dots .$$

The operations of differentiation and integration are defined by the equalities

$$A'(x) = \sum_{n=0}^{\infty} a_{n+1} \frac{x^n}{n!}, \qquad \int_0^x A(t)\,dt = \sum_{n=1}^{\infty} a_{n-1} \frac{x^n}{n!}$$

and obey the known rules.

3.1.3 Incidence algebras

A very general approach to the construction of generating functions was presented in (Doubilet, Rota and Stanley, 1972). Some ideas related to this approach will be given in this section.

Let X be a partially ordered set with the relation of partial order denoted by $\preccurlyeq$. A set of elements $z \in X$ satisfying the condition $x \preccurlyeq z \preccurlyeq y$, $x, y \in X$, is called a segment and is denoted by $[x, y]$. A partially ordered set X is called locally finite if every one of its segments is finite.

For some field $\mathscr{P}$ we define the incidence algebra $I(X)$. The elements of $I(X)$ are the functions $f(x, y)$ taking values from $\mathscr{P}$ such that $f(x, y) = 0$ if the ordering $x \preccurlyeq y$ does not hold. The sum of two such functions and the multiplication of a function by an element from $\mathscr{P}$ are defined in the usual way. The multiplication $*$ of these functions is defined as follows: if $f * g = h$, then

$$h(x, y) = \sum_{x \preccurlyeq z \preccurlyeq y} f(x, z) g(z, y).$$

It is not difficult to verify that multiplication as defined above is associative and distributive with respect to addition. The neutral element of $I(X)$ with respect to addition is Kronecker's symbol

$$\delta = \delta(x, y) = \begin{cases} 1 & \text{if } x = y, \\ 0 & \text{if } x \neq y. \end{cases}$$

The equivalence relation $\sim$ defined on the segments of a locally finite partially ordered set X is called congruent if the condition $f(x, y) =$

$f(u,v)$, $g(x,y) = g(u,v)$ for all pairs of segments such that $[x,y] \sim [u,v]$, $f, g \in I(X)$, implies that

$$(f * g)(x,y) = (f * g)(u,v).$$

Let us fix attention on some congruent equivalence relation. The equivalence classes of segments will be referred to as the types. We consider a set of functions defined on the set of types $\alpha, \beta, \gamma, \ldots$ and define the multiplication of functions $f * g = h$ as

$$h(\alpha) = \sum \begin{bmatrix} \alpha \\ \beta, \gamma \end{bmatrix} f(\beta)g(\gamma),$$

where the summation is taken over all pairs of types and the symbol $\left[\begin{smallmatrix} \alpha \\ \beta,\gamma \end{smallmatrix}\right]$ means the number of distinct elements z from a segment $[x,y]$ of type α such that $[x,z]$ is a segment of type β and $[z,y]$ of type γ. This symbol is called the incidence coefficient. If $h_\lambda \in I(X)$ and

$$h_\lambda(x,y) = \begin{cases} 1, & [x,y] \text{ is of type } \lambda, \\ 0 & \text{otherwise}, \end{cases}$$

then

$$(h_\beta * h_\gamma)(u,v) = \begin{bmatrix} \alpha \\ \beta, \gamma \end{bmatrix}.$$

Since $\sim$ is a congruent relation, the left-hand side of this last equality does not depend on a particular interval $[u,v]$ of type α.

The set of all functions defined on the types forms the reduced incidence algebra $R(X, \sim)$. We define the equivalence relation in such a way that $[x,y] \sim [u,v]$ if these two segments are isomorphic. In this case $R(X, \sim)$ is called the standard reduced incidence algebra. Now let $X = \mathbf{N}_0 = [0, 1, \ldots]$. In this case $I(X)$ is the algebra of upper triangular infinite matrices. The reduced incidence algebra $R(X)$ is isomorphic to the algebra of formal power series.

Indeed, an element from $R(X)$ is uniquely determined by the sequence $[a_n]$ of real numbers if we put $f(i,j) = a_{j-i}$, $i \leqslant j$. Multiplication of elements is defined by the equality

$$h(i,j) = \sum_{i \leqslant k \leqslant j} f(i,k)g(k,i) = \sum_{i \leqslant k \leqslant j} a_{k-i} b_{j-k}.$$

Setting $r = k - i$, $n = j - i$, we obtain

$$h(i,j) = \sum_{r=0}^{n} a_r b_{n-r} = c_n.$$

Hence it follows that the mapping of the set of power series into $R(X)$, defined as

$$F(t) = \sum_{n=0}^{\infty} a_n t^n \longrightarrow f(i,j) = a_{j-i}, \qquad j \geqslant i,$$

is an isomorphism.

3.2 The basic numbers, polynomials and relations

3.2.1 The Nörlund and Vandermonde relations

Generating functions are useful for obtaining various identities. For example, let us consider the well-known binomial formula for positive integers m:

$$(1+x)^m = \sum_{k=0}^{m} \binom{m}{k} x^k, \tag{3.2.1}$$

which can easily be proved by induction on m. Using formula (3.2.1) and calculating the coefficient of x^k in $(1+x)^m(1+x)^n$ in two different ways, we obtain the identities

$$\sum_{s=0}^{k} \binom{m}{s}\binom{n}{k-s} = \binom{m+n}{k}, \qquad k = 0, 1, \ldots. \tag{3.2.2}$$

The functions $(\alpha)_n = \alpha(\alpha-1)\cdots(\alpha-n+1)$ and $[\alpha]_n = \alpha(\alpha+1)\cdots(\alpha+n-1)$ defined for all real α and natural n are called the falling and rising factorials. These factorials satisfy the relations

$$(\alpha+\beta)_k = \sum_{j=0}^{k} \binom{k}{j} (\alpha)_j (\beta)_{k-j}, \tag{3.2.3}$$

$$[\alpha+\beta]_k = \sum_{j=0}^{k} \binom{k}{j} [\alpha]_j [\beta]_{k-j}. \tag{3.2.4}$$

The first relation is called the Vandermonde relation, and the second one is called the Nörlund relation. In classical analysis it is proved that for any $\alpha \neq 0$ and $|x| < 1$

$$(1+x)^\alpha = \sum_{k=0}^{\infty} \binom{\alpha}{k} x^k, \tag{3.2.5}$$

where

$$\binom{\alpha}{k} = \frac{(\alpha)_k}{k!} = \frac{\alpha(\alpha-1)\cdots(\alpha-k+1)}{k!}.$$

Using relation (3.2.5), we generalize equality (3.2.2):

$$\sum_{j=0}^{k} \binom{\alpha}{j}\binom{\beta}{k-j} = \binom{\alpha+\beta}{k}, \qquad k = 0, 1, \ldots .$$

Multiplying both sides of this equality by $k!$ and carrying out simple transformations, we prove the Vandermonde relation (3.2.3). From formula (3.2.5) it follows that

$$(1-x)^{-\alpha} = \sum_{k=0}^{\infty} \binom{\alpha+k-1}{k} x^k. \tag{3.2.6}$$

This equality yields the relation

$$\sum_{j=0}^{k} \binom{\alpha+j-1}{j}\binom{\beta+k-j-1}{k-j} = \binom{\alpha+\beta+k-1}{k}, \qquad k = 0, 1, \ldots .$$

Multiplying both sides of this equality by $k!$ and carrying out simple transformations, we prove the Nörlund relation (3.2.4).

3.2.2 Inversion formulae

Let two generating functions $a(x) = \sum_{n=0}^{\infty} a_n x^n$, $b(x) = \sum_{n=0}^{\infty} b_n x^n$ be related by the equality

$$a(x) = b(x)(1-x)^{\alpha}. \tag{3.2.7}$$

Hence it follows that

$$a_n = \sum_{k=0}^{n} (-1)^k \binom{\alpha}{k} b_{n-k}. \tag{3.2.8}$$

From the relation

$$b(x) = a(x)(1-x)^{-\alpha},$$

we obtain

$$b_n = \sum_{k=0}^{n} \binom{\alpha+k-1}{k} a_{n-k}. \tag{3.2.9}$$

Formula (3.2.9) is the inversion of formula (3.2.8), and vice versa.

Consider the exponential generating functions

$$A(x) = \sum_{n=0}^{\infty} a_n \frac{x^n}{n!}, \qquad B(x) = \sum_{n=0}^{\infty} b_n \frac{x^n}{n!}$$

related by the equality

$$B(x)=\sum_{n=0}^{\infty}\frac{x^n}{n!}A(x), \tag{3.2.10}$$

which means that

$$b_n=\sum_{k=0}^{n}\binom{n}{k}a_{n-k}. \tag{3.2.11}$$

From equality (3.2.10) we deduce

$$A(x)=\sum_{n=0}^{\infty}(-1)^n\frac{x^n}{n!}B(x). \tag{3.2.12}$$

Hence it follows that

$$a_n=\sum_{k=0}^{n}(-1)^k\binom{n}{k}b_{n-k}. \tag{3.2.13}$$

Formula (3.2.13) is the inversion of formula (3.2.11), and vice versa.

Now let $B(x)$ be expressed in terms of $A(x)$ as follows:

$$B(x)=\sum_{n=0}^{\infty}\frac{x^n}{n!}A(-x). \tag{3.2.14}$$

Hence we obtain

$$A(x)=\sum_{n=0}^{\infty}\frac{x^n}{n!}B(-x). \tag{3.2.15}$$

The following inversion formulae correspond to (3.2.14) and (3.2.15):

$$b_n=\sum_{k=0}^{n}(-1)^k\binom{n}{k}a_k, \tag{3.2.16}$$

$$a_n=\sum_{k=0}^{n}(-1)^k\binom{n}{k}b_k. \tag{3.2.17}$$

These inversion formulae can be extended to the two-dimensional case. If the generating functions

$$A(x,y)=\sum_{m,n=0}^{\infty}a_{m,n}\frac{x^my^n}{m!\,n!},\qquad B(x,y)=\sum_{m,n=0}^{\infty}b_{m,n}\frac{x^my^n}{m!\,n!}$$

are related by the equality

$$B(x,y)=\sum_{m,n=0}^{\infty}\frac{x^my^n}{m!\,n!}A(-x,-y), \tag{3.2.18}$$

then

$$A(x,y) = \sum_{m,n=0}^{\infty} \frac{x^m y^n}{m!\,n!} B(-x,-y). \tag{3.2.19}$$

Equalities (3.2.18) and (3.2.19) imply the inversion formulae (Carlitz, 1969)

$$a_{m,n} = \sum_{j=0}^{m}\sum_{k=0}^{n}(-1)^{j+k}\binom{m}{j}\binom{n}{k} b_{j,k},$$

$$b_{m,n} = \sum_{j=0}^{m}\sum_{k=0}^{n}(-1)^{j+k}\binom{m}{j}\binom{n}{k} a_{j,k}.$$

3.2.3 The Appell, Bernoulli, Euler, Hermite polynomials

The set of polynomials $A_n(x)$ satisfying the condition

$$A_n'(x) = nA_{n-1}(x), \qquad n = 1, 2, \ldots, \tag{3.2.20}$$

where $A_n'(x)$ means the derivative of $A_n(x)$, is called the Appell set. It can easily be seen that such a set satisfies the condition

$$\sum_{n=0}^{\infty} A_n(x)\frac{t^n}{n!} = \sum_{n=0}^{\infty} a_n \frac{t^n}{n!} \sum_{k=0}^{\infty} \frac{(xt)^k}{k!}, \tag{3.2.21}$$

where a_n does not depend on x. Hence it follows that

$$A_n(x) = \sum_{k=0}^{n} \binom{n}{k} a_k x^{n-k}.$$

Inverting this relation, we have

$$a_n = \sum_{k=0}^{n} (-1)^k \binom{n}{k} x^k A_{n-k}(x).$$

We consider the Bernoulli polynomials $B_n(x)$ which are determined by the generating function

$$\frac{te^{tx}}{e^t - 1} = \sum_{n=0}^{\infty} B_n(x)\frac{t^n}{n!}. \tag{3.2.22}$$

We also introduce the Bernoulli numbers B_n by the generating function

$$\frac{t}{e^t - 1} = \sum_{n=0}^{\infty} B_n \frac{t^n}{n!}. \tag{3.2.23}$$

Note that $B_n = B_n(0)$ and, in addition,

$$B_n(x) = \sum_{k=0}^{n} \binom{n}{k} B_k x^{n-k}.$$

The Bernoulli numbers satisfy the relation

$$B_n = \sum_{k=0}^{n} \binom{n}{k} B_{n-k}, \qquad n = 2, 3, \ldots, \tag{3.2.24}$$

and the Bernoulli polynomials form the Appell set, since

$$B_n'(x) = nB_{n-1}(x).$$

The Euler polynomials $E_n(x)$, determined by the generating function

$$\frac{2e^{xt}}{e^t + 1} = \sum_{n=0}^{\infty} E_n(x)\frac{t^n}{n!},$$

are closely related to the Bernoulli polynomials. We define the one-parameter Euler numbers E_n by means of the generating function

$$\frac{2e^t}{e^{2t} + 1} = \sum_{n=0}^{\infty} E_n \frac{t^n}{n!}. \tag{3.2.25}$$

It is easy to verify that $E_n = 2^n E_n(1/2)$ and, in addition,

$$E_n(x) = \sum_{k=0}^{n} \binom{n}{k} E_k 2^{-k}(x - 1/2)^{n-k}.$$

Hence it follows that

$$E_n'(x) = nE_{n-1}(x),$$

that is, the Euler polynomials form the Appell set.

The Hermite polynomials $H_n(x)$ are determined by the generating function

$$e^{2xt-t^2} = \sum_{n=0}^{\infty} H_n(x)\frac{t^n}{n!}. \tag{3.2.26}$$

Differentiating both sides of this equality with respect to x and equating the coefficients of $t^n/n!$, we see that

$$H_n'(x) = 2nH_{n-1}(x),$$

that is, the Hermite polynomials with a minor modification satisfy the property determining the Appell set.

It follows from the generating function that

$$H_n(x) = \sum_{k=0}^{[n/2]} (-1)^k \frac{n!}{k!\,(n-2k)!}(2x)^{n-2k}. \tag{3.2.27}$$

3.2.4 The finite difference operators; de Morgan and Stirling numbers

Let $f(x)$ be a function defined at the point $x+1$ if it is defined at a point x. We define the operator Δ by the equality

$$\Delta f(x) = f(x+1) - f(x).$$

It can easily be seen that the operator Δ is linear:

$$\Delta(cf(x)) = c\Delta f(x), \qquad \Delta(f(x) + \varphi(x)) = \Delta f(x) + \Delta\varphi(x).$$

Now let us define $\Delta^n f(x)$ by the equality $\Delta^n f(x) = \Delta(\Delta^{n-1} f(x))$. Using induction on n, using the properties of the operator Δ, we obtain the formula

$$\Delta^n f(x) = \sum_{k=0}^{n} (-1)^k \binom{n}{k} f(x+n-k). \tag{3.2.28}$$

The inversion of this formula has the form

$$f(x+n) = \sum_{k=0}^{n} \binom{n}{k} \Delta^k f(x). \tag{3.2.29}$$

Put $f(x) = x^m$, where m is a natural number. From formulae (3.2.28) and (3.2.29) we obtain

$$\Delta^n x^m = \sum_{k=0}^{n} (-1)^k \binom{n}{k} (x+n-k)^m, \tag{3.2.30}$$

$$(x+n)^m = \sum_{k=0}^{n} \binom{n}{k} \Delta^k x^m. \tag{3.2.31}$$

Introduce the notation $\Delta^n 0^m = \Delta^n x^m|_{x=0}$. The numbers $\Delta^n 0^m$ are called Morgan numbers. From formulae (3.2.30) and (3.2.31) it follows that

$$\Delta^n 0^m = \sum_{k=0}^{n} (-1)^k \binom{n}{k} (n-k)^m, \tag{3.2.32}$$

$$n^m = \sum_{k=0}^{n} \binom{n}{k} \Delta^k 0^m. \tag{3.2.33}$$

Let us consider the function $(x)_n = x(x-1)\cdots(x-n+1)$, which we recall is usually called the falling factorial function. An interesting property of this function consists of the fact that its behavior with respect to the operation of taking the finite difference is the same as the behavior of the power function with respect to differentiation. The formula

$$\Delta^k (x)_n = \begin{cases} (n)_k (x)_{n-k}, & k < n, \\ n!, & k = n, \\ 0, & k > n, \end{cases} \tag{3.2.34}$$

holds. If we put $(x)_0 = 1$, $(n)_k = 0$, $k > n$, then the formula can be simplified to $\Delta^k (x)_n = (n)_k (x)_{n-k}$.

The Stirling numbers $s(n,k)$ and $\sigma(n,k)$ of the first and second kind, respectively, can be defined by the equalities

$$(x)_n = \sum_{k=0}^{n} s(n,k) x^k, \tag{3.2.35}$$

$$x^n = \sum_{k=0}^{n} \sigma(n,k)(x)_k. \tag{3.2.36}$$

Here we set $s(0,0) = \sigma(0,0) = 1$ and $s(n,k) = \sigma(n,k) = 0$ for $n < k$.

From equalities (3.2.35) and (3.2.36) it follows that

$$(1+t)^x = \sum_{n=0}^{\infty} \sum_{k=0}^{n} s(n,k) \frac{t^n}{n!} x^k, \tag{3.2.37}$$

$$e^{xt} = \sum_{n=0}^{\infty} \sum_{k=0}^{n} \sigma(n,k) \frac{t^n}{n!} (x)_k. \tag{3.2.38}$$

From equality (3.2.37) we obtain

$$\frac{1}{k!} (\ln(1+t))^k = \sum_{n=k}^{\infty} s(n,k) \frac{t^n}{n!}. \tag{3.2.39}$$

Take the kth finite difference of both sides of equality (3.2.38). Using the obvious relation

$$\Delta^k e^{xt} = (e^t - 1)^k e^{xt}$$

and formula (3.2.34), we obtain

$$\frac{1}{k!} (e^t - 1)^k = \sum_{n=k}^{\infty} \sigma(n,k) \frac{t^n}{n!}. \tag{3.2.40}$$

Let us deduce the representation of Stirling numbers of the second kind in terms of Morgan numbers. To this end we take the kth finite difference of both sides of equality (3.2.36) and put $x = 0$; then

$$\sigma(n,k) = \frac{\Delta^k 0^n}{k!}, \qquad k = 0, 1, \ldots, n. \tag{3.2.41}$$

Now we take the kth finite difference of both sides of equality (3.2.35):

$$\Delta^k (x)_n = \sum_{j=k}^{n} s(n,j) \Delta^k x^j.$$

Setting $x = 0$ and using formula (3.2.41), we obtain

$$\sum_{j=k}^{n} s(n,j)\sigma(j,k) = \begin{cases} 1, & k = n, \\ 0, & k \neq n. \end{cases} \tag{3.2.42}$$

This is the orthogonality relation for Stirling numbers. This relation implies that the equalities

$$a_n = \sum_{k=0}^{n} s(n,k) b_k, \qquad b_n = \sum_{k=0}^{n} \sigma(n,k) a_k \tag{3.2.43}$$

hold for any sequences $\{a_n\}$ and $\{b_n\}$. In conclusion, we give the recurrence relations for Stirling numbers

$$s(n,k) = s(n-1,k-1) - (n-1)s(n-1,k), \tag{3.2.44}$$

$$\sigma(n,k) = \sigma(n-1,k-1) + k\sigma(n-1,k), \tag{3.2.45}$$

which follow from equalities (3.2.35) and (3.2.36).

3.2.5 A combinatorial interpretation of Stirling numbers

(1) Denote by $C(n,k)$ the number of substitutions of degree n with k cycles. The numbers $C(n,k)$ satisfy the recurrence relation

$$C(n,k) = C(n-1,k-1) + (n-1)C(n-1,k), \qquad C(0,0) = 1,$$

which can be obtained if we divide the substitutions with k cycles into two classes depending on whether or not a fixed element of the set X, which these substitutions act on, is placed in a cycle of length 1. From the recurrence relation it follows that

$$\sum_{k=0}^{n} C(n,k) x^k = x(x+1)\cdots(x+n-1).$$

Comparing this generating function with the generating function for Stirling numbers of the first kind, we finally obtain

$$C(n,k) = (-1)^{n+k} s(n,k).$$

(2) Let $T(m,k)$ be the number of partitions of a set X of m elements into k blocks. We divide the partitions into two classes depending on whether or not a fixed element $x \in X$ forms a block consisting of one element. The calculation of the number of partitions in these classes gives us the recurrence relation

$$T(m,k) = T(m-1,k-1) + kT(m-1,k), \qquad T(0,0) = 1.$$

Hence it follows that

$$x^m = \sum_{k=0}^{m} T(m,k)(x)_k.$$

Thus, the numbers $T(m,k)$ coincide with Stirling numbers of the second kind:

$$T(m,k) = \sigma(m,k), \qquad k = 0, 1, \ldots, m.$$

(3) Assume that m distinct particles are allocated at random to n distinct cells of unlimited capacity, and let $D_0(n,m)$ be the number of allocations with no empty cells. The relation

$$n^m = \sum_{k=0}^{n} \binom{n}{k} D_0(k,m)$$

is obvious. Inverting this relation, we obtain

$$D_0(n,m) = \sum_{k=0}^{n} (-1)^k \binom{n}{k} (n-k)^m, \tag{3.2.46}$$

which allows us to clear up the combinatorial meaning of Morgan numbers and to give a new interpretation of Stirling numbers of the second kind:

$$D_0(n,m) = \Delta^n 0^m = \sigma(m,n)n!.$$

(4) An obvious generalization of formula (3.2.46) can be obtained by the use of the inclusion–exclusion method. Indeed, if $D_r(n,m)$ is the number

of allocations of m particles into n cells under the condition that exactly r cells are empty, then

$$D_r(n,m) = \sum_{k=r}^{n}(-1)^{k-r}\binom{k}{r}\binom{n}{k}(n-k)^m.$$

Carrying out the appropriate transformations, we obtain

$$D_r(n,m) = \binom{n}{r}\sum_{k=0}^{n-r}(-1)^k\binom{n-r}{k}(n-r-k)^m.$$

This last equality means that

$$D_r(n,m) = \binom{n}{r}\Delta^{n-r}0^m = \frac{n!}{r!}\sigma(m,n-r).$$

Note that the interpretation given of Morgan numbers and Stirling numbers of the second kind allows us to apply the Bonferroni equalities to alternating sums of these numbers.

3.2.6 The Euler–Maclaurin summation formula

Consider the problem of finding a function $f(x)$ satisfying the equation

$$\Delta f(x) = \varphi(x). \tag{3.2.47}$$

This problem is equivalent to finding an operator Δ^{-1} such that $\Delta^{-1}\varphi(x) = f(x)$. To determine this operator, we consider the translation operator E defined by the condition

$$Ef(x) = f(x+1)$$

and the differentiation operator

$$Df(x) = \frac{d}{dx}f(x).$$

It is obvious that the operators E and D, as well as Δ, are linear. The operators E and Δ are related by the equality $\Delta = E - 1$. Let us find how they relate to the operator D. We assume that the function $f(x+1)$ is expanded into a convergent Taylor series and represent this expansion using the operator D:

$$f(x+1) = \left\{1 + \sum_{k=1}^{\infty}\frac{D^k}{k!}\right\}f(x).$$

By convention, this equality can be rewritten as

$$Ef(x) = e^D f(x),$$

that is, the equalities

$$E = e^D, \qquad \Delta = e^D - 1$$

hold. Thus, we can formally write

$$D\Delta^{-1} = \frac{D}{e^D - 1}.$$

If we apply the operator $D\Delta^{-1}$ to the left-hand side of equation (3.2.47) and apply the operator, obtained from $(D/(e^D - 1)$ by the expansion of form (3.2.23), to the right-hand side of that equation, then we have

$$Df(x) = \sum_{\nu=0}^{\infty} \frac{B_\nu}{\nu!} \varphi^{(\nu)}(x).$$

Hence, integrating and taking into account the formula

$$\sum_{j=m}^{n-1} \varphi(j) = f(n) - f(m),$$

we finally obtain the Euler–Maclaurin formula

$$\sum_{j=m}^{n-1} \varphi(j) = \int_m^n \varphi(x)\,dx + \sum_{\nu=1}^{\infty} \frac{B_\nu}{\nu!} \{\varphi^{(\nu-1)}(n) - \varphi^{(\nu-1)}(m)\}. \quad (3.2.48)$$

The trouble with this form of the Euler–Maclaurin formula is that we know nothing about the convergence of the series on the right-hand side. The Euler–Maclaurin formula with remainder term, whose proof we omit,

$$\sum_{j=m}^{n-1} \varphi(j) = \int_m^n \varphi(x)\,dx + \sum_{\nu=1}^{k-1} \frac{B_\nu}{\nu!} \{\varphi^{(\nu-1)}(n) - \varphi^{(\nu-1)}(m)\}$$
$$- \frac{1}{k!} \int_0^1 (B_k(x) - B_k) \sum_{j=m}^{n-1} \varphi^{(k)}(j - x + 1)\,dx, \quad (3.2.49)$$

where $B_k(x)$ is the Bernoulli polynomial, is more convenient in practice. From formula (3.2.48), it follows, in particular, that

$$\sum_{j=1}^{n-1} j^l = \frac{1}{l+1}(B_{l+1}(n) - B_{l+1}).$$

This formula can also be obtained by inspection. To this end we take the finite difference with respect to x of both sides of equality (3.2.22). From the resulting equality,

$$\sum_{k=0}^{\infty} \Delta B_k(x) \frac{t^k}{k!} = te^{xt}$$

it follows that

$$\Delta B_k(x) = kx^{k-1}, \qquad k = 0, 1, \ldots .$$

Setting $x = j$ and summing both sides of the last equality, we arrive at the required formula.

From formula (3.2.49) we can also derive the equality

$$\sum_{j=1}^{n} \frac{1}{j} = \ln n + C + \gamma_n,$$

where $C = 0.5772\ldots$ is the Euler constant and $\gamma_n \to 0$ as $n \to \infty$.

3.2.7 The Bruno and Lagrange formulae

In a number of cases the formula

$$(xD)^n f(x) = \sum_{k=0}^{n} \sigma(n,k) x^k D^k f(x), \qquad n \geqslant 0, \tag{3.2.50}$$

where $\sigma(n,k)$ are Stirling numbers of the second kind, may be useful. This formula can easily be proved by induction on n. Now let us generalize the notion of the finite difference operator, setting

$$\underset{x,h}{\Delta} f(x) = \frac{f(x+h) - f(x)}{h}.$$

The relation

$$\underset{x,h}{\Delta}^m f(x+h) = h \underset{x,h}{\Delta}^{m+1} f(x) + \underset{x,h}{\Delta}^m f(x)$$

is valid. By induction on n we can prove the analog of formula (3.2.50) (Gould, 1973):

$$\left(x \underset{x,h}{\Delta}\right)^n f(x) = \sum_{j=0}^{n} \sigma(n,j) \prod_{k=0}^{j-1} (x+kh) \underset{x,h}{\Delta}^j f(x), \tag{3.2.51}$$

where

$$\prod_{k=0}^{-1} (x+kh) = 1.$$

For $h = 1$ formula (3.2.51) was proved by Jordan. It can easily be seen that

$$\mathop{\Delta^n}_{x,h}\binom{x/h+b}{m} = h^{-n}\binom{x/h+b}{m-n}.$$

Using this equality and relation (3.2.51), we can (Gould, 1973) derive the expansion

$$\left(x\mathop{\Delta}_{x,h}\right)^n\binom{x/h+b}{m} = \sum_{j=0}^{n}\sigma(n,j)\binom{x/h+b}{m-j}h^{-j}\prod_{k=0}^{j-1}(x+kh). \tag{3.2.52}$$

Now we consider the composite function $F(t) = f(g(t))$, where f and g are infinitely differentiable. Let us introduce the notation

$$F_n = D_t^n F(t), \quad f_n = D_u^n f(u)|_{u=g(t)}, \quad g_n = D_t^n g(t).$$

By sequential differentiation we obtain the relation

$$F_n = \sum_{k=1}^{n} F_{nk} f_k, \tag{3.2.53}$$

where F_{nk}, $k = 1, 2, \ldots, n$, depend only on $g_1, g_2, \ldots, g_n$ and not on $f_1, f_2, \ldots, f_n$. Therefore, in order to evaluate the coefficients F_{nk}, we can set $f(u) = e^{au}$, where a is a constant. Then $F(t) = e^{ag(t)}$, $F_n = F_n(a)$ and $f_k = a^k e^{ag(t)}$, and relation (3.2.53) takes the form

$$e^{-ag(t)}F_n(a) = e^{-ag(t)}(D^n e^{ag(t)}) = \sum_{k=1}^{n} F_{nk}a^k. \tag{3.2.54}$$

Let us define the Bell polynomials by the formula

$$Y_n = Y_n(y_1, \ldots, y_n) = e^{-y(t)}(D^n e^{y(t)}), \tag{3.2.55}$$

where we put $y_k = D^k y(t)$. From equality (3.2.55) it follows that

$$Y_{n+1} = e^{-y(t)}(D^n(y_1 e^{y(t)})). \tag{3.2.56}$$

Using the Leibniz formula

$$D^n(uv) = \sum_{k=0}^{n}\binom{n}{k}D^k u \cdot D^{n-k}v,$$

we deduce from equality (3.2.56) that

$$Y_{n+1} = \sum_{k=0}^{n}\binom{n}{k} y_{k+1} Y_{n-k}. \tag{3.2.57}$$

ıen any function $f(y)$, analytic in the neighborhood of the point $y = 0$, expanded into the power series

$$f(y) = f(0) + \sum_{k=1}^{\infty} \frac{x^k}{k!} \left\{ \left(\frac{d}{dy}\right)^{k-1} [f'(y)\varphi^k(y)] \right\}\Bigg|_{y=0}.$$

r example, if y satisfies the functional equation $x = ye^{-y}$, then the last rmula with $f(y) = y$, $\varphi(y) = e^y$ yields

$$y = \sum_{k=1}^{\infty} \frac{k^{k-1}}{k!} x^k.$$

.8 Asymptotic formulae

t us introduce the symbols O and o which will be used frequently in ıat follows. If f and φ are real or complex functions defined on a set then the formula

$$f(z) = O(\varphi(z)), \qquad z \in Z, \tag{3.2.59}$$

eans that there exists a positive number c, independent of z, such that

$$|f(z)| \leqslant c|\varphi(z)|, \qquad z \in Z. \tag{3.2.60}$$

$|z| \to \infty$, then equality (3.2.59) means that there exists an a such that und (3.2.60) holds for $a < |z| < \infty$. Equality (3.2.59) can be treated nilarly as $|z|$ tends to other values.

The formula $f(z) = o(\varphi(z))$, $|z| \to \infty$, means that the ratio $f(z)/\varphi(z)$ ıds to zero as $|z| \to \infty$.

We now introduce the gamma function which is usually defined as an ıler integral of the second kind

$$\Gamma(a) = \int_0^{\infty} x^{a-1} e^{-x}\, dx,$$

d which converges for $a > 0$ and diverges for $a \leqslant 0$. The gamma nction satisfies the recurrence relation

$$\Gamma(a+1) = a\Gamma(a), \qquad \Gamma(1) = 1.$$

$a = n$, where n is a natural number, then $\Gamma(n+1) = n!$.

The gamma function can be analytically continued to the complex ane, and $\Gamma(z)$ as a function of the complex variable z is regular erywhere except for the points $z = 0, -1, -2, \ldots$, which are simple les.

Introducing the generating functions

$$\varphi(u) = \sum_{k=0}^{\infty} y_k \frac{u^k}{k!}, \qquad \Phi(u) = \sum_{n=0}^{\infty}$$

and using relation (3.2.57), we obtain the different

$$\frac{\Phi'(u)}{\Phi(u)} = \varphi'(u).$$

Integrating this equation with the initial conditic $\Phi(0) = Y_0 = 1$, we obtain

$$\Phi(u) = e^{\varphi(u)-y} = \exp\left\{\sum_{k=1}^{\infty} y_k \frac{u^k}{k!}\right.$$

Setting $y(t) = ag(t)$, we deduce from equalities (3.

$$e^{-ag(t)} F_n(a) = Y_n(ag_1, ag_2, \ldots, a$$

therefore

$$\sum_{n=0}^{\infty} e^{-ag(t)} F_n(a) \frac{u^n}{n!} = \exp\left\{a \sum_{k=1}^{\infty} g_k\right.$$

Calculating the coefficients of $a^k u^n/n!$ on both sid we obtain

$$F_{nk} = n! \sum_{K} \prod_{j=1}^{n} \frac{g_j^{k_j}}{(j!)^{k_j} k_j!},$$

where

$$K = \left\{k_1, \ldots, k_n : \sum_{j=1}^{n} k_j = k, \sum_{j=1}^{n} jk\right.$$

Now from equality (3.2.53) we obtain the Bruno f

$$D^n f[g(t)] = n! \sum_{k=1}^{n} f_k \sum_{K} \prod_{j=1}^{n} \frac{g_j}{(j!)^k}$$

In conclusion we give the Lagrange formula fr which allows us to derive the generating functio equation. Let $\varphi(u)$ be an analytic function in a point $y = 0$, $\varphi(0) \neq 0$, such that

$$x = \frac{y}{\varphi(y)}.$$

The gamma function satisfies the asymptotic Stirling formula

$$\Gamma(z+1) = \sqrt{2\pi} z^{z+1/2} e^{-z}(1 + O(1/z)), \tag{3.2.61}$$

which is valid for $-\pi/2 \leqslant \arg z \leqslant \pi/2$ and $|z| \to \infty$. As $n \to \infty$, the Stirling formula for factorials follows:

$$n! = \sqrt{2\pi} n^{n+1/2} e^{-n}(1 + O(1/n)). \tag{3.2.62}$$

The inequalities

$$(n/e)^n < n! < e(n/2)^n \tag{3.2.63}$$

are sometimes helpful in the estimation of $n!$. Using the Euler–Maclaurin formula, we can obtain the asymptotic expansion of the logarithm of the gamma function

$$\ln \Gamma(z+1) = \ln \sqrt{2\pi} + \left(z + \frac{1}{2}\right) \ln z - z + \frac{B_2}{1 \cdot 2} \frac{1}{z} + \frac{B_4}{3 \cdot 4} \frac{1}{z^3} + \cdots + \frac{B_{2k}}{(2k-1)2k} \frac{1}{z^{2k-1}} + O\left(\frac{1}{z^{2k+1}}\right), \tag{3.2.64}$$

where B_{2k} are the Bernoulli numbers. This formula is true for all the complex plane with an arbitrary small sector cut out, whose bisectrix is directed along the negative real axis. The series (3.2.64) diverges for all z. For fixed k and $|z| \to \infty$ the remainder term tends to zero faster than the last term included in the formula.

3.3 Non-regenerative substitutions, inversions and ascents in permutations

Now let us demonstrate how to obtain generating functions for solving the enumerative problems of combinatorial analysis. As examples we consider several problems connected with the enumeration of permutations of degree n with given values of some characteristics. Let us begin with a problem where formal power series with zero radius of convergence are used as generating functions.

3.3.1 Non-regenerative substitutions

Let a substitution s of degree n act on the set $N_n = \{1, 2, \ldots, n\}$. The substitution s is called k-regenerative if there exists a minimum number k such that $s(N_k) = N_k$, $1 \leqslant k \leqslant n$; if $k = n$, then the substitution s is called non-regenerative. On the set N_k the substitution s acts as a non-regenerative substitution of degree k. We denote by R_n the number of non-regenerative substitutions of degree n. The recurrence relation

$$n! = \sum_{k=1}^{n} R_k (n-k)!$$

holds; see (Stam, 1985).

Introducing the generating functions as formal power series

$$R(x) = \sum_{n=1}^{\infty} R_n x^n, \qquad \pi(x) = \sum_{n=0}^{\infty} n!\, x^n,$$

and using the recurrence relation for the numbers R_n, we get

$$R(x) = 1 - \pi^{-1}(x).$$

Hence, by formula (3.1.10) for the elements of the sequence corresponding to $\pi^{-1}(x)$, we obtain

$$R_n = \sum_{k=1}^{n} (-1)^{k-1} \sum_{\substack{n_1+\cdots+n_k=n,\\ n_1,\ldots,n_k \geqslant 1}} n_1! \cdots n_k!$$

The recurrence relation also implies

$$1 - \frac{4}{n} \leqslant \frac{R_n}{n!} \leqslant 1 - \frac{1}{n}.$$

Hence it follows that

$$\lim_{n\to\infty} \frac{R_n}{n!} = 1.$$

3.3.2 Inversions

Let $(a_1, a_2, \ldots, a_n)$ be a permutation of the numbers $\{1, 2, \ldots, n\}$. A pair of elements a_i and a_j forms an inversion if $a_i > a_j$ for $i < j$. The identity permutation $(1, 2, \ldots, n)$ contains no inversions; the permutation $(n, n-1, \ldots, 1)$ contains the maximum number of inversions $\binom{n}{2}$.

We denote by $B(n, r)$ the number of permutations of n elements with r inversions. The numbers $B(n, r)$ satisfy the recurrence relation

$$B(n, r) = \sum_{s=0}^{r} B(n-1, r-s), \qquad (3.3.1)$$

which can be obtained if we partition the permutations of n elements with r inversions into $l+1$ classes, $l = \min(n-1, r)$, where the ith class contains the permutations with the element n placed in the $(n-i+1)$th position.

Consider the generating function

$$b_n(x) = \sum_{r=0}^{\binom{n}{2}} B(n,r)x^r. \qquad (3.3.2)$$

From formula (3.3.1) we obtain

$$b_n(x) = (1 + x + \cdots + x^{n-1})b_{n-1}(x).$$

Since $b_1(x) = 1$, we get the final expression for the generating function

$$b_n(x) = \frac{(1-x)(1-x^2)\cdots(1-x^n)}{(1-x)^n}. \qquad (3.3.3)$$

3.3.3 Ascents

In a permutation $(a_1, a_2, \ldots, a_n)$ of the numbers $\{1, 2, \ldots, n\}$, the elements a_j and a_{j+1} form an ascent if $a_j < a_{j+1}$. It is convenient to assume that an ascent precedes the element a_1.

We denote by $A_{n,k}$ the number of permutations of n elements with k ascents. The numbers $A_{n,k}$ satisfy the recurrence relation

$$A_{n,k} = (n-k+1)A_{n-1,k-1} + kA_{n-1,k}. \qquad (3.3.4)$$

This relation can be derived by the following reasoning. If in a permutation $(a_1, a_2, \ldots, a_n)$ we place the element n before the element a_1 or between elements a_j and a_{j+1} where $a_j < a_{j+1}$, then the total number of ascents in the permutation does not change. If $a_j > a_{j+1}$, then the number of ascents is increased by one; the same occurs if n is placed at the rightmost position.

It is clear that $A_{n,1} = A_{n,n} = 1$. In addition, we set $A_{0,0} = 1$. By virtue of symmetry, $A_{n,k} = A_{n,n-k+1}$. From the recurrence relation (3.3.4) we find

$$A_{n,2} = 2^n - (n+1),$$
$$A_{n,3} = 3^n - (n+1)2^n + \binom{N+1}{2}.$$

Let us introduce the generating function

$$A_n(x) = \sum_{k=1}^{\infty} A_{n,k}x^k, \qquad A_0(x) = 1.$$

Multiplying both sides of equality (3.3.4) by x^n and summing over k, we obtain, after obvious transformations, that

$$A_n(x) = nxA_{n-1}(x) + x(1-x)A'_{n-1}(x), \qquad (3.3.5)$$

where the prime means the derivative with respect to x. By direct calculation we verify that a generating function of the form

$$A_n(x) = (1-x)^{n+1} \sum_{k=0}^{\infty} k^n x^k \qquad (3.3.6)$$

satisfies the difference-differential equation (3.3.5) with boundary conditions $A_0(x) = 1$, $A_1(x)$. From formula (3.3.6) we obtain

$$A_{n,k} = \sum_{j=0}^{k} (-1)^j \binom{n+1}{j} (k-j)^n, \qquad k = 1, 2, \ldots, n. \qquad (3.3.7)$$

The numbers $A_{n,k}$ are usually referred to as the Eulerian numbers.

We consider the following generating function of two variables:

$$A(x,t) = \sum_{n=0}^{\infty} A_n(x) \frac{t^n}{n!} = \sum_{n=0}^{\infty} \sum_{k=0}^{n} A_{n,k} \frac{t^n}{n!} x^k,$$

where we assume that $A_{0,0} = 1$, $A_{n,k} = 0$ for $n < k$. Multiplying both sides of equality (3.3.6) by $t^n/n!$ and summing, we obtain the expression

$$A(x,t) = \frac{1-x}{1-xe^{t(1-x)}}, \qquad x \neq 1. \qquad (3.3.8)$$

3.3.4 Ascents and descents

In a permutation $(a_1, a_2, \ldots, a_n)$ a pair of elements a_j, a_{j+1} obeying $a_j > a_{j+1}$ is called a descent. In contrast to the preceding case, we assume that there is no ascent to the left of a_1 and no descent to the right of a_n. If a permutation of degree n has r ascents and s descents, then $r + s = n - 1$.

We denote by $A(r,s)$ the number of permutations of degree n with r ascents and s descents. The recurrence relation

$$A(r,s) = (r+1)A(r,s-1) + (s+1)A(r-1,s) \qquad (3.3.9)$$

is valid. Indeed, let us insert the element $n+1$ into a permutation of degree n with r ascents and s descents. If this element falls between the elements forming an ascent, the number of ascents remains the same, and the number of descents increases by one; if $n+1$ falls between the elements forming a descent, the situation is the opposite. Note that $A(r,s) = A_{r+s+1,r+1}$, and relation (3.3.9), after the proper transformation, becomes (3.3.4) for the Eulerian numbers.

Let us put the permutation $(b_1, b_2, \dots, b_n)$ into one-to-one correspondence with a permutation $(a_1, a_2, \dots, a_n)$ in such a way that $b_i = n - a_i + 1$, $i = 1, 2, \dots, n$. The ascents in one permutation correspond to the descents in the other permutation, and vice versa. Therefore,

$$A(r,s) = A(s,r).$$

To obtain an expression for the generating function, it is more convenient to use the following recurrence relation (Carlitz, Roselle and Scoville, 1966):

$$A(r,s) = A(r,s-1) + A(r-1,s) + \sum_{j<r}\sum_{k<s}\binom{r+s}{j+k+1}A(j,k)A(r-j-1,s-k-1). \tag{3.3.10}$$

To derive (3.3.10), we partition all permutations of degree $r+s+1$ into $r+s+1$ classes depending on the position of the element $r+s+1$. The first two summands on the right-hand side of equality (3.3.10) give the sizes of the classes corresponding to the leftmost and rightmost positions of $r+s+1$. If $r+s+1$ is not placed at those extreme positions, then the elements to the left and to the right of $r+s+1$ are independent permutations with the numbers of ascents j and $r-j-1$ and the numbers of descents k and $s-k-1$ respectively; the number of ways of choosing the elements for the left permutation is $\binom{r+s}{j+k+1}$.

We consider the generating function

$$F(z) = \sum_{r,s=0}^{\infty} A(r,s)\frac{x^r y^s z^{r+s+1}}{(r+s+1)!}. \tag{3.3.11}$$

From equality (3.3.10) it follows that

$$\begin{aligned}\sum_{r,s=0}^{\infty} A(r,s)\frac{x^r y^s z^{r+s}}{(r+s)!} &= 1 + \sum_{r,s=0}^{\infty} A(r,s)\frac{x^r y^{s+1} z^{r+s+1}}{(r+s+1)!} \\ &+ \sum_{r,s=0}^{\infty} A(r,s)\frac{x^{r+1} y^s z^{r+s+1}}{(r+s+1)!} \\ &+ \sum_{j,k=0}^{\infty} A(j,k)\frac{x^j y^k z^{j+k+1}}{(j+k+1)!}\sum_{r,s=0}^{\infty} A(r,s)\frac{x^{r+1} y^{s+1} z^{r+s+1}}{(r+s+1)!}.\end{aligned}$$

Therefore,

$$F'(z) = 1 + (x+y)F(z) + xyF^2(z).$$

Integrating this differential equation with the initial condition $F(0) = 0$, we obtain

$$F(z) = \frac{e^{xz} - e^{yz}}{xe^{yz} - ye^{xz}}.$$

Setting $z = 1$, we have

$$F(x,y) = \sum_{r,s=0}^{\infty} A(r,s)\frac{x^r y^s}{(r+s+1)!} = \frac{e^x - e^y}{xe^y - ye^x}. \qquad (3.3.12)$$

By virtue of the equality

$$A(r,s) = A_{r+s+1,r+1},$$

from (3.3.12) we obtain the expression for the generating function of the Eulerian numbers

$$1 + \sum_{n=1}^{\infty}\frac{x^n}{n!}\sum_{k=1}^{n} A_{n,k}y^k = \frac{1-y}{1-ye^{x(1-y)}}, \qquad y \neq 1,$$

which is equivalent to relation (3.3.8).

3.3.5 The Andre problem

At the end of the nineteenth century Andre considered the following problem (Andre, 1881). For $n \geqslant 2$ we denote by C_n the number of permutations $(a_1, a_2, \ldots, a_n)$ such that $a_j > a_{j-1}$ if j is even and $a_j < a_{j-1}$ if j is odd. We call these permutations E-permutations. Similarly, a permutation $(a_1, a_2, \dots, a_n)$ is called an O-permutation if $a_j > a_{j-1}$ for odd j and $a_j < a_{j-1}$ for even j. Note that if $(a_1, a_2, \dots, a_n)$ is an O-permutation, then $(n-a_1+1, n-a_2+1, \dots, n-a_n+1)$ is an E-permutation, and vice versa. Thus, there exists a one-to-one correspondence between the O-permutations and the E-permutations. Let us evaluate the numbers C_n, $n = 2, 3, \dots$.

The number of O-permutations with $a_1 = n$ is C_{n-1}; there exist no such E-permutations. The number of E-permutations with $a_2 = n$ is $(n-1)C_{n-2}$; there exist no such O-permutations. Similarly, for $a_3 = n$ the number of O-permutations is equal to $\binom{n-1}{2}C_2C_{n-3}$ and the number of E-permutations is zero. Looking over the cases $a_1 = n$, $a_2 = n$, $\dots$, $a_n = n$ and assuming that $C_0 = C_1 = 1$, we arrive at the recurrence relation

$$2C_n = \sum_{j=0}^{n-1} \binom{n-1}{j} C_j C_{n-j-1}, \qquad n = 2, 3, \dots. \tag{3.3.13}$$

Consider the generating function

$$C(x) = \sum_{n=0}^{\infty} C_n \frac{x^n}{n!}.$$

Multiplying both sides of equality (3.3.13) by $x^{n-1}/(n-1)!$ and summing over n, we obtain the differential equation

$$2C'(x) = C^2(x) + 1, \qquad C(0) = 1. \tag{3.3.14}$$

It is not difficult to see that the generating function

$$C(x) = \tan x + \sec x$$

satisfies equation (3.3.14); here $\tan x$ contains only odd powers of x, and $\sec x$ only even ones.

Up to this point we have considered the generating functions as formal power series. Assuming x to be a real variable such that $|x| < \pi/2$, from the generating function (3.2.23) for the Bernoulli numbers we obtain

$$\frac{x}{e^x - 1} + \frac{x}{2} = 1 + \sum_{k=1}^{\infty} B_{2k} \frac{x^{2k}}{(2k)!}.$$

Replacing x by ix, $i = \sqrt{-1}$, we obtain the equality

$$\frac{ix}{2}\left(\frac{e^{ix}+1}{e^{ix}-1}\right) = \frac{x}{2}\cot\frac{x}{2} = 1 + \sum_{k=1}^{\infty} (-1)^k B_{2k} \frac{x^{2k}}{(2k)!}.$$

Hence it follows that

$$\cot x - \frac{1}{x} = 1 + \sum_{k=1}^{\infty} (-1)^k B_{2k} 2^{2k} \frac{x^{2k}}{(2k)!}.$$

Using the elementary identity $\tan\theta = \cot\theta - 2\cot 2\theta$, we obtain

$$\tan x = \sum_{k=1}^{\infty} (-1)^{k-1} B_{2k} 2^{2k} (2^{2k} - 1) \frac{x^{2k-1}}{(2k)!}.$$

Hence it follows that

$$C_{2k-1} = (-1)^{k-1} B_{2k} \frac{2^{2k-1}}{k} (2^{2k} - 1). \tag{3.3.15}$$

The first few values of the Bernoulli numbers are:

$$B_0 = 1, \quad B_1 = -\frac{1}{2}, \quad B_2 = \frac{1}{6}, \quad B_4 = -\frac{1}{30}, \quad B_6 = \frac{1}{42};$$

therefore from formula (3.3.15) we get the following values of the numbers C_{2k-1}:

$$C_1 = 1, \quad C_3 = 2, \quad C_5 = 16.$$

Further, from the equality

$$\sec x = \sum_{k=0}^{\infty} C_{2k} \frac{x^{2k}}{(2k)!}$$

and the expansion of $\cos x$ we obtain

$$\sum_{k=0}^{\infty} C_{2k} \frac{x^{2k}}{(2k)!} \sum_{k=0}^{\infty} \frac{(-1)^k x^{2k}}{(2k)!} = 1.$$

Hence it follows that $C_0 = 1$ and

$$\sum_{j=0}^{k} (-1)^j \binom{2k}{2j} C_{2(k-j)} = 0, \qquad k = 1, 2, \ldots.$$

These equalities yield some values of the numbers C_{2k}:

$$C_0 = 1, \quad C_2 = 1, \quad C_4 = 5.$$

Using the expansions of the tangent and secant, we can derive the following asymptotic formula as $n \to \infty$:

$$C_n = \left(\frac{2}{\pi}\right)^{n+1} n!\,(1 + (-3)^{-n-1} + 5^{-n-1} + (-7)^{-n-1} + \cdots).$$

3.4 Gaussian coefficients and polynomials

In this section we consider those elements of combinatorial analysis concerning finite-dimensional vector spaces over Galois fields. Gaussian coefficients, Gaussian polynomials and Galois numbers play an important role in these investigations.

3.4.1 Gaussian coefficients and Galois numbers

Let $w_{n,k}$ be the number of linearly independent systems of vectors $v_0, v_1, \ldots, v_{k-1}$ in an n-dimensional vector space V_n over the Galois field GF(q). There exist $q^n - q^i$ ways of choosing a vector v_i, $0 \leqslant i \leqslant k-1$, which is not a linear combination of the vectors $v_0, v_1, \ldots, v_{i-1}$. Hence

$$w_{n,k} = (q^n - 1)(q^n - q) \cdots (q^n - q^{k-1}), \qquad 1 \leqslant k \leqslant n. \qquad (3.4.1)$$

In particular, the number of ordered bases of V_n is equal to

$$w_{n,n} = (q^n - 1)(q^n - q) \cdots (q^n - q^{n-1}). \qquad (3.4.2)$$

Let $\begin{bmatrix} n \\ k \end{bmatrix}_q$ be the number of k-dimensional subspaces in the space V_n. Since linearly independent vectors $v_1, \ldots, v_k \in V_n$ generate a k-dimensional subspace, the equality

$$(q^n - 1)(q^n - q) \cdots (q^n - q^{k-1}) = \begin{bmatrix} n \\ k \end{bmatrix}_q (q^k - 1)(q^k - q) \cdots (q^k - q^{k-1})$$

holds. Hence we obtain the formula

$$\begin{bmatrix} n \\ k \end{bmatrix}_q = \frac{(q^n - 1)(q^n - q) \cdots (q^n - q^{k-1})}{(q^k - 1)(q^k - q) \cdots (q^k - q^{k-1})}, \qquad 1 \leqslant k \leqslant n. \qquad (3.4.3)$$

The numbers $\begin{bmatrix} n \\ k \end{bmatrix}_q$ are referred to as Gaussian coefficients.

From the obvious equality

$$\begin{bmatrix} n \\ k \end{bmatrix}_q = \begin{bmatrix} n \\ n-k \end{bmatrix}_q \qquad (3.4.4)$$

it follows that we may set

$$\begin{bmatrix} n \\ 0 \end{bmatrix}_q = \begin{bmatrix} n \\ n \end{bmatrix}_q = 1. \qquad (3.4.5)$$

From formula (3.4.3) we immediately derive the recurrence relations for the Gaussian coefficients:

$$\begin{bmatrix} n \\ k \end{bmatrix}_q = \begin{bmatrix} n-1 \\ k \end{bmatrix}_q + q^{n-k} \begin{bmatrix} n-1 \\ k-1 \end{bmatrix}_q, \tag{3.4.6}$$

$$\begin{bmatrix} n \\ k \end{bmatrix}_q = \begin{bmatrix} n-1 \\ k-1 \end{bmatrix}_q + q^{k} \begin{bmatrix} n-1 \\ k \end{bmatrix}_q. \tag{3.4.7}$$

By induction on n, using relations (3.4.7) and (3.4.5), we can easily see that the Gaussian coefficient $\left[\begin{smallmatrix} n \\ k \end{smallmatrix}\right]_q$ as a function of q is a polynomial of degree $k(n-k)$ and, therefore, the Gaussian coefficients are determined for all real values of k by relation (3.4.3).

Note that

$$\lim_{q \to 1} \begin{bmatrix} n \\ k \end{bmatrix}_q = \binom{n}{k}, \tag{3.4.8}$$

that is, for $q = 1$ the Gaussian coefficients coincide with the binomial coefficients, and relations (3.4.6) and (3.4.7) take the form

$$\binom{n}{k} = \binom{n-1}{k} + \binom{n-1}{k-1}, \tag{3.4.9}$$

that is, they determine the Pascal triangle for the binomial coefficients. From relation (3.4.6) and the equality

$$\begin{bmatrix} n \\ k-1 \end{bmatrix}_q = \frac{q^n - 1}{q^{n-k+1} - 1} \begin{bmatrix} n-1 \\ k-1 \end{bmatrix}_q$$

we derive one more relation for the Gaussian coefficients:

$$\begin{bmatrix} n+1 \\ k \end{bmatrix}_q = \begin{bmatrix} n \\ k \end{bmatrix}_q + \begin{bmatrix} n \\ k-1 \end{bmatrix}_q + (q^n - 1) \begin{bmatrix} n-1 \\ k-1 \end{bmatrix}_q. \tag{3.4.10}$$

Consider the generating function

$$F_n(x) = F_n(x, q) = \sum_{k=0}^{n} \begin{bmatrix} n \\ k \end{bmatrix}_q x^k, \qquad n = 0, 1, \ldots. \tag{3.4.11}$$

Multiplying both sides of equality (3.4.10) by x^n and summing over n, we obtain the recurrence relation for this generating function:

$$F_{n+1}(x) = (1 + x)F_n(x) + (q^n - 1)xF_{n-1}(x), \qquad F_0(x) = 1. \tag{3.4.12}$$

Let us introduce the Galois numbers G_n by the equality

$$G_n = G_n(q) = \sum_{k=0}^{n} \begin{bmatrix} n \\ k \end{bmatrix}_q, \qquad G_0 = 1. \tag{3.4.13}$$

The Galois numbers $G_n(q)$ determine the total number of subspaces of an n-dimensional vector space V_n over the Galois field GF(q) if $q = p^\alpha$, where p is a prime number and α is a natural number. Substituting $x = 1$ into both sides of equality (3.4.12) and taking into account the equality $G_n = F_n(1)$, we obtain the recurrence relation for the Galois numbers (Goldman and Rota, 1970)

$$G_{n+1} = 2G_n + (q^n - 1)G_{n-1}, \quad n = 0, 1, \ldots, \quad G_0 = 1, \quad G_1 = 2. \tag{3.4.14}$$

Further, substituting $x = -1$ into both sides of equality (3.4.12), we obtain the relation

$$\sum_{k=0}^{n+1} (-1)^k \begin{bmatrix} n+1 \\ k \end{bmatrix}_q = -(q^n - 1) \sum_{k=0}^{n-1} (-1)^k \begin{bmatrix} n-1 \\ k \end{bmatrix}_q, \qquad n = 1, 2, \ldots .$$

Using this relation repeatedly, we obtain

$$\sum_{k=0}^{n} (-1)^k \begin{bmatrix} n \\ k \end{bmatrix}_q = \begin{cases} (1 - q^{n-1})(1 - q^{n-3}) \cdots (1 - q), & n \text{ is even}, \\ 0, & n \text{ is odd}. \end{cases} \tag{3.4.15}$$

Let us consider the Eulerian generating function for the Galois numbers:

$$G(x) = 1 + \sum_{n=1}^{\infty} \frac{G_n x^n}{(1-q)(1-q^2) \cdots (1-q^n)}. \tag{3.4.16}$$

From formula (3.4.3) and equality (3.4.13) it follows that the equality

$$G(x) = g^2(x) \tag{3.4.17}$$

holds with

$$g(x) = 1 + \sum_{n=1}^{\infty} \frac{x^n}{(1-q)(1-q^2) \cdots (1-q^n)}. \tag{3.4.18}$$

From relation (3.4.14) for the Galois numbers we derive

$$q^n G_n = G_n - 2(1 - q^n)G_{n-1} + (1 - q^n)(1 - q^{n-1})G_{n-2}.$$

Multiplying both sides of this equality by $x^n/((1-q)(1-q^2) \cdots (1-q^n))$, we obtain the functional relation

$$G(x) = \frac{1}{(1-x)^2} G(qx).$$

Using this relation repeatedly, we obtain

$$G(x) = \prod_{k=0}^{\infty} \frac{1}{(1 - q^k x)^2}. \tag{3.4.19}$$

By virtue of relation (3.4.17) and formulae (3.4.16) and (3.4.18), the representation of $G(x)$ in form (3.4.19) is equivalent to the validity of the equality

$$\prod_{k=0}^{\infty} \frac{1}{(1-q^k x)^2} = \sum_{n=0}^{\infty} \frac{x^n}{(1-q)(1-q^2)\cdots(1-q^n)}, \tag{3.4.20}$$

which will be proved independently in Section 5.5.

We note that for $q = 1$

$$G_n(1) = 2^n, \qquad F_n(x, 1) = (1+x)^n,$$

and equalities (3.4.15) yield

$$\sum_{k=0}^{n} (-1)^k \binom{n}{k} = \begin{cases} 1, & n = 0, \\ 0, & n > 0. \end{cases}$$

3.4.2 Gaussian polynomials

The Gaussian polynomials $g_n(x)$ in a real variable x are defined by the equalities

$$g_0(x) = 1, \qquad g_n(x) = (x-1)(x-q)\cdots(x-q^{n-1}), \quad n = 1, 2, \ldots. \tag{3.4.21}$$

Let $q = p^\alpha$, where p is a prime number and α is a natural number, let V_n be a vector space of dimension n over the Galois field GF(q), and let X be a vector space consisting of x elements over the same field. We consider a linear operator $A: V_n \to X$. If $v_1, v_2, \ldots, v_n$ is a basis of the space V_n, then the linear operator is uniquely determined by the totality of the images of this basis $A(v_1), A(v_2), \ldots, A(v_n)$. Hence it follows that the Gaussian polynomial $g_n(x)$ gives the number of non-singular linear operators of the form under consideration.

The Gaussian polynomials satisfy the relation (Goldman and Rota, 1970)

$$x^n = \sum_{k=0}^{n} \begin{bmatrix} n \\ k \end{bmatrix}_q g_k(x). \tag{3.4.22}$$

Let $v_1, v_2, \ldots, v_n$ be a basis of the space V_n and

$$W_k = \{v : A(v) = 0, \ v \in V_n\}$$

be the kernel of the operator A, which is a subspace of V_n of dimension k. If $k = 0$, the operator A is non-singular, and if $k = n$, then all vectors of V_n

are mapped onto the zero vector. Without loss of generality, we assume that $v_1, \dots, v_k$ is a basis of W_k and therefore, $A(v_1) = \cdots = A(v_k) = 0$, $A(v_{k+1}) \neq 0, \dots, A(v_n) \neq 0$, and the vectors $A(v_{k+1}), \dots, A(v_n)$ are linearly independent. The number of linear operators A with kernel W_k is equal to the number of ways of choosing the vectors $A(v_{k+1}), \dots, A(v_n)$, that is, it is equal to $g_{n-k}(x)$. The total number of linear operators of the form $A\colon V_n \to X$ with kernels of dimension k is equal to $\begin{bmatrix} n \\ k \end{bmatrix}_q g_{n-k}(x)$, $0 \leqslant k \leqslant n$. Summing these values over k, we obtain the total number of linear operators, which, on the other hand, is equal to x^n. Thus, equality (3.4.22) is proved for infinitely many natural values $x = q^s$, $s = 1, 2, \dots$. Since both sides of this equality are polynomials, its validity for all real x follows. Setting $x = 0$ in both sides of this equality and assuming that $0^0 = 1$, we obtain the orthogonality relation for the Gaussian coefficients

$$\sum_{k=0}^{n} (-1)^k \begin{bmatrix} n \\ k \end{bmatrix}_q q^{\binom{k}{2}} = \delta_{n,0}, \qquad n = 0, 1, \dots, \tag{3.4.23}$$

where $\delta_{n,0}$ is Kronecker's symbol, that is, $\delta_{0,0} = 1$, $\delta_{n,0} = 0$, $n > 0$. Since the left-hand side of relation (3.4.23) is a polynomial in q, this relation holds for all real values of q.

Now let the relations

$$u_n = \sum_{k=0}^{n} \begin{bmatrix} n \\ k \end{bmatrix}_q v_k, \qquad n = 0, 1, \dots, \tag{3.4.24}$$

hold for arbitrary numerical sequences $u_0, u_1, \dots; v_0, v_1, \dots$. Then by using the orthogonality relation (3.4.23) we obtain the inversion formula

$$v_n = \sum_{k=0}^{n} (-1)^k \begin{bmatrix} n \\ k \end{bmatrix}_q q^{\binom{k}{2}} u_{n-k}, \qquad n = 0, 1, \dots. \tag{3.4.25}$$

Conversely, formulae (3.4.25) yield (3.4.24).

Applying the inversion formula (3.4.25) to relation (3.4.22) for the Gaussian polynomials, we obtain the expression

$$g_n(x) = \sum_{k=0}^{n} (-1)^k \begin{bmatrix} n \\ k \end{bmatrix}_q q^{\binom{k}{2}} u_{n-k}, \qquad n = 0, 1, \dots. \tag{3.4.26}$$

Substituting $x = 1, q, \dots, q^{n-1}$ into (3.4.26) and taking (3.4.21) into account, we obtain, in addition to (3.4.23), $n-1$ orthogonality relations for the Gaussian coefficients

$$\sum_{k=0}^{n}(-1)^k \begin{bmatrix} n \\ k \end{bmatrix}_q q^{k(k-2j-1)/2} = \delta_{n,0}, \quad n = 0, 1, \ldots, \quad 1 \leqslant j \leqslant n-1.$$

Applying the inversion formula (3.4.25) to equality (3.4.13), we obtain one more recurrence relation for the Galois numbers

$$\sum_{k=0}^{n}(-1)^k \begin{bmatrix} n \\ k \end{bmatrix}_q q^{\binom{k}{2}} G_{n-k} = 1,$$

which, of course, is less convenient than relation (3.4.14).

3.4.3 The Rota method

Now let us derive relation (3.4.14) for the Galois numbers by method due to Rota (Rota, 1964a) which uses the properties of linear functionals. This derivation was suggested in paper (Goldman and Rota, 1970).

We consider the infinite-dimensional vector space V whose elements are the polynomials with rational coefficients. As bases of V we can choose the sequences of polynomials

$$1, x, x^2, \ldots, x^n, \ldots; \tag{3.4.27}$$

$$g_0(x), g_1(x), \ldots, g_n(x), \ldots, \tag{3.4.28}$$

where $g_n(x)$ is the Gaussian polynomial. A linear functional L is a linear mapping $L\colon V \to \mathbf{R}$, where $\mathbf{R}$ is the set of real numbers. From the linearity of the functional it follows that it is completely determined by its values at the elements of a basis of V. We take the sequence (3.4.28) as a basis and consider the linear functional determined by the system of equations

$$L(g_n(x)) = 1, \qquad n = 0, 1, \ldots. \tag{3.4.29}$$

Calculating the values of the functional L for both sides of equality (3.4.22) and taking formula (3.4.13) into account, we obtain

$$G_n = L(x^n). \tag{3.4.30}$$

By virtue of (3.4.30), relation (3.4.14) will be proved if we demonstrate that

$$L(x^{n+1}) = 2L(x^n) + (q^n - 1)L(x^{n-1}). \tag{3.4.31}$$

Let us introduce the linear operator $\mathscr{D}_q$:

$$\mathscr{D}_q P(x) = \frac{1}{x}(P(qx) - P(x)), \qquad P(x) \in V.$$

It is clear that $\mathscr{D}_q x^n = (q^n - 1)x^{n-1}$. Then relation (3.4.31) can be rewritten in terms of the operator $\mathscr{D}_q$ as

$$L(x \cdot x^n) = 2L(x^n) + L(\mathscr{D}_q x^n). \tag{3.4.32}$$

Since the sequence (3.4.27) is a basis of V and the operator $\mathscr{D}_q$ is linear, we can write

$$L(xP(x)) = 2L(P(x)) + L(\mathscr{D}_q P(x)) \tag{3.4.33}$$

for an arbitrary polynomial $P(x) \in V$. Setting $P(x) = g_n(x)$, from (3.4.33) we obtain

$$L(xg_n(x)) = 2L(g_n(x)) + L(\mathscr{D}_q g_n(x)). \tag{3.4.34}$$

Since the sequence (3.4.28) is a basis of V, the validity of equality (3.4.34) implies that of (3.4.33) and, therefore, of relation (3.4.32). Thus, (3.4.14) will be proved if we demonstrate that (3.4.34) holds.

Formula (3.4.21) yields

$$xg_n(x) = g_{n+1}(x) + q^n g_n(x).$$

Letting the operator L act on both sides of this equality bearing in mind formula (3.4.29), we obtain

$$L(xg_n(x)) = 1 + q^n.$$

On the other hand, since $\mathscr{D}_q g_n(x) = (q^n - 1)g_{n-1}(x)$, the relation

$$2L(g_n(x)) + L(\mathscr{D}_q g_n(x)) = 1 + q^n$$

holds. Thus, equality (3.4.34) is proved.

3.4.4 Partitions of numbers with constraints

A representation of a natural number n as the sum $n = 1 \cdot d_1 + 2 \cdot d_2 + \cdots + nd_n$, where $d_1, d_2, \ldots, d_n$ are non-negative integers, is called a partition of n, where the sum $d_1 + d_2 + \cdots + d_n$ determines the number of summands of the partition. We denote by $R(n, \alpha, \beta)$ the number of partitions of n such that $d_1 + d_2 + \cdots + d_n \leqslant \alpha$ and $d_{\beta+1} = \cdots = d_n = 0$, that is, with the number of summands at most α and no summands greater than β. It is clear that

$$R(\alpha\beta, \alpha, \beta) = 1, \qquad R(n, \alpha, \beta) = 0 \quad \text{for} \quad n > \alpha\beta. \tag{3.4.35}$$

In addition, we set

$$R(n, 0, \beta) = R(n, \alpha, 0) = \begin{cases} 1, & n = \alpha = \beta = 0, \\ 0 & \text{otherwise.} \end{cases} \tag{3.4.36}$$

The recurrence relation

$$R(n, \alpha, \beta) - R(n, \alpha - 1, \beta) = R(n - \alpha, \alpha, \beta - 1) \tag{3.4.37}$$

holds. Indeed, the left-hand side of this equality is equal to the number of partitions of n into exactly α summands not exceeding β. If we subtract one from every summand of such a partition, then, as a result, we obtain a partition into no more than α summands not exceeding $\beta - 1$. The number of such partitions coincides with the right-hand side of equality (3.4.37). Since there exists a bijection between the partitions of both these types, equality (3.4.37) is proved.

We consider the generating function for the numbers $R(n, \alpha, \beta)$:

$$f(q; \alpha, \beta) = \sum_{n=0}^{\alpha\beta} R(n, \alpha, \beta) q^n.$$

Taking equalities (3.4.35), (3.4.36) and

$$R(n, \alpha, \beta) = R(n, \alpha - 1, \beta), \qquad n < \alpha,$$

into account, we find from relation (3.4.37) that

$$f(q; \alpha, \beta) = f(q; \alpha - 1, \beta) + q^\alpha f(q; \alpha, \beta - 1), \tag{3.4.38}$$

where

$$f(q; 0, \beta) = f(q; \alpha, 0) = 1. \tag{3.4.39}$$

From equalities (3.4.5) and (3.4.7) it follows that (Andrews, 1976)

$$f(q; \alpha, \beta) = \begin{bmatrix} \alpha + \beta \\ \beta \end{bmatrix}_q, \tag{3.4.40}$$

since $f(q; \alpha, \beta)$ satisfies the same recurrence relation as the Gaussian coefficients $\begin{bmatrix} \alpha+\beta \\ \beta \end{bmatrix}_q$ with the same initial conditions.

3.4.5 Inversions in binary sequences

Let us consider a sequence of length $\alpha+\beta$ containing α units and β zeros. The number of zeros situated to the right of some unit will be referred to as the number of inversions generated by this unit. The sum of inversions of all α units is the number of inversions of the whole sequence. We denote by $J(n, \alpha, \beta)$ the number of sequences consisting of α units and β zeros which have n inversions. Let us prove that (Andrews, 1976)

$$J(n, \alpha, \beta) = R(n, \alpha, \beta), \tag{3.4.41}$$

where $R(n, \alpha, \beta)$ is the number of partitions of n into no more than α summands not exceeding β. Any such partition can be represented as

$$n = \gamma_1 d_{\gamma_1} + \cdots + \gamma_\nu d_{\gamma_\nu},$$
$$1 \leqslant \gamma_1 < \cdots < \gamma_\nu \leqslant \beta, \qquad 1 \leqslant d_{\gamma_1} + \cdots + d_{\gamma_\nu} \leqslant \alpha. \tag{3.4.42}$$

There exists a bijection between the partitions (3.4.42) and the sequences with α units and β zeros of the form

$$\underbrace{0\ldots0}_{\beta-\gamma_\nu}\underbrace{1\ldots1}_{d_{\gamma_\nu}}\underbrace{0\ldots0}_{\gamma_\nu-\gamma_{\nu-1}}\underbrace{1\ldots1}_{d_{\gamma_{\nu-1}}}\ldots\underbrace{0\ldots0}_{\gamma_2-\gamma_1}\underbrace{1\ldots1}_{d_{\gamma_1}}\underbrace{0\ldots0}_{\gamma_1}\ \underbrace{1\ldots1}_{\alpha-d_{\gamma_1}-\cdots-\gamma_\nu}\ . \tag{3.4.43}$$

The number of inversions generated by the ith series of units in sequence (3.4.43) is equal to $\gamma_i d_{\gamma_i}$, $1 \leqslant i \leqslant \nu$. Therefore, the total number of inversions is equal to $n = \gamma_1 d_{\gamma_1} + \gamma_2 d_{\gamma_2} + \cdots + \gamma_\nu d_{\gamma_\nu}$, and equality (3.4.41) is proved.

By virtue of formula (3.4.40), equality (3.4.41) yields

$$\sum_{n=0}^{\alpha\beta} J(n, \alpha, \beta) q^n = \sum_{n=0}^{\alpha\beta} R(n, \alpha, \beta) q^n$$
$$= \frac{(1-q^{\alpha+\beta})(1-q^{\alpha+\beta-1})\cdots(1-q^{\alpha+1})}{(1-q)(1-q^2)\cdots(1-q^\beta)}.$$

3.5 The Dirichlet generating functions

3.5.1 The Möbius inversion formula

In number theory and combinatorial analysis the generating functions defined by the Dirichlet series of the form

$$d(s) = \sum_{n=1}^{\infty} \frac{a_n}{n^s}$$

are widely used. If generating functions

$$d_1(s) = \sum_{n=1}^{\infty} \frac{b_n}{n^s}, \qquad d_2(s) = \sum_{n=1}^{\infty} \frac{c_n}{n^s}$$

are given, then the condition

$$d(s) + d_1(s) = d_2(s)$$

means that

$$a_n + b_n = c_n, \qquad n = 1, 2, \ldots .$$

On the other hand, the condition

$$d(s)d_1(s) = d_2(s)$$

is equivalent to the equality

$$c_n = \sum_{sr=n} a_s b_r,$$

where the summation is over all numbers r and s whose product is equal to n.

Thus, we have defined two composition laws, addition and multiplication, on the set of the generating functions determined by the Dirichlet series; both of these laws are commutative and associative, and the second law is connected with the first by the distributive rule. The validity of these properties follows from the corresponding relations between the elements of the sequences. For example, the associativity of multiplication means that

$$\sum_{d|n} a_{n/d} \sum_{\delta|d} b_{d/\delta} c_\delta = \sum_{d|n} \left(\sum_{\delta|d} a_\delta b_{d/\delta} \right) c_{n/d},$$

where the summation is taken over all divisors of n. This equality can easily be verified.

Thus, the set of all Dirichlet series forms a ring with the unit element

$$e(s) = \frac{1}{1^s} + \frac{0}{2^s} + \cdots .$$

If $a_1 \neq 0$, then for a generating function

$$d(s) = \sum_{n=1}^{\infty} \frac{a_n}{n^s}$$

there exists an inverse one

$$d^{-1}(s) = \sum_{n=1}^{\infty} \frac{a_n^{(-1)}}{n^s}$$

satisfying the condition

$$d(s)d^{-1}(s) = e(s).$$

This inverse generating function is determined by the system of equations

$$\sum_{d|n} a_d a_{n/d}^{(-1)} = \begin{cases} 1, & n = 1, \\ 0, & n > 1. \end{cases}$$

If the canonical expansion of a number n is

$$n = p_1^{s_1} p_2^{s_2} \cdots p_r^{s_r},$$

then the Möbius function can be defined as follows:

$$\mu(n) = \begin{cases} 1, & n = 1, \\ (-1)^r, & s_1 = s_2 = \cdots = s_r = 1, \\ 0 & \text{otherwise.} \end{cases} \tag{3.5.1}$$

The Möbius function satisfies the relation

$$\sum_{d|n} \mu(d) = \begin{cases} 1, & n = 1, \\ 0, & n > 1. \end{cases} \tag{3.5.2}$$

Indeed, put $n^* = p_1 p_2 \cdots p_r$ for $n > 1$; then

$$\sum_{d|n} \mu(d) = \sum_{d|n^*} \mu(d) = \sum_{k=0}^{r} (-1)^k \binom{r}{k} = 0.$$

The generating function

$$\zeta(s) = \sum_{n=1}^{\infty} \frac{1}{n^s}$$

is called the zeta function. It can easily be seen that the generating function

$$\zeta^{-1}(s) = \sum_{n=1}^{\infty} \frac{\mu(n)}{n^s}$$

is inverse to $\zeta(s)$. We consider two Dirichlet generating functions

$$d(s) = \sum_{n=1}^{\infty} \frac{f(n)}{n^s}, \qquad d_1(s) = \sum_{n=1}^{\infty} \frac{g(n)}{n^s},$$

related by the equality

$$d(s) = d_1(s)\zeta(s).$$

This last relation means that

$$f(n) = \sum_{d|n} g(d). \tag{3.5.3}$$

The obvious equality

$$d_1(s) = d(s)\zeta^{-1}(s)$$

yields the Möbius inversion formula

$$g(n) = \sum_{d|n} \mu(d) f(n/d). \tag{3.5.4}$$

It is easy to verify that

$$\zeta(s) = \sum_{n=1}^{\infty} \frac{1}{n^s} = \prod_p \frac{1}{1 - \frac{1}{p^s}}, \qquad \Re s > 1,$$

where the product is taken over all prime numbers. This equality is known in number theory as the Euler identity. Since $\zeta(s) \to \infty$ as $s \to 1$, this identity, in particular, implies that there are infinitely many prime numbers.

3.5.2 The problem on cyclic sequences

We consider n^m mappings of the set $X = \{1, 2, \ldots, m\}$ into the set $A = \{a_1, a_2, \ldots, a_n\}$. Every such mapping determines a linear sequence of length m consisting of elements of A. By definition, we say that the set of all sequences obtained from a linear one by cyclic shift is called the cyclic sequence corresponding to the given linear sequence. If d is the minimum period of a given cyclic sequence, then there are d linear sequences corresponding to this sequence. We denote by $N_{nm}(d)$ the number of cyclic sequences of the minimum period d. Since $dN_{nm}(d)$ linear sequences correspond to them and d divides m, the equality

$$\sum_{d|m} dN_{nm}(d) = n^m$$

is true. Using the Möbius inversion formula, we obtain

$$N_{nm}(m) = \frac{1}{m} \sum_{d|m} \mu(d) n^{m/d},$$

where $N_{nm}(m)$ is the number of cyclic sequences of the maximum period m. Hence it follows that the total number of cyclic sequences is equal to

$$N_{nm} = \sum_{d|m} \frac{1}{d} \sum_{d'|d} \mu(d') n^{d/d'}. \tag{3.5.5}$$

Let us consider the Euler totient function $\varphi(n)$ equal to the number of elements of the set $0, 1, \ldots, n-1$ which are relatively prime with n. For this function the Gaussian formula is well known:

$$\sum_{d|m} \varphi(d) = m.$$

Applying the Möbius inversion formula to this last equality, we obtain

$$\varphi(m) = \sum_{d|m} \mu(d) \frac{m}{d}. \tag{3.5.6}$$

Note that

$$\begin{aligned} \sum_{d|m} \frac{1}{d} \sum_{d'|d} \mu(d') n^{d/d'} &= \sum_{d|m} n^d \sum_{d'|m/d} \frac{1}{dd'} \mu(d') \\ &= \frac{1}{m} \sum_{d|m} n^d \sum_{d'|m/d} \mu(d') \frac{(m/d)}{d'}. \end{aligned}$$

Then formulae (3.5.5) and (3.5.6) yield

$$N_{nm} = \frac{1}{m} \sum_{d|m} \varphi(d) n^{m/d}. \tag{3.5.7}$$

If $m = p$, where p is a prime number, then by virtue of the fact that $\varphi(1) = 1$, $\varphi(p) = p - 1$, the formula takes the form

$$N_{np} = \frac{n^p - n}{p} + n.$$

Let n be fixed and $m \to \infty$. Formula (3.5.7) can be rewritten as follows:

$$N_{nm} = \frac{n^m}{m} + R_{nm}, \tag{3.5.8}$$

where

$$mR_{nm} = \sum_{d|m,\, d>1} \varphi(d) n^{m/d}.$$

Note that

$$mR_{nm} \leqslant n^{m/2} \sum_{d|m} \varphi(d) = n^{m/2} m.$$

Using this estimate, we find from formula (3.5.8) that

$$N_{nm} = \frac{n^m}{m}\left(1 + O\left(\frac{m}{(\sqrt{n})^m}\right)\right).$$

It is not difficult to see that the formula obtained remains meaningful under more general conditions, namely, if $m/(\sqrt{n})^m \to 0$.

3.5.3 The number of irreducible polynomials over the Galois field

A polynomial $f(x) = x^n + a_1x^{n-1} + \cdots + a_n$ with coefficients $a_1, a_2, \ldots, a_n$ from the Galois field GF(q) and with the leading coefficient equal to 1 is called unitary. We denote by J_n the number of irreducible unitary polynomials over the field GF(q). If $f_1(x), f_2(x), \ldots, f_N(x)$ are all irreducible unitary polynomials whose degrees divide n, then the equality

$$x^{q^n} - x = f_1(x)f_2(x)\cdots f_N(x)$$

holds. Equating the degrees of the polynomials on the left and right-hand sides of this equality, we obtain

$$q^n = \sum_{d|n} dJ_d. \tag{3.5.9}$$

Using the Möbius inversion formula, we obtain

$$J_n = \frac{1}{n}\sum_{d|n} \mu(d)q^{n/d}.$$

From relation (3.5.9) it follows that

$$\frac{q^n - q^{n/2+1}}{n} \leqslant J_n \leqslant \frac{q^n}{n}.$$

These inequalities imply that

$$\lim_{n\to\infty} \frac{nJ_n}{q^n} = 1.$$

3.6 Asymptotic behavior of Stirling numbers

In this section we consider the asymptotic formulae for Stirling numbers of the first kind. These formulae are of considerable importance due to the role played by Stirling numbers in combinatorial analysis. In addition, the methods used to obtain these asymptotic formulae have wide use in mathematics and are thus of independent interest.

3.6.1 The basic definitions and results

Three asymptotic formulae for Stirling numbers of the first kind $s(n,m)$ are known. These formulae are valid as $n \to \infty$ in three different overlapping regions covering the whole range of values of m.

For $m = o(\ln n)$ the Jordan formula

$$|s(n,m)| = \frac{(n-1)!}{(m-1)!}(\ln n + C)^{m-1}(1+o(1)) \tag{3.6.1}$$

is valid; here $C = 0.5772\ldots$ is the Euler constant.

In the case where $n, m \to \infty$ in such a way that $m \leqslant n - cn^{\alpha}$, $c > 0$, $0 < \alpha < 1$, it was proved by Moser and Wyman that

$$s(n,m) = (-1)^{n+m}\frac{\Gamma(n+R)}{R^m\Gamma(R)\sqrt{2\pi H}}(1+o(1)), \tag{3.6.2}$$

where R is the unique solution of the equation

$$\sum_{h=0}^{n-1}\frac{R}{R+h} = m \tag{3.6.3}$$

and

$$H = m - \sum_{h=0}^{n-1}\frac{R^2}{(R+h)^2}. \tag{3.6.4}$$

Finally, Moser and Wyman also proved that for $n, m \to \infty$ in such a way that $n - m = o(\sqrt{n})$:

$$s(n,m) = (-1)^{n+m}\binom{n}{m}\left(\frac{m}{2}\right)^{n-m}(1+o(1)). \tag{3.6.5}$$

In order to derive the asymptotic formulae, we introduce some notation and formulate several assertions from mathematical analysis. We set

$$M(m,n) = \sum_{k=1}^{n-1}\frac{1}{k^m}, \qquad M(n) = M(1,n).$$

Using the zeta function and the generalized Riemann's zeta function,

$$\zeta(z) = \sum_{k=1}^{\infty}\frac{1}{k^z}, \qquad \zeta(z,n) = \sum_{k=n}^{\infty}\frac{1}{k^z},$$

we can write

$$M(m,n) = \zeta(m) - \zeta(m,n).$$

For the logarithmic derivative of the gamma function $\psi(z) = \frac{d}{dz}\ln\Gamma(z)$ the expression

$$\psi(n+1) = -C + \sum_{l=1}^{n} \frac{1}{l}$$

is true, where n is a natural number and C is the Euler constant defined by the equality

$$C = \lim_{n\to\infty}\left(\sum_{l=1}^{n}\frac{1}{l} - \ln n\right) = 0.5772\ldots.$$

Using the Euler–Maclaurin formula, we find that

$$M(n) = \psi(n) + C = \ln n + C - \frac{1}{2n} + O\left(\frac{1}{n^2}\right). \qquad (3.6.6)$$

Taking the obvious formula

$$\psi(R+n) - \psi(R) = \sum_{h=0}^{n-1}\frac{1}{R+h}$$

into account, we introduce the following notation

$$P(R,n) = R(\psi(R+n) - \psi(R)) = \sum_{h=0}^{n-1}\frac{R}{R+h}, \qquad (3.6.7)$$

$$H(R,n) = R\frac{d}{dR}P(R,n) = \sum_{h=0}^{n-1}\frac{Rh}{R+h}. \qquad (3.6.8)$$

3.6.2 The Jordan formula

Introducing the notation

$$D_0^{m-1}f(x) = \left(\frac{d}{dx}\right)^{m-1} f(x)|_{x=0},$$

from the equality

$$(x)_n = \sum_{m=1}^{n} s(n,m)x^m, \qquad (3.6.9)$$

we see that

$$s(n,m) = \frac{1}{(m-1)!}D_0^{m-1}(x-1)_{n-1}$$

$$= \frac{(-1)^{n-1}(n-1)!}{(m-1)!}D_0^{m-1}\exp\left\{\sum_{k=1}^{n-1}\ln\left(1-\frac{x}{k}\right)\right\}.$$

Since the coefficient of x^{m-1} in $\ln(1 - x/k)$ is equal to the corresponding coefficient in the sum

$$-\sum_{p=1}^{m-1} \frac{x^p}{pk^p},$$

and, in addition,

$$\sum_{k=1}^{n-1} \ln\left(1 - \frac{x}{k}\right) = -xM(n) - \sum_{p=2}^{\infty} \frac{x^p M(p,n)}{p},$$

we see that

$$s(n,m) = \frac{(-1)^{n-1}(n-1)!}{(m-1)!} D_0^{m-1} \exp\left\{-xM(n) - \sum_{p=2}^{\infty} \frac{x^p M(p,n)}{p}\right\}.$$

We carry out the change of variable $z = -xM(n)$ and define the function $F(z)$ as

$$F(z) = \exp\left\{\sum_{p=2}^{m-1} (-1)^{p+1} \frac{M(p,n)}{p(M(n))^p} z^p\right\}.$$

As a result we obtain

$$s(n,m) = \frac{(-1)^{m+n}(n-1)!}{(m-1)!} (M(n))^{m-1} D_0^{m-1}(e^z F(z)).$$

Using the Leibniz formula for the kth derivative of a function, we find that

$$s(n,m) = \frac{(-1)^{m+n}(n-1)!}{(m-1)!} (M(n))^{m-1} \left(1 + \sum_{k=2}^{m-1} \binom{m-1}{k} F^{(k)}(0)\right), \tag{3.6.10}$$

where $F^{(k)}(0)$ is the kth derivative of $F(z)$ at the point $z = 0$.

For s, $2 \leqslant s \leqslant m-1$, we introduce the notation

$$E_s = \sum_{k=s+1}^{m-1} \binom{m-1}{k} F^{(k)}(0).$$

Then

$$\sum_{k=2}^{m-1} \binom{m-1}{k} F^{(k)}(0) = \sum_{k=2}^{s} \binom{m-1}{k} F^{(k)}(0) + E_s.$$

Using the Cauchy integral formula, we obtain

$$b_k = \frac{1}{k!} F^{(k)}(0) = \frac{1}{2\pi i} \oint_C F(z) \frac{dz}{z^{k+1}},$$

where C is a circle with its center at the point $z = 0$ of the complex plane.

Since we want to calculate the kth derivative at the point $z = 0$, we can substitute $F_k(z)$ for $F(z)$, where

$$F_k(z) = \exp\left\{\sum_{p=2}^{k}(-1)^{p+1}\frac{M(p,n)}{p(M(n))^p}z^p\right\}.$$

Grouping the terms in the sum $M(p,n)$, we can obtain the bound

$$M(p,n) \leqslant \frac{2^{p-1}}{2^{p-1}-1} \leqslant p,$$

which is valid for $p \geqslant 2$. Using this and the bound

$$|F_k(z)| \leqslant e^k,$$

which is valid for $|z| = M(n)$, we see that

$$|b_k| \leqslant \left(\frac{e}{M(n)}\right)^k.$$

Thus,

$$|E_s| \leqslant \sum_{k=s+1}^{m-1}\left(\frac{e(m-1)}{M(n)}\right)^k.$$

Hence we obtain for $m = o(M(n))$ and $s = 2, 3, \ldots, m-1$ that

$$|E_s| \leqslant 2\left(\frac{e(m-1)}{M(n)}\right)^{s+1}.$$

Note that for $m = o(M(n))$

$$\binom{m-1}{2}F^{(2)}(0) = -\binom{m-1}{2}\frac{M(2,n)}{(M(n))^2} = o(1).$$

Now, substituting the asymptotic representation (3.6.6) of $M(n)$, we derive the Jordan formula (3.6.1) from (3.6.10).

3.6.3 The first Moser and Wyman formula

In the second domain of variation of m we assume that

$$h(n) \leqslant m \leqslant n - cn^{\alpha}, \qquad 0 < \alpha < 1, \tag{3.6.11}$$

where c is a positive constant and $h(n)$ is an arbitrary function such that $h(n) \to \infty$ as $n \to \infty$.

Lemma 3.6.1 *For all m satisfying condition* (3.6.11), *equation* (3.6.3), *written in the form*

$$P(R,n) = m, \tag{3.6.12}$$

has a unique solution $R > 0$ *such that the function* $H = H(R,n)$ *defined by equality* (3.6.4) *satisfies the condition*

$$\lim_{n\to\infty} H(R,n) = \infty. \tag{3.6.13}$$

Proof It can easily be seen that $P(R,n)$ is an increasing continuous function of R for $R > 0$ and fixed n; moreover, $\lim_{R\to 0} P(R,n) = 1$ and $\lim_{R\to\infty} P(R,n) = n$. Therefore, equation (3.6.12) has a unique solution for $1 < m < n$.

From (3.6.7) and (3.6.8) it follows that

$$H = P - \sum_{h=0}^{n-1} \frac{R^2}{(R+h)^2}.$$

If R is a solution of equation (3.6.12), then

$$H = m - \sum_{h=0}^{n-1} \frac{R^2}{(R+h)^2}. \tag{3.6.14}$$

Since the series $\sum_{h=0}^{\infty}(R+h)^{-2}$ converges, it is clear that $H \to \infty$ as $n \to \infty$, R is bounded and m satisfies condition (3.6.11).

Now let us consider the behavior of H as $R \to \infty$. From formulae (3.6.6) and (3.6.7) with large R we deduce

$$P(R,n) = R \ln\left(1 + \frac{n}{R}\right) + \frac{n}{2(R+n)} + o(1).$$

Since the second summand is less than $1/2$ for $R > 0$ and we are interested only in those values of R which yield large $P(R,n)$, it is sufficient to consider the equation

$$R \ln\left(1 + \frac{n}{R}\right) = m. \tag{3.6.15}$$

Substituting

$$R = \frac{n}{e^v - 1}$$

into equality (3.6.15), we obtain

$$\frac{nv}{e^v - 1} = m. \tag{3.6.16}$$

From formula (3.6.14) we deduce that

$$H = m - \frac{Rn}{R+n} + o(1) = m\left(1 - \frac{1-e^{-v}}{v} + o(1)\right).$$

The function $(1-e^{-v})/v$ decreases for $v \geqslant 0$; therefore $H \to \infty$ provided $\lim_{n\to\infty} v \neq 0$. If v is small enough, then $H = mv(1+o(1))/2$, and $H \to \infty$ provided $v \geqslant cm^{\alpha-1}$, where $c > 0$, $\alpha > 0$. Since

$$\frac{1-e^{-v}}{v} \geqslant 1 + \frac{v}{2},$$

relation (3.6.16) implies the validity of the lemma for

$$n \geqslant m\left(1+\frac{v}{2}\right) \geqslant m + cm^{\alpha}.$$

Since $m \leqslant n$, the lemma is valid for $m \leqslant n - cn^{\alpha}$. □

Lemma 3.6.2 *Let R be a positive variable, h be a positive constant and $\theta = R\frac{d}{dR}$. Then for any non-negative integer s the functions*

$$a_s = \frac{1}{s!}\theta^s\left(\frac{R}{(R+h)^2}\right)$$

satisfy the inequality

$$|a_s| \leqslant \frac{R}{(R+h)^2}.$$

Proof We prove the lemma by induction on s. It is clear that the assertion of the lemma is true for $s = 0$. Using the formulae

$$(s+1)!\,a_{s+1} = \theta^{s+1}\left(\frac{R}{(R+h)^2}\right) = \theta^s\left(\frac{R}{(R+h)^2}\frac{h-R}{R+h}\right),$$

$$\theta^s(uv) = \sum_{k=0}^{s}\binom{s}{k}\theta^k(u)\theta^{s-k}(v),$$

we obtain

$$(s+1)!\,a_{s+1} = \frac{h-R}{R+h}s!\,a_s - 2hs!\sum_{k=0}^{s-1}\frac{a_k a_{s-k-1}}{s-k}.$$

Now, taking the induction assumption into account, we obtain the inequality

$$|a_{s+1}| \leqslant \frac{R}{(R+h)^2(s+1)} + \frac{2h}{s+1}\frac{R^2}{(R+h)^4}\sum_{k=1}^{s}\frac{1}{k}. \qquad (3.6.17)$$

Since R and h are positive, $(2hR)/(R+h)^2 \leqslant 1$ and $\sum_{k=1}^{s} \frac{1}{k} \leqslant s$. These bounds, together with inequality (3.6.17), yield the assertion of the lemma.

□

Lemma 3.6.3 *Under the hypotheses of Lemma* 3.6.2 *define* $C_k = C_k(R, n)$ *for* $k \geqslant 2$ *by the equalities*

$$C_k = \frac{1}{k!}\theta^{k-2} H(R,n) = \frac{1}{k!}\theta^{k-2} \sum_{h=1}^{n-1} \frac{Rh}{(R+h)^2}.$$

Then the inequality

$$|C_k| \leqslant H(R,n) = \sum_{h=1}^{n-1} \frac{Rh}{(R+h)^2}$$

holds.

The assertion of the lemma follows from Lemma 3.6.2.

Lemma 3.6.4 *Let a function* $f(z)$ *of a complex variable* z *be regular in a neighborhood of the point* $z=0$. *If* $f(0)=0$ *and*

$$f(z) = \sum_{k=1}^{\infty} a_k z^k, \qquad e^{f(z)} = 1 + \sum_{k=1}^{\infty} b_k z^k,$$

where $|a_k| \leqslant K\sigma^k$, *with positive constants* K *and* σ, *then*

$$|b_k| \leqslant K\sigma^k (1+K)^{k-1}.$$

Proof Since $b_1 = a_1$, the assertion of the lemma is true for $k=1$. From the equality

$$b_{k+1} = \frac{1}{(k+1)!} \frac{d^k}{dz^k} \left(e^{f(z)} \frac{df(z)}{dz} \right) \bigg|_{z=0},$$

using the Leibniz formula, we derive the equality

$$b_{k+1} = \frac{1}{k+1} \sum_{j=0}^{k} b_j a_{k-j+1}(k-j+1), \qquad b_0 = 1.$$

Hence the inequality

$$|b_{k+1}| \leqslant \sum_{j=0}^{k} |b_j||a_{k-j+1}|$$

follows, from which, using the hypothesis of the lemma, we obtain the bound

$$|b_{k+1}| \leqslant K\sigma^{k+1}\left(1+\sum_{j=1}^{k}|b_j|\sigma^{-j}\right).$$

Hence the assertion of the lemma follows by induction. □

Now let us turn to the derivation of the asymptotic formula. From equality (3.6.9) we deduce that

$$s(n,m) = \frac{1}{2\pi i}\oint_{|z|=R}\frac{(z)_n}{z^{m+1}}\,dz. \tag{3.6.18}$$

Substituting $-z$ for z, we obtain

$$s(n,m) = \frac{(-1)^{n+m}}{2\pi i}\oint_{|z|=R}\frac{\Gamma(z+n)}{\Gamma(z)z^{m+1}}\,dz.$$

Set $z = Re^{i\theta}$, then this last formula yields

$$s(n,m) = B\int_{-\pi}^{\pi} F(\theta)\,d\theta,$$

where

$$B = \frac{(-1)^{n+m}\Gamma(R+n)}{2\pi R^m\Gamma(R)},$$

$$F(\theta) = \frac{\Gamma(Re^{i\theta}+n)\Gamma(R)}{e^{im\theta}\Gamma(R+n)\Gamma(Re^{i\theta})}.$$

It is not difficult to calculate the square of the absolute value of $F(\theta)$:

$$|F(\theta)|^2 = \prod_{h=1}^{n-1}\left(1-\frac{4Rh}{(R+h)^2}\sin^2\frac{\theta}{2}\right).$$

It can be seen that $|F(\theta)|$ is a decreasing function of θ for $0 \leqslant \theta \leqslant \pi$. For all θ such that $0 < \varepsilon \leqslant \theta \leqslant \pi$ the inequality $|F(\theta)| \leqslant |F(\varepsilon)|$ holds. If ε is sufficiently small, then there exists a constant $c_1 > 0$ such that

$$|F(\theta)| \leqslant e^{-c_1\varepsilon^2 H},$$

where H is the same as in Lemma 3.6.1, and therefore $H \to \infty$ as $n \to \infty$. We choose $\varepsilon = H^{-3/8}$; then we obtain the bound

$$|F(\theta)| \leqslant e^{-c_1 H^{1/4}}.$$

Since $|F(\theta)| = |F(-\theta)|$, the same bound is valid for $-\pi \leqslant \theta \leqslant -\varepsilon$. Hence it follows that

$$s(n,m) = B \int_{-\varepsilon}^{\varepsilon} F(\theta)\, d\theta (1 + o(1)),$$

where the error is of an exponential character.

Since $F(0) = 1$ and $F(\theta)$ is regular in a neighborhood of the point $\theta = 0$, the logarithm $\ln F(\theta)$ can be expanded as a Maclaurin series

$$\ln F(\theta) = (P - m)i\theta - \frac{1}{2}H\theta^2 + \sum_{k=3}^{\infty} C_k (i\theta)^k.$$

Supposing R to be the solution of the equation $P = m$, we obtain

$$\ln F(\theta) = -\frac{1}{2}H\theta^2 + \sum_{k=3}^{\infty} C_k (i\theta)^k.$$

Let us carry out the change of variable $\theta = \varphi\sqrt{2/H}$. Then

$$s(n,m) = D \int_{-\delta}^{\delta} G(\varphi) e^{-\varphi^2}\, d\varphi, \tag{3.6.19}$$

where

$$D = B\sqrt{\frac{2}{H}},$$

$$\delta = \varepsilon\sqrt{\frac{H}{2}} = \frac{1}{\sqrt{2}} H^{1/8} \to \infty, \qquad n \to \infty,$$

$$G(\varphi) = \exp\left\{ \sum_{k=3}^{\infty} \frac{C_k \left(i\varphi\sqrt{2}\right)^k}{H^{k/2}} \right\}.$$

Set $\bar{C}_k = C_k/H$, then

$$G(\varphi) = \exp\left\{ \sum_{k=1}^{\infty} \frac{\bar{C}_{k+2} \left(i\varphi\sqrt{2}\right)^{k+2}}{H^{k/2}} \right\}.$$

Now consider the function

$$G(\varphi, z) = \exp\left\{ \sum_{k=1}^{\infty} \bar{C}_{k+2} \left(i\varphi\sqrt{2}\right)^{k+2} z^k \right\},$$

where $\bar{C}_{k+2}$ are introduced above and z is an independent complex variable. From Lemma 3.6.3 it follows that $|\bar{C}_{k+2}| \leqslant 1$ and $|\varphi| \leqslant$

$\delta = H^{1/8}/\sqrt{2}$. Hence $G(\varphi, z)$ is a regular function of z in the region $|z| < H^{-1/8}$. In particular, the point $z = H^{-1/2}$ is regular. We expand $G(\varphi, z)$ as a Maclaurin series in a neighborhood of the point $z = 0$, and substitute $z = H^{-1/2}$. Then we obtain

$$G(\varphi) = \sum_{k=0}^{\infty} g_k(\varphi) H^{-k/2}, \qquad g_0 = 1. \tag{3.6.20}$$

Using the Bruno formula, we can prove that $g_{2k}(\varphi)$ and $g_{2k+1}(\varphi)$ are polynomials in φ, where $g_{2k}(\varphi)$ contains only even powers of φ, and $g_{2k+1}(\varphi)$ only odd ones. Using Lemma 3.6.4 with $a_k = \bar{C}_{k+2}\left(i\varphi\sqrt{2}\right)^{k+2}$ and $b_k = g_k(\varphi)$, we obtain the inequality

$$|g_k(\varphi)| \leqslant \left(|\varphi|\sqrt{2}(1 + 2\varphi^2)\right)^2. \tag{3.6.21}$$

Thus, from equalities (3.6.19) and (3.6.20) it follows that, for any fixed s,

$$s(n, m) = D\left(\sum_{k=0}^{s} \frac{1}{H^{k/2}} \int_{-\delta}^{\delta} g_k(\varphi) e^{-\varphi^2}\, d\varphi + R_s\right), \tag{3.6.22}$$

where

$$R_s = \int_{-\delta}^{\delta} \sum_{k=s+1}^{\infty} g_k(\varphi) e^{-\varphi^2} H^{-k/2}\, d\varphi.$$

From inequality (3.6.21) we can derive the bound

$$|R_s| \leqslant H^{-(s+1)/2} \int_{-\infty}^{\infty} P_s(|\varphi|) e^{-\varphi^2}\, d\varphi, \tag{3.6.23}$$

where $P_s(|\varphi|)$ is a polynomial in $|\varphi|$ which does not depend on H. Since the integral in (3.6.23) exists, $|R_s| \to 0$ as $H \to \infty$. Therefore, equality (3.6.22) implies the asymptotic formula

$$s(n, m) = D \sum_{k=0}^{\infty} \frac{1}{H^k} \int_{-\infty}^{\infty} g_{2k}(\varphi) e^{-\varphi^2}\, d\varphi.$$

Writing out the first two terms of the asymptotic expansion we finally obtain

$$s(n, m) = \frac{(-1)^{n+m}\Gamma(n + R)}{(2\pi H)^{1/2} R^m \Gamma(R)} \left(1 + \frac{3C_4}{H^2} - \frac{15C_3^2}{2H^3} + \cdots\right), \tag{3.6.24}$$

where, as can easily be seen, the second and third terms are of order $O(1/H)$.

3.6.4 The second Moser and Wyman formula

We now assume that $n-m = o(\sqrt{n})$. We consider the generating function

$$\frac{1}{m!}(\ln(1+z))^m = \sum_{n=m}^{\infty} s(n,m)\frac{z^n}{n!}.$$

Using the Cauchy integral formula, we can write

$$s(n,m) = \frac{n!}{2\pi i m!}\oint_C (\ln(1+z))^m \frac{dz}{z^{n+1}}, \tag{3.6.25}$$

where C is a closed contour in the complex plane containing the origin of coordinates and lying inside the unit circle. Substituting

$$f(z) = \frac{1}{z}(\ln(1+z)-z), \quad q = \frac{2}{m}, \quad z = qw$$

into formula (3.6.25), we obtain

$$s(n,m) = \frac{q^{n-m}n!}{2\pi i m!}\oint_C (1+f(qw))^{2/q}\frac{dw}{w^{n-m+1}}. \tag{3.6.26}$$

Define the function $T(q,w)$ as

$$T(q,w) = e^w(1+f(qw))^{2/q}. \tag{3.6.27}$$

It is not difficult to show that $T(q,w)$ has the Maclaurin expansion at the point $z=0$

$$T(q,w) = 1 + \sum_{k=1}^{\infty} T_k(w)q^k,$$

where $T_k(w)$ are polynomials in w with the smallest power of w no less than $k+1$. Carrying out simple calculations, we find that

$$T(q,w) = 1 + \frac{5}{12}w^2q - \cdots.$$

From formula (3.6.26) it follows that

$$s(n,m) = q^{n-m}\binom{n}{m}\operatorname{coef}_{w^{n-m}/(n-m)!}\, e^w T(q,w).$$

Thus,

$$s(n,m) = \binom{n}{m}\left(-\frac{m}{2}\right)^{n-m}\left(1+\frac{5(n-m)^2}{6m}+\cdots\right).$$

Hence the asymptotic formula (3.6.5) follows for all m such that $n-m = o(\sqrt{n})$.

3.7 The saddle point method and asymptotic behavior of Stirling numbers

3.7.1 The saddle point method

The saddle point method is intended for the derivation of asymptotic formulae for expressions presented as contour integrals of certain forms of functions of complex variable. Let us consider how to apply this method to integrals of the form

$$F(\lambda) = \int_C \varphi(z) e^{\lambda f(z)}\, dz, \tag{3.7.1}$$

where $\varphi(z)$ and $f(z)$ are functions of a complex variable z which are analytic in a region $\mathscr{G}$ containing the curve C, and λ is a large positive number. It is assumed that integral (3.7.1) exists, and we need to find the asymptotic expansion of $F(\lambda)$ in powers of $1/\lambda$.

Let us first describe the general idea of the saddle point method (Sveshnikov and Tikhonov, 1967). By virtue of Cauchy's theorem, for analytic functions $\varphi(z)$ and $f(z)$ in the region $\mathscr{G}$ the value of integral (3.7.1) is determined solely by the initial point z_1 and the end point z_2 of the integration curve. Therefore, we try to choose the contour C in the region $\mathscr{G}$ in such a way that the absolute value of the integrand is close to its maximum on a small portion of the contour and then decreases quickly. If such a choice is possible, then integral (3.7.1) can be approximately evaluated by means of the maximum absolute value of the integrand function corrected for the rate of its decrease along the contour.

Since we are interested in the values of integral (3.7.1) for large λ, it is natural to expect that the main contribution will be made by those portions of the contour C where the real part $u(x, y)$ of the function $f(z) = u(x, y) + iv(x, y)$ takes the largest values. The function $u(x, y)$, which is harmonic in the region $\mathscr{G}$, has no points where it increases or decreases along all the directions, but, rather, has only saddle points.

Let $z_0 = x_0 + iy_0$ be the only saddle point in $\mathscr{G}$ and let the trivial case where $f(z)$ is a constant be excluded from consideration. In the theory of analytic functions it is known that the curves $u(x, y) = u(x_0, y_0)$ cannot be closed and either meet the bounds of the region $\mathscr{G}$ or go to infinity. These curves partition the region near the point z_0 into sectors where the values of $u(x, y)$ are either less or greater than $u(x_0, y_0)$. The first sectors are called negative, and the latter, positive. A saddle point with two positive and two negative sectors is presented in Figure 3.7.1.

If the boundary points z_1 and z_2 of the contour C lie in the same sector, then the contour can be deformed in such a way that $u(x, y)$ changes

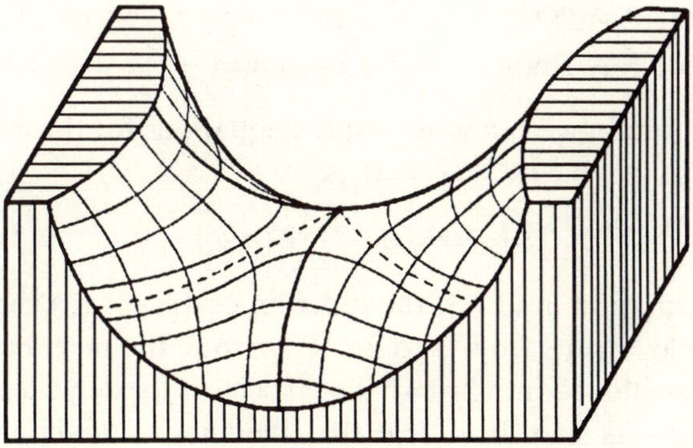

Fig. 3.7.1. A saddle point with four sectors

monotonically, and the main contribution to the value of the integral is made in the neighborhood of that boundary point at which $u(x, y)$ takes the greatest value. The same situation takes place where the points z_1 and z_2 lie in sectors of distinct kinds. The saddle point method is used in the case where the points z_1 and z_2 lie in distinct negative sectors. In this case we can choose the contour C including z_0 in such a manner that $u(x, y)$ attains its maximum at $z_0 = (x_0, y_0)$ and decreases quickly in the directions of the boundary points, and the imaginary part $v(x, y)$ of the function $f(z)$ remains constant.

If the saddle point is not unique, then we should consider the contributions of several points to the value of the integral. It frequently occurs that this contribution is small in comparison with the contribution of one of the saddle points.

Now let us determine the directions of the steepest descent of $u(x, y)$ in a neighborhood of the point z_0. Since $f'(z_0) = 0$, in a neighborhood of the point z_0 the function $f(z)$ can be represented as

$$f(z) = f(z_0) + (z - z_0)^p(a_0 + a_1(z - z_0) + \cdots),$$

where $p \geqslant 2$, $a_0 \neq 0$. Setting $a_k = \rho_k e^{i\omega_k}$, $k = 0, 1, \ldots$, $z - z_0 = \rho e^{i\omega}$, we obtain

$$f(z) - f(z_0) = \rho^p\{\rho_0 e^{i(p\omega+\omega_0)} + \rho\rho_1 e^{i((p+1)\omega+\omega_1)} + \cdots\}.$$

The real and imaginary parts of the expression in braces satisfy the relations

$$U(\rho,\omega) = \rho_0 \cos(p\omega + \omega_0) + \rho\rho_1 \cos((p+1)\omega + \omega_1) + \cdots,$$
$$V(\rho,\omega) = \rho_0 \sin(p\omega + \omega_0) + \rho\rho_1 \sin((p+1)\omega + \omega_1) + \cdots.$$

Using this notation we can write out the equations for the curves $u(x,y) - u(x_0,y_0) = 0$, $v(x,y) - v(x_0,y_0) = 0$ as

$$U(\rho,\omega) = 0, \qquad V(\rho,\omega) = 0. \tag{3.7.2}$$

In the first equation in (3.7.2) the function $\cos(p\omega + \omega_0)$ changes its sign $2p$ times while ω runs from 0 to 2π. Therefore, the neighborhood of the point z_0 is partitioned into $2p$ curvilinear sectors inside which the function $U(\rho,\omega)$ retains its sign. It is clear that the lines of directions of steepest descent go through distinct negative sectors such that $U(\rho,\omega) < 0$, $V(\rho,\omega) = 0$. The corresponding angles ω are determined from the condition $\cos(p\omega + \omega_0) = -1$ and therefore take the values $\omega = (-\omega_0 + (2m+1)\pi)/p$, $m = 0, 1, \ldots, p-1$.

In the case where $p = 2$, $f''(z_0) \neq 0$, $a_0 = f''(z_0)/2$, $\omega_0 = \arg f''(z_0)$ there exist only two negative sectors; these contain the line of steepest descent of $u(x,y)$. The tangent direction for this line is determined by the angle

$$\omega = \frac{\pi}{2} - \frac{1}{2} \arg f''(z_0).$$

3.7.2 Asymptotic expansions

Let us apply the saddle point method to the derivation of the asymptotic expansions for Stirling numbers of the second kind. From the generating function

$$\frac{1}{r!}(e^z - 1)^r = \sum_{n=r}^{\infty} \sigma(n,r) \frac{z^n}{n!}$$

of Stirling numbers of the second kind $\sigma(n,r)$, using the Cauchy integral formula, we obtain

$$\sigma(n,r) = \frac{n!}{2\pi i r!} \oint_C (e^z - 1)^r \frac{dz}{z^{n+1}}, \tag{3.7.3}$$

where C is a closed contour on the complex plane containing the origin of coordinates. Setting $f(z) = \ln(e^z - 1) - t \ln z$, $t = (n+1)/r$, from the condition $f'(z) = 0$ we obtain the equation

$$(t - z)e^z = t \tag{3.7.4}$$

determining the saddle points of $f(z)$. The main contribution is made by the saddle point $z_0 = u$ lying on the positive real axis and satisfying the condition $t-1 < u < t$. Note that, for $|z-t| \leqslant t-u$, equation (3.7.4) has no other roots, and we can use the Lagrange inversion formula:

$$u = t - \sum_{m=1}^{\infty} \frac{m^{m-1}(te^{-t})^m}{m!}. \tag{3.7.5}$$

This series converges for $t > 1$. Since the second derivative at the saddle point $z_0 = u$ takes a real value, the direction of steepest descent is perpendicular to the real axis, and for a portion of the contour C passing through the point u we can choose the straight line $z = u + iy$; denote it by C_1. This straight line can be completed to a closed contour by the semicircle C_2 containing the origin of coordinates, with center at the saddle point and radius tending to infinity. The integral along C_2 is equal to zero, and integral (3.7.4) can be represented in the form

$$\sigma(n,r) = \frac{n!\,(e^u-1)^r}{2\pi r!\,u^{n+1}} \int_{-\infty}^{\infty} \exp\psi(u+iy)\,dy, \tag{3.7.6}$$

$$\psi(u+iy) = r\ln\left(\frac{e^{u+iy}-1}{e^u-1}\right) - (n+1)\ln\left(\frac{u+iy}{u}\right). \tag{3.7.7}$$

We now divide the contour C_1 into three parts determined by the conditions $-\infty < y \leqslant -\pi$, $-\pi \leqslant y \leqslant \pi$, $\pi \leqslant y < \infty$. It is clear that

$$\Re\psi(u+iy) = r\ln((e^{2u}-2e^u\cos y+1)^{1/2}/(e^u-1)) - (n+1)\ln((1+y^2u^{-2})^{1/2}).$$

Taking into account the estimate $1+y^2u^{-2} \geqslant 1+\pi(2y-\pi)u^{-2}$, we obtain

$$\frac{(e^u-1)^r}{u^{n+1}}\left|\int_{\pi}^{\infty} \exp\psi(u+iy)\,dy\right| \leqslant \frac{(e^u+1)^r}{\pi(n-1)u^{n-1}(1+\pi^2u^{-2})^{(n-1)/2}}.$$

This estimate is of order $O((2e/\ln n)^n/n)$ for r such that $r < n/\ln n$. Hence it follows that the corresponding part of integral (3.7.6) for $\pi \leqslant y < \infty$ tends to zero exponentially. The same estimate occurs for $-\infty < y \leqslant -\pi$. Hence, from formula (3.7.6) we obtain the relation

$$\sigma(n,r) = \frac{n!\,(e^u-1)^r}{2\pi r!\,u^{n+1}} \int_{-\pi}^{\pi} \exp\psi(u+iy)\,dy(1+o(1)), \tag{3.7.8}$$

where the error is of an exponential character. We expand the function $\psi(u+iy)$ into the Taylor series in a neighborhood of the point $y=0$:

$$\psi(u+iy) = -\frac{n+1}{2u}\left(\frac{1}{u}-\frac{1}{e^u-1}\right)y^2 + (n+1)\sum_{j=3}^{\infty}\frac{(iy)^j}{j!}\left(\frac{d}{dz}\right)^{j-1}\left(\frac{1-e^{-u}}{u(e^z-1)}-\frac{1}{z}\right)\Bigg|_{z=u}.$$

After the change of variable,

$$w = w(y) = \sqrt{\frac{n+1}{2}}\left(1-\frac{u}{e^u-1}\right)^{1/2}\frac{y}{u}$$

formula (3.7.8) becomes

$$\sigma(n,r) = B(n,r)\int_{-\delta}^{\delta}\exp\left\{-w^2+f\left(\frac{1}{\sqrt{n+1}}\right)\right\}dw(1+o(1)), \qquad (3.7.9)$$

where $\pm\delta = w(\pm\pi)$ and

$$B(n,r) = \frac{n!\,(e^u-1)^r}{\pi u^n r!\,\sqrt{2(n+1)}\left(1-\frac{u}{e^u-1}\right)^{1/2}}, \qquad (3.7.10)$$

$$f(\alpha) = \sum_{j=1}^{\infty} a_j\alpha^j, \qquad (3.7.11)$$

$$a_j = \frac{(iwu)^{j+2}\left(\frac{d}{dz}\right)^{j+1}\left(\frac{1-e^{-u}}{u(e^z-1)}-\frac{1}{z}\right)\Big|_{z=u}}{(j+2)!\left(\frac{1}{2}-\frac{u}{2(e^u-1)}\right)^{j/2+1}}. \qquad (3.7.12)$$

Again introducing the error of exponential character, we rewrite formula (3.7.9) as follows:

$$\sigma(n,r) = B(n,r)\int_{-\infty}^{\infty}\exp\left\{-w^2+f\left(\frac{1}{\sqrt{n+1}}\right)\right\}dw(1+o(1)). \qquad (3.7.13)$$

We now obtain an upper bound for $|a_j|$, $j=1,2,\ldots$. Using the expansion

$$(e^z-1)^{-1} = \sum_{k=1}^{\infty} e^{-kz},$$

which is valid for $\Re z > 0$, we obtain

$$\left(\frac{d}{dz}\right)^k\left(\frac{1}{e^u-1}\right) = (-1)^k\sum_{j=1}^{\infty} j^k e^{-uj}.$$

To calculate the sum on the right-hand side of the last equality, we use the Euler–Maclaurin formula

$$\sum_{j=0}^{\infty} j^k e^{-uj} = \int_0^{\infty} x^k e^{-ux}\,dx + R_k.$$

Using the known estimate of the remainder term of the Euler–Maclaurin formula and the integral form of the gamma function, we obtain

$$\left(\frac{d}{dz}\right)^k \left(\frac{1}{e^u - 1}\right) = (-1)^k \left(\frac{k!}{u^{k+1}} + R_k\right),$$

where R_k is small in comparison with $k!/u^{k+1}$. Now formula (3.7.12) yields the bound

$$|a_j| < 2\beta^{j+2}/j, \tag{3.7.14}$$

where

$$\beta = w\sqrt{2}\left(1 - \frac{u}{e^u - 1}\right)^{-1/2} = \frac{y}{u}(n+1)^{1/2}.$$

Hence it follows that, by virtue of the inequality $|y| \leqslant \pi$, the series (3.7.11) converges at the point $\alpha = 1/\sqrt{n+1}$ if $u > \pi$. We consider the expansion

$$\exp f\left(\frac{1}{\sqrt{n+1}}\right) = \sum_{j=0}^{\infty} b_j \frac{1}{(n+1)^{j/2}}, \tag{3.7.15}$$

where $b_0 = 1$ and b_j is a polynomial in w. By virtue of Lemma 3.6.4, inequality (3.7.14) yields

$$|b_j| \leqslant \beta^{j+2}(1+\beta^2)^{j-1}. \tag{3.7.16}$$

Now formula (3.7.13) can be written in the form

$$\sigma(n,r) = B(n,r)\left(\sum_{j=0}^{s-1} \frac{1}{(n+1)^j} \int_{-\infty}^{\infty} b_{2j} e^{-w^2}\,dw + R_s\right). \tag{3.7.17}$$

From inequality (3.7.16) we see that

$$|R_s| \leqslant \frac{1}{(n+1)^s} \int_{-\infty}^{\infty} \frac{\beta^{2(s+1)}(1+\beta^2)^{2s-1}}{1 - \beta^2(1+\beta^2)^2/(n+1)} e^{-w^2}\,dw. \tag{3.7.18}$$

The denominator in the integrand function differs from zero if

$$\frac{\pi}{u}\left(1 + \left(\frac{\pi}{u}\right)^2 (n+1)\right) < 1. \tag{3.7.19}$$

Since $(n+1)/r = t < u+1$, inequality (3.7.19) is satisfied provided

$$r < (n+1)^{2/3}(\pi + (n+1)^{-1/3}). \tag{3.7.20}$$

It is clear that $\beta^{2(s+1)}(1+\beta^2)^{2s-1}$ is a polynomial in w; therefore under condition (3.7.19) the integral in inequality (3.7.18) is bounded and formula (3.7.17) can be rewritten as

$$\sigma(n,r) = B(n,r)\left(\sum_{j=0}^{s-1}\frac{1}{(n+1)^j}\int_{-\infty}^{\infty} b_{2j}e^{-w^2}\,dw + O\left(\frac{1}{(n+1)^s}\right)\right). \tag{3.7.21}$$

This complete asymptotic expansion can be applied for sufficiently large n to the calculation of the numbers $\sigma(n,r)$ provided that inequality (3.7.20) is satisfied. Since $b_0 = 1$, we can derive the following asymptotic formula from equality (3.7.21):

$$\sigma(n,r) = \frac{(e^u-1)^r n!}{(2\pi(n+1))^{1/2} r!\, u^n (1-G(u))^{1/2}}(1+o(1)), \tag{3.7.22}$$

where $G(u)$ is the generating function for the Bernoulli numbers:

$$G(u) = \frac{u}{e^u-1} = 1 - \frac{u}{2} + \sum_{k=1}^{\infty} B_{2k}\frac{u^{2k}}{(2k)!}.$$

Using Stirling's formula, from formula (3.7.22) we finally obtain (Bleick and Wang, 1974)

$$\sigma(n,r) = \frac{1}{r!}(e^u-1)^r\left(\frac{n}{eu}\right)^n(1-G(u))^{-1/2}(1+o(1)),$$

where u is the root of equality (3.7.4) satisfying the condition $(n+1)/r-1 < u < (n+1)/r$. If the value $n-r$ is small as $n\to\infty$, then for Stirling numbers of the second kind (Hsu, 1948) the asymptotic expansion

$$\sigma(n,r) = \frac{(r^2/2)^{n-r}}{(n-r)!}\left(1+\sum_{j=1}^{s} f_j(n-r)\frac{1}{r^j} + O\left(\frac{1}{r^{s+1}}\right)\right) \tag{3.7.23}$$

is valid, where $f_j(x)$ is a polynomial such that $f_j(0) = 0$.

For $r < n/\ln n$ the asymptotic formula

$$\sigma(n,r) = \frac{r^n}{r!}\exp\left\{\left(\frac{n}{2r}-r\right)e^{-n/r}\right\}(1+o(1)), \tag{3.7.24}$$

due to Erdős and Szekeres, is valid. We omit the proofs of formulae (3.7.23) and (3.7.24).

4
Graphs and mappings

In this chapter we consider enumerative problems of graph theory, that is, problems arising when counting graphs with specific properties. We also consider similar problems about mappings of finite sets with various constraints. Particular attention is given to mappings of bounded height h, whose cycle lengths are the elements of a given sequence A. For $h = 0$ these mappings become substitutions whose cycle lengths belong to a given sequence A.

The method of generating functions can be effectively applied to these problems. As a result, we obtain either explicit formulae or some expressions for generating functions which allow us to find the asymptotic expressions of the corresponding coefficients, for instance by using the saddle point method. The results of the application of the saddle point method to the derivation of such asymptotic expressions are not given here. They can be found, for example, in papers (Sachkov, 1972; Sachkov, 1971b; Sachkov, 1971a).

4.1 The generating functions for graphs

4.1.1 Basic definitions

Let us formulate the basic definitions about the graphs; we follow the most common terminology of graph theory (Berge, 1958; Ore, 1962; Harary, 1969). A graph $\Gamma = \Gamma(X, W)$ consists of a set X containing $n \geqslant 1$ elements called vertices and of a set W of unordered pairs of vertices called edges. Usually a graph is geometrically represented on the plane by points corresponding to the vertices, and lines corresponding to the edges which join pairs of vertices from W; the intersections of the lines at points that differ from the vertices are not taken into account.

It is clear that every graph Γ with a set of vertices X can be associated with a symmetric antireflexive binary relation on this set, and vice versa. The matrix of this binary relation is called the adjacency matrix of the corresponding graph.

We say that vertices x and y are incident to the edge (x, y), or that an edge (x, y) joins the vertices x and y; in its turn, the edge (x, y) is incident to the vertices x and y. The number of edges incident to a vertex x is called the degree of this vertex and is denoted by $\deg x$. A vertex x is called an end vertex if $\deg x = 1$, and it is called isolated if $\deg x = 0$. The identity

$$\sum_{x \in X} \deg X = 2|W|$$

goes back to Euler. This identity implies that in any graph the number of vertices of odd degree is even. A graph whose vertices are all of the same degree r is called regular. If $r = 0$, then such a graph contains no edges and consists of isolated vertices only.

If two edges are incident to the same vertex, then they are called adjacent.

An alternating sequence $x_1, w_1, x_2, w_2, \dots, x_{l-1}, w_{l-1}, x_l$, of vertices and edges where $x_i \in X$, $w_i \in W$, $w_i \neq w_j$, $i \neq j$, and w_i is incident to x_i and x_{i+1}, is called a path connecting the vertices x_1 and x_l. A path is called a simple path if $x_i \neq x_j$, $i \neq j$. A closed path with $x_1 = x_l$ is called a cycle. A cycle whose vertices are all distinct is called a simple cycle.

A graph is connected if any pair of its vertices is connected by a simple path. The number of edges in a simple path is called the length of this path. The distance between two vertices x and x' of a connected graph Γ is the length of the shortest simple path connecting x and x'. A graph Γ' is a subgraph of a graph Γ if all vertices and all edges of Γ' belong to Γ. A maximal connected subgraph of a graph Γ is called a connected component of the graph Γ.

A connected graph containing no cycles is called a tree. Any two vertices of a tree are connected by a unique path. A tree with n vertices has $n - 1$ edges. A graph whose connected components are all trees is called a forest.

A directed graph, or digraph, $\bar{\Gamma} = \Gamma(X, \bar{W})$ consists of a finite nonempty set X whose elements are called vertices, and a set $\bar{W}$ of ordered pairs of distinct vertices called arcs. We say that an arc $\overline{(x, x')}$ emerges from the vertex x and terminates at the vertex x', and that the vertex x is adjacent to x' and x' is adjacent to x. An arc $\overline{(x, x')}$ is incident to x and x'. An alternating sequence of vertices and arcs of a digraph $x_1, \bar{w}_1, x_2, \bar{w}_2, \dots, x_{l-1}, \bar{w}_{l-1}, x_l$, where $x_i \in X$, $\bar{w}_i \in \bar{W}$, $x_i \neq x_j$, $i \neq j$, and $\bar{w}_i = (x_i, x_{i+1})$, $1 \leqslant i \leqslant l-1$, is called a walk connecting x_1 and x_l. If in the alternating sequence of vertices and arcs $x_1 = x_l$, and all other vertices are distinct, then this sequence is called a contour.

A digraph is said to be strongly connected if for any two vertices x and x' there exist walks connecting x with x' and x' with x. A digraph $\bar{\Gamma}'$ is called a subgraph of a digraph $\bar{\Gamma}$ if all vertices and all arcs of $\bar{\Gamma}'$ belong to $\bar{\Gamma}$. A maximal strongly connected subgraph of a digraph $\bar{\Gamma}$ is called a strongly connected component of the digraph $\bar{\Gamma}$.

It is clear that a digraph $\Gamma(X, \bar{W})$ can be associated with an antireflexive binary relation R such that xRx' if $\overline{(x, x')} \in \bar{W}$. Any digraph $\bar{\Gamma} = \Gamma(X, \bar{W})$ can be represented geometrically on the plane by the points corresponding to the vertices of $\bar{\Gamma}$ and the arrows corresponding to its arcs. If $\overline{(x, x')} \in \bar{W}$, then the vertices x and x' are joined by the arrow emerging from x and terminating at x'. The points of intersection of those arrows which do not coincide with vertices are not taken into account.

From the given definitions of graphs (digraphs) it follows that in their geometric representations the loops, that is, those edges joining a vertex to itself, are not allowed, neither are the multiple, or parallel, edges (arcs) which join the same pair of vertices (with regard to their directions). If the presence of loops and multiple edges in a graph (digraph) is allowed, then the graph is sometimes called a multigraph (directed multigraph). In what follows we do not introduce special terminology and speak only of graphs (digraphs), pointing out whether loops and multiple edges are allowed.

In the geometric representation of a graph $\Gamma = \Gamma(X, W)$, its vertices are labeled with the corresponding elements of the set X. We may consider the graph obtained from Γ by taking off all the labels at the vertices. In the first case the graph is called labeled, and in the second it is referred to as unlabeled. If only some of the vertices have labels, then such a graph is called partially labeled. A tree with one chosen vertex is called a rooted tree, and the vertex chosen is referred to as the root. A tree with no chosen vertex is called an unrooted tree.

4.1.2 Graphs with labeled vertices

Let us consider the set of all graphs with m edges and n vertices labeled by the numbers $1, 2, \dots, n$ such that every connected component of a graph from the set possesses a property Λ. This property may relate to the vertex degrees, cycle lengths, etc.

We denote by $T_{nm}^{(k)}(\Lambda)$ the number of graphs in the set with k connected components, by $T_{nm}(\Lambda)$, the total number of the graphs, and set $C_{nm}(\Lambda) = T_{nm}^{(1)}(\Lambda)$. For convenience, we set $T_{00}^{(0)} = 1$ and $T_{nm}^{(k)}(\Lambda) = 0$ if $nmk = 0$ and at least one of the indices n, m, k differs from zero. The recurrence relation

$$T_{n+1,m}^{(k)}(\Lambda) = \sum_{i=0}^{n}\sum_{j=0}^{m}\binom{n}{j} C_{i+1,j}(\Lambda) T_{n-i,m-j}^{(k-1)}(\Lambda) \tag{4.1.1}$$

holds. To prove it, we divide all graphs of the set under consideration with $n+1$ vertices, m edges and k connected components, into classes depending on the numbers of vertices and edges in the connected component containing the vertex labeled by $n+1$. If this component has $i+1$ vertices and j edges, then the number of ways of constructing the component is equal to $\binom{n}{i}C_{i+1,j}(\Lambda)$. The number of ways of constructing the other $k-1$ components is equal to $T_{n-i,m-j}^{(k-1)}(\Lambda)$. Multiplying and summing these values, we obtain the expression for $T_{n+1,m}^{(k)}(\Lambda)$.

Now we consider the generating functions

$$T_n(y, z; \Lambda) = \sum_{m=0}^{\infty}\sum_{k=0}^{n} T_{nm}^{(k)}(\Lambda) y^m z^k,$$

$$C_n(y; \Lambda) = \sum_{m=0}^{\infty} C_{nm}(\Lambda) y^m.$$

Relation (4.1.1) yields the equality

$$T_{n+1}(y, z; \Lambda) = z \sum_{i=0}^{n} \binom{n}{i} C_{i+1}(y; \Lambda) T_{n-i}(y, z; \Lambda). \tag{4.1.2}$$

Let us introduce the generating functions

$$T(x, y, z; \Lambda) = \sum_{n=0}^{\infty} T_n(y, z; \Lambda) \frac{x^n}{n!},$$

$$C(x, y; \Lambda) = \sum_{n=0}^{\infty} C_n(y; \Lambda) \frac{x^n}{n!}.$$

Then equality (4.1.2) means that

$$T(x, y, z; \Lambda) = \exp\{zC(x, y; \Lambda)\}. \tag{4.1.3}$$

For $z = 1$, this equality turns into the well-known relation (Gilbert, 1959)

$$T(x, y, 1; \Lambda) = \exp\{C(x, y; \Lambda)\}, \tag{4.1.4}$$

where

$$T(x, y, 1; \Lambda) = \sum_{n=0}^{\infty} T_{nm}(\Lambda)\frac{x^n}{n!}y^m.$$

Let us consider the case where Λ is the empty set, that is, there are no constraints on the connected components. We find the numbers of graphs and digraphs, including the case where loops and multiple edges are allowed.

Theorem 4.1.1 (Austin (1960)) *The number of distinct graphs (digraphs) with n labeled vertices and m edges is equal to*

$$T_{nm} = \binom{\alpha(n) + \beta(m) - 1}{m}, \tag{4.1.5}$$

where $\alpha(n)$ and $\beta(m)$ depend on the kind of graph under consideration and are defined by the equalities

$$\alpha(n) = \begin{cases} \binom{n}{2} & \textit{for ordinary graphs with no loops;} \\ \binom{n+1}{2} & \textit{for ordinary graphs with loops allowed;} \\ n(n-1) & \textit{for directed graphs with no loops;} \\ n^2 & \textit{for directed graphs with loops allowed;} \end{cases}$$

and

$$\beta(m) = \begin{cases} 1 & \textit{if there are no multiple edges;} \\ m & \textit{if multiple edges are allowed.} \end{cases}$$

The generating functions for the enumeration of graphs (digraphs) and connected graphs (digraphs) are of the form

$$\sum_{n=0}^{\infty}\sum_{m=0}^{\infty} T_{nm}\frac{x^n}{n!}y^m = \sum_{n=0}^{\infty}(1+\gamma y)^{\gamma\alpha(n)}\frac{x^n}{n!}, \tag{4.1.6}$$

$$\sum_{n=0}^{\infty}\sum_{m=0}^{\infty} C_{nm}\frac{x^n}{n!}y^m = \ln\left(1 + \sum_{k=1}^{\infty}(1+\gamma y)^{\gamma\alpha(k)}\frac{x^k}{k!}\right), \tag{4.1.7}$$

where

$$\gamma = \begin{cases} 1 & \textit{if there are no multiple edges,} \\ -1 & \textit{if multiple edges are allowed,} \end{cases}$$

and the expansion of the logarithm is carried out according to the common rules for formal power series.

Proof To prove formula (4.1.5), we notice that $\alpha(n)$ is the number of distinct pairs of vertices depending on the kind of graph (digraph); $\beta(m)$ is equal to 1 or to m depending on whether combinations without repetitions or with repetitions from the set of distinct pairs are used to construct n edges (arcs) of the graph (digraph). The structure of generating function (4.1.6) follows from formula (4.1.5). Equality (4.1.7) follows from formula (4.1.6) and relation (4.1.4). □

4.1.3 Graphs with labeled edges; graphs with labeled vertices and edges

Let us consider graphs whose n vertices have no labels but whose edges are labeled by the numbers $1, 2, \ldots, m$. If $\tilde{T}_{nm}^{(k)}(\Lambda)$ is the number of such graphs with k connected components, whose every component possesses a property Λ, and $\tilde{C}_{nm}(\Lambda)$ is the number of the corresponding connected graphs, then the following recurrence relation, which is similar to relation (4.1.1), is valid:

$$\tilde{T}_{n,m+1}^{(k)}(\Lambda) = \sum_{i=0}^{n} \sum_{j=0}^{m} \binom{m}{j} \tilde{C}_{i,j+1}(\Lambda) \tilde{T}_{n-i,m-j}^{(k-1)}(\Lambda). \tag{4.1.8}$$

Note the validity of all results similar to those from Subsection 4.1.2 that follow from relation (4.1.1). In particular, the generating functions

$$\tilde{T}(x, y, z; \Lambda) = \sum_{n=0}^{\infty} \sum_{m=0}^{\infty} \sum_{k=0}^{n} \tilde{T}_{nm}^{(k)}(\Lambda) x^n \frac{y^m}{m!} z^k,$$

$$\tilde{C}(x, y; \Lambda) = \sum_{n=0}^{\infty} \sum_{m=0}^{\infty} \tilde{C}_{nm}(\Lambda) x^n \frac{y^m}{m!}$$

are related by the equality

$$\tilde{T}(x, y, z; \Lambda) = \exp\{z\tilde{C}(x, y; \Lambda)\}. \tag{4.1.9}$$

Further, let us consider graphs whose vertices and edges are all labeled. If $\hat{T}_{nm}^{(k)}(\Lambda)$ is the number of such graphs with k connected components, every one of which possesses a property Λ, and $\hat{C}_{nm}(\Lambda)$ is the number of the corresponding connected graphs, then

$$\hat{T}_{n+1,m}^{(k)}(\Lambda) = \sum_{i=0}^{n} \sum_{j=0}^{m} \binom{n}{i} \binom{m}{j} \hat{C}_{i+1,j}(\Lambda) \hat{T}_{n-i,m-j}^{(k-1)}(\Lambda). \tag{4.1.10}$$

By virtue of this relation, the generating functions

$$\hat{T}(x, y, z; \Lambda) = \sum_{n=0}^{\infty} \sum_{m=0}^{\infty} \sum_{k=0}^{n} \hat{T}_{nm}^{(k)}(\Lambda) \frac{x^n}{n!} \frac{y^m}{m!} z^k,$$

$$\hat{C}(x, y; \Lambda) = \sum_{n=0}^{\infty} \sum_{m=0}^{\infty} \hat{C}_{nm}(\Lambda) \frac{x^n}{n!} \frac{y^m}{m!}$$

satisfy the equality

$$\hat{T}(x, y, z; \Lambda) = \exp\{z\hat{C}(x, y; \Lambda)\}. \tag{4.1.11}$$

4.1.4 Colored graphs

A graph Γ is referred to as s-colorable if its n vertices can be divided into s non-intersecting classes containing $\alpha_1, \alpha_2, \ldots, \alpha_s$ elements, respectively, $\alpha_1 + \alpha_2 + \cdots + \alpha_s = n$, in such a way that the edges of Γ may only join vertices of different classes. For the sake of convenience we can suppose that the vertices belonging to one class are colored with the same color, and the vertices belonging to different classes are of different colors, so that the edges join vertices of different colors. In particular, if $s = 2$, the corresponding graphs are called bicolorable or bipartite.

We denote by $N_m^{(k)}(\alpha_1, \alpha_2, \ldots, \alpha_s)$ the number of s-colorable graphs with α_i vertices of the ith color, with m edges and with k connected components, and denote by $C_m(\alpha_1, \alpha_2, \ldots, \alpha_s)$ the number of the corresponding connected graphs. The recurrence relation

$$\begin{aligned} N_m^{(k)}&(\alpha_1 + 1, \alpha_2, \ldots, \alpha_s) \\ &= \sum_{i_1=0}^{\alpha_1} \cdots \sum_{i_s=0}^{\alpha_s} \sum_{j=0}^{m} \binom{\alpha_1}{i_1} \cdots \binom{\alpha_s}{i_s} C_j(i_1 + 1, i_2, \ldots, i_s) \\ &\qquad \times N_{m-j}^{(k-1)}(\alpha_1 - i_1, \alpha_2 - i_2, \ldots, \alpha_s - i_s) \end{aligned} \tag{4.1.12}$$

is valid. Hence, the generating functions

$$\begin{aligned} N(x_1, x_2, \ldots, x_s; y, z) &= \sum_{\alpha_1=0}^{\infty} \cdots \sum_{\alpha_s=0}^{\infty} \sum_{m=0}^{\infty} \sum_{k=0}^{\alpha_1+\cdots+\alpha_s} N_m^{(k)}(\alpha_1, \ldots, \alpha_s) \\ &\qquad \times \frac{x_1^{\alpha_1}}{\alpha_1!} \cdots \frac{x_s^{\alpha_s}}{\alpha_s!} y^m z^k, \\ C(x_1, x_2, \ldots, x_s; y) &= \sum_{\alpha_1=0}^{\infty} \cdots \sum_{\alpha_s=0}^{\infty} \sum_{m=0}^{\infty} C_m(\alpha_1, \ldots, \alpha_s) \frac{x_1^{\alpha_1}}{\alpha_1!} \cdots \frac{x_s^{\alpha_s}}{\alpha_s!} y^m \end{aligned}$$

satisfy the equality

$$\frac{\partial}{\partial x_1}N(x_1,x_2,\dots,x_s;y,z) = z\frac{\partial}{\partial x_1}C(x_1,x_2,\dots,x_s;y)N(x_1,x_2,\dots,x_s;y,z),$$

where $\frac{\partial}{\partial x}$ means as usual the operation of partial differentiation with respect to x. This equality yields the relation

$$N(x_1,x_2,\dots,x_s;y,z) = \exp\{zC(x_1,x_2,\dots,x_s;y)\}, \qquad (4.1.13)$$

which generalizes relation (4.1.3).

The number of possible edges in an s-colorable graph is equal to

$$\frac{1}{2}\sum_{i=1}^{s}(n-\alpha_i)\alpha_i = \frac{1}{2}\left(n^2 - \sum_{i=1}^{s}\alpha_i^2\right).$$

Therefore the number of s-colorable graphs with m edges is equal to

$$\binom{\frac{1}{2}\left(n^2 - \sum_{i=1}^{s}\alpha_i^2\right)}{m}. \qquad (4.1.14)$$

For the corresponding graph with multiple edges this number is equal to

$$\binom{\frac{1}{2}\left(n^2 - \sum_{i=1}^{s}\alpha_i^2\right) + m - 1}{m}. \qquad (4.1.15)$$

It is not difficult to derive the corresponding formulae for directed graphs and the corresponding graphs (digraphs) with multiple edges (arcs) allowed.

4.2 Trees and forests

The main aim of this section is to obtain the formulae for determining the number of trees with a given number of vertices. We will consider rooted, unrooted and partially labeled trees and also forests whose connected components are such trees. The possible applications of these formulae are illustrated by solving some problems.

4.2.1 Rooted trees

We denote by r_n the number of rooted trees with n labeled vertices, and by $\tilde{r}_n$ the number of unrooted trees. The following formulae, due to Cayley, are valid:

$$r_n = n^{n-1}, \qquad n = 1,2,\dots, \qquad (4.2.1)$$

$$\tilde{r}_n = n^{n-2}, \qquad n = 1,2,\dots. \qquad (4.2.2)$$

Note that formula (4.2.2) follows from formula (4.2.1). Indeed, with every unrooted tree we can associate n rooted trees obtained by choosing any of n vertices as roots. Hence it follows that

$$r_n = n\tilde{r}_n.$$

Let us establish the validity of formula (4.2.1). Denote by $r_n(k)$ the number of rooted trees with n vertices and k edges adjacent to the root. Then

$$r_n(k) = \frac{n}{k!} \sum_{j_1+\cdots+j_k=n-1} \frac{(n-1)!}{j_1!\, j_2! \cdots j_k!} r_{j_1} r_{j_2} \cdots r_{j_k}, \tag{4.2.3}$$

where the summation is taken over all positive integer solutions of the equation $j_1 + j_2 + \cdots + j_k = n - 1$.

Indeed, the polynomial coefficient gives the number of ways of allocating the elements of the set X, $|X| = n - 1$, for the construction of k rooted trees with $j_1, j_2, \ldots, j_k$ vertices; the total number of such trees is equal to $r_{j_1} r_{j_2} \cdots r_{j_k}$. Multiplication by n corresponds to the ways of choosing the label of the root; division by $k!$ eliminates the variants differing by permutation of trees at the root, since these variants give the same rooted tree with n vertices.

The formula

$$r_n = \sum_{k=1}^{n-1} r_n(k), \qquad n = 2, 3, \ldots, \tag{4.2.4}$$

is obvious. We consider the generating function for rooted trees:

$$r(x) = \sum_{n=1}^{\infty} r_n \frac{x^n}{n!}. \tag{4.2.5}$$

It can easily be seen that

$$xr^k(x) = \sum_{n=k+1}^{\infty} \frac{x^n}{n!} \sum_{j_1+\cdots+j_k=n-1} \frac{n!}{j_1!\, j_2! \cdots j_k!} r_{j_1} r_{j_2} \cdots r_{j_k}.$$

Therefore, by virtue of formula (4.2.3),

$$r(x;k) = \sum_{n=k+1}^{\infty} r_n(k) \frac{x^n}{n!} = \frac{x}{k!} r^k(x), \qquad k = 1, 2, \ldots. \tag{4.2.6}$$

In addition, since $r_1 = r_1(0) = 1$,

$$r(x;0) = x. \tag{4.2.7}$$

Summing equalities (4.2.6) and taking (4.2.4), (4.2.5) and (4.2.7) into account, we obtain

$$r(x) = x\sum_{k=0}^{\infty} \frac{1}{k!} r^k(x),$$

or, what amounts to the same,

$$r(x) = xe^{r(x)}. \tag{4.2.8}$$

To derive $r(x)$ from this expression, we use the Lagrange inversion formula and obtain

$$r(x) = \sum_{n=1}^{\infty} n^{n-1} \frac{x^n}{n!}. \tag{4.2.9}$$

This last formula and (4.2.5) imply the validity of formula (4.2.1).

From formulae (4.2.1), (4.2.3) and (4.2.4) we easily derive the identity

$$\frac{n^{n-1}}{n} = \sum_{k=1}^{n-1} \frac{1}{k!} \sum_{\substack{j_1+\cdots+j_k=n-1,\\ j_1,\ldots,j_k>0}} \frac{j_1^{j_1-1} j_2^{j_2-1} \cdots j_k^{j_k-1}}{j_1!\, j_2! \cdots j_k!}. \tag{4.2.10}$$

4.2.2 Forests with k disjoined vertices

We denote by $\tilde{r}_{nk}$ the number of forests which have $n+k-1$ vertices and which consist of k trees such that the fixed vertices $x_1, x_2, \ldots, x_k$ belong to different ones. The following Cayley formula holds:

$$\tilde{r}_{nk} = k(n+k-1)^{n-2}. \tag{4.2.11}$$

To prove this formula, we first establish the validity of the identity (Rényi, 1959):

$$\frac{kn^{n-k-1}}{(n-k)!} = \sum_{\substack{j_1+\cdots+j_k=n,\\ j_1,\ldots,j_k>0}} \frac{j_1^{j_1-1} j_2^{j_2-1} \cdots j_k^{j_k-1}}{j_1!\, j_2! \cdots j_k!}. \tag{4.2.12}$$

We set $y = r(x)$ and notice that

$$ky^{k-1}y' = ky^{k-2}y' - \frac{ky^{k-1}}{x}, \tag{4.2.13}$$

where the prime denotes the derivative with respect to x. Indeed, equality (4.2.13) is equivalent to

$$(1-y)y' = y/x,$$

which is the result of differentiation of the relation $y = xe^y$ equivalent to (4.2.8).

Integrating both sides of relation (4.2.13) from 0 to x, we obtain

$$y^k = \frac{k}{k-1} y^{k-1} - k \int_0^x \frac{y^{k-1}(t)}{t}\, dt. \tag{4.2.14}$$

Using this relation, let us prove the equality

$$y^k = \sum_{n=k}^{\infty} k n^{n-k-1} (n)_k \frac{x^n}{n!}. \tag{4.2.15}$$

For $k = 1$ equality (4.2.15) is valid since it is equivalent to formula (4.2.9). By induction, using equality (4.2.14), we see that formula (4.2.15) holds. From (4.2.1) and (4.2.5) it follows that

$$y^k = \sum_{n=k}^{\infty} \frac{x^n}{n!} \sum_{\substack{j_1+\cdots+j_k=n,\\ j_1,\ldots,j_k>0}} \frac{n!\, j_1^{j_1-1} j_2^{j_2-1} \cdots j_k^{j_k-1}}{j_1!\, j_2! \cdots j_k!}. \tag{4.2.16}$$

Equating the coefficients of $x^n/n!$ on the right-hand sides of equalities (4.2.15) and (4.2.16) and carrying out obvious transformations, we see that (4.2.12) is valid.

Since none of the k fixed vertices $x_1, x_2, \ldots, x_k$ belong to the same tree and the remaining $n-1$ vertices can be allocated over k trees with j_1+1, $j_2+1, \ldots, j_k+1$ vertices in $(n-1)!\,(j_1!\, j_2! \cdots j_k!)^{-1}$ ways, we have

$$\tilde{r}_{n,k} = \sum_{j_1+\cdots+j_k=n-1} \frac{(n-1)!\,(j_1+1)^{j_1-1}(j_2+1)^{j_2-1}\cdots(j_k+1)^{j_k-1}}{j_1!\, j_2! \cdots j_k!}.$$

We carry out the change of variable $j_i' = j_1 + 1$ on the right-hand side and obtain

$$\tilde{r}_{n,k} = (n-1)! \sum_{\substack{j_1+\cdots+j_k=n+k-1,\\ j_1,\ldots,j_k>0}} \frac{j_1^{j_1-1} j_2^{j_2-1} \cdots j_k^{j_k-1}}{j_1!\, j_2! \cdots j_k!}. \tag{4.2.17}$$

Using identity (4.2.12), we obtain the equality

$$\tilde{r}_{n,k} = (n-1)! \frac{k(n+k-1)^{n-2}}{(n-1)!},$$

which implies formula (4.2.11).

4.2.3 Forests of unrooted trees

We denote by $\tilde{r}_n^{(k)}$ the number of forests with n vertices which consist of k unrooted trees. The formula

$$\tilde{r}_n^{(k)} = \frac{1}{k!} \sum_{j_1+\cdots+j_k=n} \frac{n!\, j_1^{j_1-2} j_2^{j_2-2} \cdots j_k^{j_k-2}}{j_1!\, j_2! \cdots j_k!} \tag{4.2.18}$$

holds. The derivation of this formula is based on partitioning the set of vertices into k subsets out of which the unrooted trees are constructed; division by $k!$ eliminates the equivalent constructions.

Let us consider the generating function for unrooted trees:

$$\tilde{r}(x) = \sum_{n=1}^{\infty} \tilde{r}_n \frac{x^n}{n!} = \sum_{n=1}^{\infty} n^{n-2} \frac{x^n}{n!}. \tag{4.2.19}$$

It can easily be seen that

$$\frac{1}{k!} (\tilde{r}(x))^k = \sum_{n=k}^{\infty} \tilde{r}_n^{(k)} \frac{x^n}{n!}. \tag{4.2.20}$$

Introducing the notation $Y = \tilde{r}(x)$ and setting, as above,

$$y = r(x) = \sum_{n=1}^{\infty} n^{n-1} \frac{x^n}{n!},$$

we see that

$$Y = \int_0^x \frac{y(t)}{t}\, dt. \tag{4.2.21}$$

Now from (4.2.14) it follows that

$$Y = y - y^2/2. \tag{4.2.22}$$

This equality means that the generating function $\tilde{r}(x)$ for unrooted trees is expressed in terms of the generating function $r(x)$ for rooted trees as

$$\tilde{r}(x) = r(x) - r^2(x)/2. \tag{4.2.23}$$

From equalities (4.2.20) and (4.2.22) we obtain

$$\frac{1}{k!} (\tilde{r}(x))^k = \sum_{n=k}^{\infty} \tilde{r}_n^{(k)} \frac{x^n}{n!} = \frac{1}{k!} \left(y - \frac{y^2}{2} \right)^k$$

$$= \frac{1}{k!} \sum_{j=0}^{k} \binom{k}{j} \left(-\frac{1}{2} \right)^j y^{k+j}. \tag{4.2.24}$$

From formula (4.2.15) it follows that

$$y^{k+j} = \sum_{n=k+j}^{\infty} (k+j)n^{n-k-j-1}(n)_{k+j}\frac{x^n}{n!}. \tag{4.2.25}$$

Now equalities (4.2.24) and (4.2.25) yield the formula

$$\tilde{r}_n^{(k)} = \frac{1}{k!}\sum_{j=0}^{k}\binom{k}{j}\left(-\frac{1}{2}\right)^j (k+j)n^{n-k-j-1}(n)_{k+j},$$
$$k = 1,2,\dots,n, \qquad n = 1,2,\dots. \tag{4.2.26}$$

Let us demonstrate that for fixed k the equality

$$\lim_{n\to\infty}\frac{\tilde{r}_n^{(k)}}{n^{n-2}} = \frac{1}{2^{k-1}(k-1)!}, \qquad k = 1,2,\dots, \tag{4.2.27}$$

holds. To this end we represent $\tilde{r}_n^{(k)}$ as

$$\tilde{r}_n^{(k)} = \sum_{l=1}^{2k}\alpha_{nk}^{(l)}n^{n-l}.$$

It can easily be seen that

$$\alpha_{nk}^{(1)} = \frac{1}{k!}\sum_{j=0}^{k}\binom{k}{j}\left(-\frac{1}{2}\right)^j (k+j) = 0,$$

$$\alpha_{nk}^{(2)} = \frac{-1}{k!}\sum_{j=0}^{k}\binom{k}{j}\left(-\frac{1}{2}\right)^j (k+j)\binom{k+j}{2} = \frac{1}{2^{k-1}(k-1)!}.$$

Hence equality (4.2.27) follows.

Now we denote by $\tilde{L}_n$ the total number of forests with n vertices composed of unrooted trees. It is clear that

$$\tilde{L}_n = \sum_{k=0}^{n}\tilde{r}_n^{(k)}. \tag{4.2.28}$$

Assuming

$$\tilde{L}_0 = \tilde{r}_0^{(0)} = 1,$$

it follows from equalities (4.2.20) and (4.2.22) that

$$\sum_{n=0}^{\infty}\tilde{L}_n\frac{x^n}{n!} = e^{r(x)-r^2(x)/2}.$$

From formulae (4.2.27) and (4.2.28) we obtain

$$\lim_{n\to\infty} \frac{\tilde{L}_n}{n^{n-2}} = \sqrt{e}. \tag{4.2.29}$$

4.2.4 End vertices in trees

Let us consider an unrooted tree with labeled vertices $V_1, V_2, \ldots, V_n$. A vertex V_i is called an end vertex if there exists only one edge joining V_i to other vertices, that is, $\deg V_i = 1$. We denote by $\tilde{r}(n,k)$ the number of unrooted trees with k labeled end vertices. Let us prove Rényi's formula

$$\tilde{r}(n,k) = \frac{n!}{k!}\sigma(n-2, n-k), \tag{4.2.30}$$

where $\sigma(n,k)$ are Stirling numbers of the second kind. To this end we use Prüfer's method, namely, we establish a one-to-one correspondence between the unrooted trees with n vertices and the arrangements of size $n-2$ with unlimited repetitions of the numbers $1, 2, \ldots, n$. The first step is to remove the end vertex V_{i_1} with the smallest index, together with the edge from the tree; let j_1 be the index of the other vertex of the tree which is incident to the eliminated edge. We repeat this process until a single edge remains. As a result, we obtain the $(n-2)$-arrangement $(j_1, j_2, \ldots, j_{n-2})$. This $(n-2)$-arrangement determines the tree uniquely. It is clear that a tree with k end vertices is associated with an $(n-2)$-arrangement $(j_1, j_2, \ldots, j_{n-2})$ where exactly $n-k$ numbers from the set $1, 2, \ldots, n$ appear at least once. The numbers for the labels of the end vertices can be chosen in $\binom{n}{k}$ ways, and the number of $(n-2)$-arrangements generated by $n-k$ distinct elements which appear in the arrangement at least once is equal to $\Delta^{n-k}0^{n-2} = (n-k)!\,\sigma(n-2, n-k)$, using the notation from Chapter 3. Hence formula (4.2.30) follows.

4.2.5 Unlabeled end vertices

In structure diagrams representing hydrocarbons, the symbol H denoting hydrogen is frequently dropped, since this element is usually placed at the end vertices of the diagram, which is a tree. These circumstances inspire the problem of determining the number $r^*(n)$ of unrooted trees whose vertices, other than the end ones, are labeled.

If $r^*(n,k)$ is the number of trees under consideration with k unlabeled end vertices, then it is obvious that

$$r^*(n) = \sum_{k=2}^{n-1} r^*(n,k). \tag{4.2.31}$$

The relation

$$r^*(n,k) = \sum_{i=2}^{k} \binom{n-i-1}{k-i} \tilde{r}(n-k,i), \quad k = 2,3,\dots,n-1, \quad n = 3,4,\dots, \tag{4.2.32}$$

holds, where $\tilde{r}(n,k)$ means the number of trees with n labeled vertices, k of which are end vertices. Indeed, removing k unlabeled end vertices in a tree we obtain a tree with $n-k$ labeled vertices, i of which are end vertices. Consider k edges corresponding to the end vertices; the number of ways of adding them to i end vertices and $n-k-i$ inner vertices is equal to the number of allocations of k identical particles into $n-k$ distinct cells, provided that i given cells are non-empty. This number is equal to

$$\operatorname{coef}_{x^k}(x+x^2+\cdots)^i(1+x+x^2+\cdots)^{n-k-i}$$
$$= \operatorname{coef}_{x^k} x^i(1-x)^{-n+k} = \binom{n-i-1}{k-i}.$$

Hence formula (4.2.32) follows.

From equalities (4.2.32) and (4.2.30) we derive the formula (Harary, Mowshowitz and Riordan, 1970)

$$r^*(n,k) = \sum_{i=2}^{k} \binom{n-i-1}{k-i} \frac{(n-k)!}{i!} \sigma(n-k-2, n-k-i). \tag{4.2.33}$$

The inversion of formula (4.2.32) yields the equality

$$\tilde{r}(n,k) = \sum_{i=2}^{k} (-1)^{k-i} \binom{n}{k-i} r^*(n+i,i). \tag{4.2.34}$$

Since

$$n^{n-2} = \sum_{k=2}^{n-1} \tilde{r}(n,k),$$

summing both sides of equality (4.2.34), we obtain the recurrence relation

$$n^{n-2} = \sum_{i=0}^{n-3} (-1)^i \binom{n-1}{i} r^*(2n-i-1, n-i-1), \qquad n = 3,4,\dots. \tag{4.2.35}$$

By somewhat different reasoning the following formula for the number of trees with unlabeled end vertices was obtained in (Moon, 1969):

$$r^*(n) = \sum_{k=2}^{n-1} \sum_{j=0}^{n-k-1} (-1)^j \binom{n-k}{j} \binom{n-j-1}{k} (n-k-j)^{n-k-2}. \tag{4.2.36}$$

Let $C(n,m)$ be the number of connected graphs with n labeled vertices and m edges, and let $H(n,m,i)$ be the number of such graphs with i end vertices each, $0 \leqslant i \leqslant n-1$. Using the inclusion–exclusion method we can prove that

$$H(n,m,i) = \sum_{j=i}^{n-1} (-1)^{j-i} \binom{j}{i} \binom{n}{j} C(n-j, m-j)(n-j)^j, \tag{4.2.37}$$

since for $n > 2$ there is no pair of end vertices which are adjacent. There exist $\binom{n+k-i-1}{k-i}$ ways of adding k unlabeled end vertices to a connected graph with n vertices, m edges and i end vertices. Therefore, the number of connected graphs with $m+k$ edges, $n+k$ vertices, n of which are labeled and k are unlabeled end vertices, is equal to

$$E(n+k, m+k, k) = \sum_{i=0}^{n-1} \binom{n+k-i-1}{k-i} H(n,m,i)$$

$$= \sum_{j=0}^{n-1} (-1)^j \binom{n}{j} \binom{n+k-j-1}{k} C(n-j, m-j)(n-j)^j.$$

Substituting n for $n+k$ in these last equalities, setting $C(n-k-j, m-j) = (n-k-j)^{n-k-j-2}$, and summing over k from 2 to $n-1$, we obtain (4.2.36).

4.2.6 A problem on car parking

Let us consider the following problem: there are n cars labeled by numbers from 1 to n and n parking places arranged in a line and numbered from left to right. Assume that the ith car out of the n arriving sequentially for parking has the initial preference of choosing the position labeled by p_i; if the position p_i is already occupied, then it occupies the first unoccupied position to the right. There are n^n ways to assign the labels determining the initial preferences $p_1, \ldots, p_n$. It is clear that only some of these assignments allow all the cars to be parked. The question is, what is the number T_n of such assignments?

In (Riordan, 1969) it is proved that

$$T_n = (n+1)^{n-1}. \tag{4.2.38}$$

Each method of assigning the labels is determined by some transformation φ of the set $X = \{1, 2, \ldots, n\}$. If $\varphi(i) = p_i$, $i = 1, 2, \ldots, n$, then the vector $(p_1, p_2, \ldots, p_n)$, referred to as the preference vector, uniquely determines the assignment of the preference labels. We consider the totality of mappings of the form $\varphi' : X \to Y$, where $X = \{1, 2, \ldots, n\}$, $Y = \{1, 2, \ldots, n+1\}$. Every mapping φ' is determined by a vector $(p'_1, p'_2, \ldots, p'_n)$, $p'_i = \varphi'(i)$, $i = 1, 2, \ldots, n$, which is referred to as the derived vector. Every derived vector is considered as the preference vector for the case where the parking places are allocated on a circle and are labeled by the numbers $1, 2, \ldots, n+1$. While using the preference vector $(p'_1, p'_2, \ldots, p'_n)$, we assume that if $p'_i = n+1$ and the place $n+1$ is occupied, then the ith car goes along the circle to the next unoccupied place. Each derived vector determines an allocation of cars with one unoccupied place. The total number of derived vectors, which is equal to $(n+1)^n$, must be divided by $n+1$, since only one of $n+1$ ways of choosing the label k of the unoccupied place, namely $k = n+1$, is suitable for us. Hence formula (4.2.38) follows.

We can construct a mapping which puts preference vectors into one-to-one correspondence with unrooted trees with $n+1$ vertices. A vector $(p_1, p_2, \ldots, p_n)$ is associated with another $(\pi_1, \pi_2, \ldots, \pi_{n-1})$, where

$$\pi_i \equiv (p_{i+1} - p_i) \pmod{n+1}.$$

Then a tree with $n+1$ vertices is put into correspondence with the vector $(\pi_1, \pi_2, \ldots, \pi_{n-1})$ by Prüfer's method described above.

4.2.7 Generation of the symmetric group by transpositions

Let S_n be the symmetric group of degree n, and let T_n be the set of transpositions of degree n (substitutions with one cycle of length two and the remainder of length one). We consider a set $R_n^{(r)} \subseteq T_n$, $R_n^{(r)} = \{\tau_1, \tau_2, \ldots, \tau_r\}$, and put the graph $\Gamma(R_n^{(r)})$, whose vertices are the elements of the set X on which the group S_n acts, into correspondence with the set $R_n^{(r)}$. Let us write a transposition $\tau_k \in R_n^{(r)}$ as $\tau_k = (i, j)$ if the cycle of length two belonging to it is (i, j). Then every transposition $\tau_k = (i, j)$ is associated with the edge in $\Gamma(R_n^{(r)})$ which joins the vertices i and j, where $i, j \in X$. The graph $\Gamma(R_n^{(r)})$ is called the Pólya graph.

Let us consider a non-empty subset M of a group G. The set of all elements of the group G which are equal to products with finite numbers of factors, being powers (positive and negative) of elements of M, is a subgroup of the group G. This subgroup is called the subgroup generated by a set M, and is denoted by $\langle M \rangle$. The general form of an element of $\langle M \rangle$ is $a_{i_1}^{\varepsilon_1} a_{i_2}^{\varepsilon_2} \cdots a_{i_\nu}^{\varepsilon_\nu}$, where each ε_i is an integer, and the $a_{i_1}, a_{i_2}, \dots, a_{i_\nu}$ are the elements of the set M, which need not all be different. It is clear that the subgroup $\langle M \rangle$ is contained in any subgroup of G which contains the whole set M, and so is the intersection of all subgroups of G which contain M.

The set M is called a generator system of the subgroup $\langle M \rangle$. If $\langle M \rangle$ coincides with the entire group G, then M is a generator system of the group G. In this case, if M is finite, G is referred to as a group with a finite number of generators.

It is clear that any substitution from the group S_n can be represented as a product of transpositions; therefore the set T_n is a generator system of S_n, and $|T_n| = \binom{n}{2}$. One can find generator systems of S_n which consist of transpositions and contain a smaller number of elements. Since $(i, j) = (1, i)(1, j)(1, i)$, then $T_n' = \{(1,2), (1,3), \dots, (1,n)\}$ is a generator system of S_n.

Lemma 4.2.1 *A set of transpositions*

$$R_n^{(r)} = \{\tau_1, \tau_2, \dots, \tau_r\}, \qquad r \geqslant n-1,$$

generates the symmetric group S_n if and only if the corresponding Pólya graph $\Gamma(R_n^{(r)})$ is connected.

Proof Let the graph $\Gamma(R_n^{(r)})$ be connected and assume that a vertex $i \in X$ can be connected with a vertex $j \in X$ by a path composed of the edges corresponding to transpositions $\tau_{i_1}, \tau_{i_2}, \dots, \tau_{i_l}$. Then the equality

$$(i, j) = \tau_{i_1} \tau_{i_2} \cdots \tau_{i_{l-1}} \tau_{i_l} \tau_{i_{l-1}} \cdots \tau_{i_2} \tau_{i_1}$$

holds, that is, a transposition $(i, j) \in T_n$ can be represented as the product of a finite number of transpositions belonging to $R_n^{(r)}$. Since (i, j) is chosen arbitrarily and T_n generates S_n, then $R_n^{(r)}$ also generates S_n.

Now let us assume that $R_n^{(r)}$ generates S_n but that the graph $\Gamma(R_n^{(r)})$ is not connected and contains components $\Gamma_1, \Gamma_2, \dots, \Gamma_k$, $k > 1$. Then an arbitrary substitution $s \in S_n$ can be represented as

$$s = \tau_{i_1}^{(1)} \tau_{i_2}^{(1)} \cdots \tau_{i_{\nu_1}}^{(1)} \tau_{i_1}^{(2)} \tau_{i_2}^{(2)} \cdots \tau_{i_{\nu_2}}^{(2)} \cdots \tau_{i_1}^{(k)} \tau_{i_2}^{(k)} \cdots \tau_{i_{\nu_k}}^{(k)},$$

where $\tau_k^{(\mu)}$ is the transposition corresponding to an edge of the component Γ_μ, $1 \leqslant \mu \leqslant k$. Since the components $\Gamma_1, \Gamma_2, \dots, \Gamma_k$ have no common vertices, s is a product of k independent cycles. The condition $k > 1$

implies that $R_n^{(r)}$ cannot generate unicyclic substitutions and therefore, cannot generate S_n. □

Corollary 4.2.1 *A set of transpositions $R_n^{(n-1)}$ generates the symmetric group S_n if and only if the corresponding Pólya graph is an unrooted tree with n labeled vertices.*

Each unrooted tree Γ_n with n labeled vertices can be associated with a set of $n-1$ transpositions $R_n^{(n-1)}$ which generate the symmetric group S_n, that is, we can establish a one-to-one correspondence between the trees with n vertices and the above-mentioned sets of $n-1$ transpositions. This reasoning yields one more corollary.

Corollary 4.2.2 *The number of generator systems of the symmetric group S_n which consist of $n-1$ transpositions is equal to n^{n-2}.*

4.2.8 The Hurwitz problem

The Hurwitz problem consists of finding the number of ways of representing a given substitution of degree n as a product of w transpositions. Let the substitutions of degree n act on a set X. The support of a set of transpositions $R_n^{(r)} = \{\tau_1, \tau_2, \ldots, \tau_r\}$ is the set of all elements of the set X belonging to the cycles of length two of the transpositions $\tau_1, \tau_2, \ldots, \tau_r$. A set of transpositions is called regular if it cannot be represented as a union of sets of transpositions with disjoint supports. The product of transpositions is called regular if the set of all transpositions from this product is regular.

Lemma 4.2.2 *A substitution $s \in S_n$ with k cycles cannot be represented as a product of fewer than $n-k$ transpositions. A product of $n-1$ transpositions acting on the set $X = \{1, 2, \ldots, n\}$ is equal to a cycle of length n if and only if this product is regular.*

Proof The second assertion of the lemma is obvious, so we confine ourselves to proving the first. Let us assume that a substitution $s \in S_n$ can be represented as a product of transpositions

$$s = \tau_1 \tau_2 \cdots \tau_{n-k-v}, \qquad v \geqslant 1.$$

It can easily be seen that the number of cycles in the substitutions $s\tau$ and τs is equal to $k+1$ and $k-1$ depending on whether or not the elements of the transposition τ belong to the same cycle of s. Hence it follows that the substitution $\tau_{n-k-v} \cdots \tau_2\tau_1 s$ has no more than $n - v$ cycles; but this substitution is the identity, and we obtain a contradiction which proves the lemma. □

Theorem 4.2.1 (Dénes (1959)) *The number of representations of a given cycle of degree n as a product of the minimum number of transpositions U_n is equal to the number of unrooted trees with n labeled vertices, that is,*

$$U_n = n^{n-2}. \tag{4.2.39}$$

Proof If R_n is a regular set of $n-1$ transpositions, then the corresponding graph $\Gamma(R_n)$ is a tree, being a connected graph with $n-1$ edges. Each of $(n-1)!$ permutations of the transpositions is associated with the same tree; therefore the number of regular products of $n-1$ transpositions is equal to $n^{n-2}(n-1)!$. On the other hand, the number of representations of all cycles of length n as products of $n-1$ transpositions is equal to $(n-1)!\, U_n$. The validity of the theorem follows from the trivial equality $n^{n-2}(n-1)! = (n-1)!\, U_n$. □

Denote by $U(\alpha_1, \alpha_2, \ldots, \alpha_n)$ the number of representations of a substitution s belonging to the cycle class $[1^{\alpha_1}2^{\alpha_2}\ldots n^{\alpha_n}]$ as a product of the minimum number $n-k$ of transpositions, where $k = \sum_{j=1}^{n} \alpha_j$. The following formula, due to Dénes, is valid:

$$U(\alpha_1, \alpha_2, \ldots, \alpha_n) = (n-k)! \prod_{j=1}^{n} \left(\frac{j^{j-2}}{(j-1)!} \right)^{\alpha_j},$$

$$\sum_{j=1}^{n} j\alpha_j = n, \qquad \sum_{j=1}^{n} \alpha_j = k. \tag{4.2.40}$$

For every cycle of the substitution s we choose one of its representations as a product of the minimum number of transpositions. This can be done in $U_1^{\alpha_1} U_2^{\alpha_2} \cdots U_n^{\alpha_n}$ ways. Two transpositions are commutative if and only if the sets of their elements do not intersect. Therefore, any permutation of $n-k$ transpositions which preserves the order of transpositions relating to the same cycle preserves their total product, which is equal to s. The number of such permutations is equal to

$$(n-k)! \left(\prod_{j=1}^{n} ((j-1)!)^{\alpha_j} \right)^{-1}.$$

Hence formula (4.2.40) follows.

4.2.9 Forests and products of transpositions

Let us prove another formula due to Dénes which gives us an expression for the number $U_k(n)$ of products of $n-k$ transpositions which are equal to a substitution of degree n with k cycles, in terms of the number of forests $\tilde{r}_n^{(k)}$ with n vertices and k unrooted trees, namely,

$$U_k(n) = \tilde{r}_n^{(k)}(n-k)!. \tag{4.2.41}$$

Indeed, using formula (4.2.40), we obtain

$$U_k(n) = \sum_{\substack{\alpha_1+2\alpha_2+\cdots+n\alpha_n=n,\\ \alpha_1+\cdots+\alpha_n=k}} C(\alpha_1,\alpha_2,\ldots,\alpha_n)(n-k)!\prod_{j=1}^{n}\left(\frac{j^{j-2}}{(j-1)!}\right)^{\alpha_j},$$

where $C(\alpha_1,\alpha_2,\ldots,\alpha_n)$ is the number of substitutions in the cycle class $[1^{\alpha_1}2^{\alpha_2}\ldots n^{\alpha_n}]$. On the other hand, formula (4.2.18) for the number of forests can be rewritten as

$$\tilde{r}_n^{(k)} = \sum_{\substack{\alpha_1+2\alpha_2+\cdots+n\alpha_n=n,\\ \alpha_1+\cdots+\alpha_n=k}} \frac{n!}{\alpha_1!\,\alpha_2!\cdots\alpha_n!\,(1!)^{\alpha_1}(2!)^{\alpha_2}\cdots(n!)^{\alpha_n}}\prod_{j=1}^{n}(j^{j-2})^{\alpha_j}.$$

Comparing these last two equalities, we obtain formula (4.2.41). From formula (4.2.26) for the number of forests of k unrooted trees it follows that

$$U_k(n) = \frac{(n-k)!}{k!}\sum_{j=0}^{k}\binom{k}{j}\left(-\frac{1}{2}\right)^{j}(k+j)n^{n-k-j-1}(n)_{k+j}. \tag{4.2.42}$$

Further, from formula (4.2.27), we obtain

$$\lim_{n\to\infty}\frac{U_k(n)}{n^{n-2}(n-k)!} = \frac{1}{2^{k-1}(k-1)!}, \qquad k=1,2,\ldots. \tag{4.2.43}$$

4.2.10 Bichromatic trees

Consider colored trees with p white vertices and q black vertices such that any pair of adjacent vertices are of different colors. Such trees will be referred to as bichromatic. We denote by R_{pq} the number of rooted trees, and by T_{pq} the number of unrooted trees of the structure under consideration. Then

$$R_{pq} = (p+q)T_{pq} = pT_{pq} + qT_{pq} = r_{pq} + r'_{pq}, \tag{4.2.44}$$

where r_{pq} and r'_{pq} are the numbers of trees with white roots and black roots respectively. Consider the generating functions

$$r(x, y) = \sum_{p=0}^{\infty} \sum_{q=0}^{\infty} r_{pq} \frac{x^p}{p!} \frac{y^q}{q!},$$

$$r'(x, y) = \sum_{p=0}^{\infty} \sum_{q=0}^{\infty} r'_{pq} \frac{x^p}{p!} \frac{y^q}{q!}.$$

The relation

$$r(x, y) = x \exp\{r'(x, y)\} \tag{4.2.45}$$

holds. The proof of this equality is similar to the proof of relation (4.2.8). We should point out that a tree with a white root, after elimination of that root, corresponds to a forest with black roots. The generating function for such forests with k trees is of the form $\{r'(x, y)\}^k/k!$. Summing these generating functions over k, we obtain relations (4.2.45). For reasons of symmetry, the equality

$$r'(x, y) = y \exp\{r(x, y)\}, \tag{4.2.46}$$

corresponding to relation (4.2.45), holds. From equalities (4.2.45) and (4.2.46) we derive that

$$r(x, y) = x \exp\{y \exp\{r(x, y)\}\}.$$

We use Lagrange's formula, which gives an expansion of the solution of the equation

$$x = \frac{z}{\varphi(z)}$$

as a series of the form

$$z = \sum_{k=1}^{\infty} \frac{x^k}{k!} \left\{ \frac{d^{k-1}}{dz^{k-1}} \varphi^k(z) \right\} \bigg|_{z=0},$$

with

$$z = r(x, y), \qquad \varphi(z) = \exp\{y \exp\{z\}\},$$

and obtain

$$\frac{d^{k-1}}{dz^{k-1}} \exp\{ky \exp\{z\}\}|_{z=0} = \sum_{j=0}^{k-1} \sigma(k-1, j)(ky)^j e^{ky},$$

where $\sigma(k,j)$ are Stirling numbers of the second kind. Therefore,

$$r(x,y) = \sum_{p=1}^{\infty}\sum_{q=0}^{\infty} p^q q^{p-1} \frac{x^p}{p!}\frac{y^q}{q!}. \tag{4.2.47}$$

Similarly, we obtain

$$r'(x,y) = \sum_{p=0}^{\infty}\sum_{q=1}^{\infty} p^{q-1} q^{p} \frac{x^p}{p!}\frac{y^q}{q!}. \tag{4.2.48}$$

Expressions (4.2.47) and (4.2.48) yield the formulae for counting unrooted and rooted bichromatic trees (Austin, 1960):

$$T_{pq} = p^{q-1}q^{p-1}, \qquad R_{pq} = p^{q-1}q^{p-1}(p+q).$$

4.3 Cycle classes

4.3.1 Cycles in substitutions

Let a substitution s of degree n act on a set X, $|X| = n$. We define the equivalence relation corresponding to the substitution s on this set X, and say that $x \sim x'$ for $x, x' \in X$ if there exists j such that $x' = s^j(x)$. The blocks of the corresponding partition $X = X_1 \cup X_2 \cup \cdots \cup X_k$ are called orbits or transitivity sets of the substitution s. The restriction s_i of the substitution s to the orbit X_i is called a cycle of this substitution. Sometimes an orbit itself is called a cycle. Two cycles are called independent if the symbols permuted by them form disjoint subsets of X. It is clear that multiplication of independent cycles is commutative. Setting $s_i(x) = x$ for $x \in X \setminus X_i$, we can extend s_i to a substitution of degree n. Then we can write $s = s_1 s_2 \cdots s_k$, that is, any substitution can be represented as a product of their independent cycles. A cycle s_i of a substitution s which acts on an orbit X_i can be associated with the representation $(x, s_i(x), \ldots, s^{l_i - 1}(x))$, called a cycle of length l of the substitution s. It is clear that this representation of the cycle s_i does not depend on a cyclic shift of the elements. The representation consisting of the same elements written in reverse order corresponds to the cycle s_i^{-1} being inverse to s_i ($s_i^{-1}s_i = s_i s_i^{-1} = e$, e is the identity substitution). We always assume that $s_i^0 = e$; by virtue of the commutativity of multiplication of independent cycles, $s^{-v} = s_1^{-v} s_2^{-v} \cdots s_k^{-v}$. Note that the graph $\Gamma(X, s)$ contains only cyclic vertices and therefore every one of its connected components is a cycle with the same number of vertices as the number of elements of the corresponding orbit.

Since the arrangement of elements in a cycle is determined up to a cyclic shift, and the arrangement of the cycles themselves is arbitrary, we obtain the following formula for the number of substitutions of degree n belonging to a cycle class $[1^{\alpha_1}2^{\alpha_2}\dots n^{\alpha_n}]$, that is, containing α_j cycles of length j, $j=1,\dots,n$:

$$C(\alpha_1,\alpha_2,\dots,\alpha_n)=\frac{n!}{1^{\alpha_1}2^{\alpha_2}\cdots n^{\alpha_n}\alpha_1!\,\alpha_2!\cdots\alpha_n!},\qquad \sum_{r=1}^{n} r\alpha_r=n. \quad (4.3.1)$$

4.3.2 Minimal and maximal cycle classes

We investigate the behavior of the function $C(\alpha_1,\alpha_2,\dots,\alpha_n)$ for $\sum_{r=1}^{n} r\alpha_r=n$ and fixed n. Let us find a cycle class which minimizes the value of this function. Such a class will be called minimal. It can easily be seen that the cycle class $[1^n2^0\dots n^0]$ is minimal. Indeed, $C(n,0,\dots,0)=1$. For $n>2$ this class is unique, since any other cycle class contains at least one cycle of length $l\geqslant 2$. The number of ways of forming this cycle is equal to

$$\binom{n}{l}(l-1)!=\frac{(n)_l}{l}\geqslant 2,$$

and the number of substitutions in this cycle class is no less than two.

The cycle class which maximizes the value of $C(\alpha_1,\alpha_2,\dots,\alpha_n)$ is called maximal. It is clear that to find the class we should evaluate

$$\min\chi(\alpha_1,\alpha_2,\dots,\alpha_n)=\min\{1^{\alpha_1}2^{\alpha_2}\cdots n^{\alpha_n}\alpha_1!\,\alpha_2!\cdots\alpha_n!\}$$

under the condition $\sum_{r=1}^{n} r\alpha_r=n$. Let $\alpha_1,\alpha_2,\dots,\alpha_n$ be the minimizing sequence, and let v be the number determined by the condition

$$\alpha_v\geqslant 1,\qquad \alpha_{v+1}=\cdots=\alpha_n=0.$$

The cases $v=1,2$ are eliminated from consideration since for $n>3$ the value $C(0,0,\dots,0,1)$ exceeds the values $C(n,0,\dots,0)$ and $C(\alpha,(n-\alpha)/2,0,\dots,0)$ with any α.

Lemma 4.3.1 *If $v\geqslant 3$, $n>2$, then $\alpha_2=\alpha_3=\cdots=\alpha_{v-1}=0$ in a maximal class.*

Proof Let us assume that $\alpha_k\geqslant 1$ for $2\leqslant k\leqslant v-1$. It is clear that

$$n=\sum_{r=1}^{n} r\alpha_r\geqslant k\alpha_k+v\alpha_v\geqslant k+v.$$

By virtue of the obvious equality

$$k\alpha_k + v\alpha_v = k(\alpha_k - 1) + v(\alpha_v - 1) + (k+v)\cdot 1,$$

the value $n = \sum_{r=1}^{n} r\alpha_r$ does not change if we substitute 1 for α_{k+v} equal to 0 and decrease each α_k and α_v by 1. Thus, a new cycle class can be constructed. Let us prove that the function χ takes a smaller value at this new cycle class. It is clear that

$$\frac{1^{\alpha_1}\cdots k^{\alpha_k}\cdots v^{\alpha_v}\alpha_1!\cdots\alpha_k!\cdots\alpha_v!}{1^{\alpha_1}\cdots k^{\alpha_k-1}\cdots v^{\alpha_v-1}(k+v)^1\alpha_1!\cdots(\alpha_k-1)!\cdots(\alpha_v-1)!}$$
$$= \frac{kv\alpha_k\alpha_v}{k+v} \leqslant \frac{kv}{k+v} \leqslant \frac{1}{1/2+1/3} > 1,$$

and the initial class is not minimal. Thus, $\alpha_2 = \alpha_3 = \cdots = \alpha_{v-1} = 0$ for $v \geqslant 3$. The lemma is proved. □

Hereafter, we assume that $v \geqslant 2$, since for $v = 1$ only α_1 is not zero and the minimal and maximal classes coincide, which is impossible for $n > 2$.

Lemma 4.3.2 *The equality $\alpha_1 = 1$ holds for a maximal class for $n > 2$.*

Proof Let us assume that $\alpha_1 \geqslant 2$. Then we decrease the number of cycles of length 1 and of length v by 1 and insert a cycle of length $v+1$. As a result, we obtain a new cycle class. Let us demonstrate that the value of the function χ at this class is smaller than that at a maximal class:

$$\frac{1^{\alpha_1}2^{\alpha_2}\cdots v^{\alpha_v}\alpha_1!\,\alpha_2!\cdots\alpha_v!}{1^{\alpha_1-1}2^{\alpha_2}\cdots v^{\alpha_nu-1}(v+1)(\alpha_1-1)!\,\alpha_2!\cdots(\alpha_v-1)!\,1!}$$
$$= \frac{v\alpha_1\alpha_v}{v+1} \geqslant \frac{2v}{v+1} \geqslant \frac{4}{3}.$$

The contradiction obtained implies that $\alpha_1 \leqslant 1$. Now assume that $\alpha_1 = 0$. Then a cycle class of the form $[1^0\ldots(v-1)^0 v^{\alpha_v}(v+1)^0\ldots n^0]$ becomes maximal. It is easy to show that the function χ takes values smaller than that at a maximal class at a cycle class of the form $[1^1, 2^0, \ldots, (n-1)^1, n^0]$ for $\alpha_v = 1$ $(v = n)$ and at a class of the form $[1^1, 2^0, \ldots, (v-1)^1, v^{\alpha_v - 1}, (v+1)^0, \ldots, n^0]$ for $\alpha_v \geqslant 2$. In the second case,

$$\frac{v^{\alpha_v}\alpha_v!}{(v-1)v^{\alpha_v-1}(\alpha_v-1)!} = \frac{v\alpha_v}{v-1} > 1.$$

Hence it follows that $\alpha_1 = 1$. □

Lemma 4.3.3 *The equality* $\alpha_\nu = 1$ *holds for a maximal class for* $n > 2$, $\nu \geqslant 2$.

Proof Let us assume that $\alpha_\nu \geqslant 2$. Then $n = \sum_{r=0}^{n} r\alpha_r \geqslant \nu\alpha_\nu \geqslant 2\nu$. We decrease the number of cycles of length ν in a maximal class by two and insert one cycle of length 2ν. At the class thus constructed, the value of χ is smaller than the value of χ at the initial class. Indeed,

$$\frac{1^{\alpha_1}\cdots\nu^{\alpha_\nu}\alpha_1!\cdots\alpha_\nu!}{1^{\alpha_1}\cdots\nu^{\alpha_\nu-2}\cdots 2\nu\cdot\alpha_1!\cdots(\alpha_\nu-2)!\,1!} = \frac{\nu\alpha_\nu(\alpha_\nu-1)}{2} \geqslant \nu > 1.$$

The contradiction obtained implies that $\alpha_\nu \leqslant 1$. Since $\alpha_\nu \neq 0$ by virtue of the definition of ν, the lemma is proved. □

Now let us find the number ν determining a maximal cycle class. It is clear that in the case where $\nu = 1$ the maximal cycle class coincides with the minimal one, which is impossible for $n > 2$. For $\nu \geqslant 2$, by virtue of the equalities $\alpha_1 = \alpha_\nu = 1$, $\alpha_2 = \cdots = \alpha_{\nu-1} = \alpha_{\nu+1} = \cdots = \alpha_n = 0$, the relation $1 \cdot 1 + \nu \cdot 1 = n$ holds. Hence it follows that $\nu = n - 1$. Thus, we have proved the following theorem due to V. L. Goncharov (Goncharov, 1944).

Theorem 4.3.1 *The cycle class* $[1^1, 2^0, \ldots, (n-2)^0, (n-1)^1, n^0]$ *is maximal for* $n > 2$.

4.4 Generating functions of cycles of substitutions

Let $\mathbf{N}_0 = \{0, 1, \ldots\}$, and let $\Lambda = \{\Lambda_1, \Lambda_2, \ldots\}$ be a sequence of subsequences Λ_j of the sequence $\mathbf{N}_0$. The generating function of an infinite number of variables

$$F(t; x_1, x_2, \ldots; \Lambda) = \prod_{j=1}^{\infty} \sum_{\alpha_j \in \Lambda_j} \left(\frac{x_j t^j}{j}\right)^{\alpha_j} \frac{1}{\alpha_j!} \tag{4.4.1}$$

is called the enumerator of the cycle classes of substitutions when the indices of the cycle classes are determined by Λ. This last condition means that if $[1^{\alpha_1} 2^{\alpha_2} \ldots n^{\alpha_n}]$ is the cycle class under consideration, then $\alpha_j \in \Lambda_j$, $j = 1, 2, \ldots$. Let us rewrite the enumerator as

$$F(t; x_1, x_2, \ldots; \Lambda) = \sum_{n=0}^{\infty} \frac{t^n}{n!} \sum_{\substack{1\alpha_1+\cdots+n\alpha_n=n,\\ \alpha_j\in\Lambda_j,\ j=1,\ldots,n}} \frac{n!}{1^{\alpha_1}\cdots n^{\alpha_n}\alpha_1!\cdots\alpha_n!} x_1^{\alpha_1}\cdots x_n^{\alpha_n}. \tag{4.4.2}$$

Taking formula (4.4.2) into account, we can say that the enumerator F enumerates the totality of the substitutions belonging to cycle classes of the form $[1^{\alpha_1}2^{\alpha_2}\dots n^{\alpha_n}]$, where $\alpha_j \in \Lambda_j$, $j = 1,2,\dots$, since the coefficient of $\frac{t^n}{n!}x_1^{\alpha_1}\cdots x_n^{\alpha_n}$ is equal to $C(\alpha_1,\alpha_2,\dots,\alpha_n)$.

A substitution for which the number α_j of cycles of length j is an element of the sequence Λ_j, $j = 1,\dots,n$, and $\Lambda = (\Lambda_1,\Lambda_2,\dots)$ is called a Λ-substitution. The generating function

$$C_n(x_1,x_2,\dots,x_n;\Lambda) = \sum_{\substack{1\alpha_1+\cdots+n\alpha_n=n,\\ \alpha_j\in\Lambda_j,\ j=1,\dots,n}} C(\alpha_1,\alpha_2,\dots,\alpha_n)x_1^{\alpha_1}\cdots x_n^{\alpha_n} \quad (4.4.3)$$

is called the cycle indicator of Λ-substitutions. It is clear that

$$\sum_{n=0}^{\infty} C_n(x_1,x_2,\dots,x_n;\Lambda)\frac{t^n}{n!} = \prod_{j=1}^{\infty}\sum_{\alpha_j\in\Lambda_j}\left(\frac{x_jt^j}{j}\right)^{\alpha_j}\frac{1}{\alpha_j!}. \quad (4.4.4)$$

We denote by $C_{nk}(\Lambda)$ the number of Λ-substitutions with k cycles, by $C_{nk}^{(l)}(\Lambda)$, the number of Λ-substitutions with k cycles of length l, and by $C_n(\Lambda)$, the total number of Λ-substitutions for a given sequence Λ. It is clear that

$$C_n(\Lambda) = C_n(1,1,\dots,1;\Lambda), \quad (4.4.5)$$

$$C_n(x;\Lambda) = \sum_{k=0}^{n} C_{nk}(\Lambda)x^k = C_n(x,x,\dots,x;\Lambda), \quad (4.4.6)$$

$$C_n^{(l)}(x;\Lambda) = \sum_{k=0}^{[n/l]} C_{nk}^{(l)}(\Lambda)x^k = C_n(1,\dots,x,\dots,1;\Lambda), \quad (4.4.7)$$

where x in the last equality is at the lth position.

From equalities (4.4.4)–(4.4.7) it follows that

$$\sum_{n=0}^{\infty} C_n(\Lambda)\frac{t^n}{n!} = \prod_{j=1}^{\infty}\sum_{\alpha_j\in\Lambda_j}\left(\frac{t^j}{j}\right)^{\alpha_j}\frac{1}{\alpha_j!}, \quad (4.4.8)$$

$$\sum_{n=0}^{\infty}\sum_{k=0}^{n} C_{nk}(\Lambda)\frac{t^n}{n!}x^k = \prod_{j=1}^{\infty}\sum_{\alpha_j\in\Lambda_j}\left(\frac{xt^j}{j}\right)^{\alpha_j}\frac{1}{\alpha_j!}, \quad (4.4.9)$$

$$\sum_{n=0}^{\infty}\sum_{k=0}^{[n/l]} C_{nk}^{(l)}(\Lambda)\frac{t^n}{n!}x^k = \left(\sum_{\alpha_l\in\Lambda_l}\left(\frac{xt^l}{l}\right)^{\alpha_l}\frac{1}{\alpha_l!}\right)\prod_{\substack{j=1,\\ j\neq l}}^{\infty}\sum_{\alpha_j\in\Lambda_j}\left(\frac{t^j}{j}\right)^{\alpha_j}\frac{1}{\alpha_j!}. \quad (4.4.10)$$

Let us turn to consideration of some specific cases.

4.4.1 The numbers of substitutions with k cycles and with k cycles of length l

We set $\Lambda_j = \mathbf{N}_0 = \{0, 1, \ldots\}$, $j = 1, 2, \ldots$, and $C_n = C_n(\mathbf{N}_0)$, $C_{nk} = C_{nk}(\mathbf{N}_0)$, $C_{nk}^{(l)} = C_{nk}^{(l)}(\mathbf{N}_0)$. Then equality (4.4.8) yields

$$\sum_{n=0}^{\infty} C_n \frac{t^n}{n!} = (1-t)^{-1}, \qquad |t| < 1.$$

Hence the obvious equality $C_n = n!$ follows. Put $C_n(x) = C_n(x, \mathbf{N}_0)$, from equality (4.4.9) we obtain

$$\sum_{n=0}^{\infty} C_n(x) \frac{t^n}{n!} = \sum_{n=0}^{\infty} \sum_{k=0}^{n} C_{nk} \frac{t^n}{n!} x^k = (1-t)^{-x}, \qquad |t| < 1.$$

This relation means that

$$C_n(x) = \sum_{k=0}^{n} C_{nk} x^k = x(x+1)\cdots(x+n-1). \tag{4.4.11}$$

Hence we obtain the well-known formula

$$C_{nk} = |s(n,k)|, \tag{4.4.12}$$

where $s(n,k)$ are Stirling numbers of the first kind.

Introducing $C_n^{(l)}(x) = C_n^{(l)}(x; \mathbf{N}_0)$, from formula (4.4.10) we find that

$$\sum_{n=0}^{\infty} C_n^{(l)}(x) \frac{t^n}{n!} = \sum_{n=0}^{\infty} \sum_{k=0} [n/l] C_{nk}^{(l)} \frac{t^n}{n!} x^k = \frac{1}{1-t} e^{(x-1)t^l/l}. \tag{4.4.13}$$

Hence after simple transformations we obtain

$$C_n^{(l)}(x) = \sum_{k=0} [n/l] C_{nk}^{(l)} x^k = n! \sum_{j=0}^{[n/l]} \frac{(x-1)^j}{j!\, l^j}. \tag{4.4.14}$$

From this generating function we deduce the formula

$$C_{nk}^{(l)} = \frac{n!}{k!\, l^k} \sum_{j=0}^{[n/l]-k} \frac{(-1)^j}{j!\, l^j}, \qquad k = 0, 1, \ldots, [n/l]. \tag{4.4.15}$$

4.4.2 The numbers of substitutions with cycles of even and odd lengths

We consider the case where

$$\Lambda_j = \begin{cases} \mathbf{N}_0, & j \in A, \\ \{0\}, & j \notin A, \end{cases}$$

where A is a fixed subsequence of the sequence $\mathbf{N} = \{1, 2, \ldots\}$ and $\mathbf{N}_0 = \{0, 1, \ldots\}$. In this case we deal with the set of substitutions whose cycle lengths are the elements of the sequence A. We denote by C_n^A, C_{nk}^A, C_{nkl}^A the total number of substitutions of the structure under consideration, the number of such substitutions with k cycles and the number of such substitutions with k cycles of length l respectively. We put

$$a(t) = \sum_{j \in A} \frac{t^j}{j}.$$

Equalities (4.4.8)–(4.4.10) imply that

$$\sum_{n=0}^{\infty} C_n^A \frac{t^n}{n!} = e^{a(t)}, \tag{4.4.16}$$

$$\sum_{n=0}^{\infty} \sum_{k=0}^{n} C_{nk}^A \frac{t^n}{n!} x^k = e^{xa(t)}, \tag{4.4.17}$$

$$\sum_{n=0}^{\infty} \sum_{k=0}^{[n/l]} C_{nkl}^A \frac{t^n}{n!} x^k = e^{(x-1)t^l/l + a(t)}, \qquad l \in A. \tag{4.4.18}$$

Now let A be the sequence of even numbers $A = \{2, 4, 6, \ldots\}$. In this case we put $C_n^A = C_n^{\mathrm{e}}$, $C_{nk}^A = C_{nk}^{\mathrm{e}}$, $C_{nkl}^A = C_{nkl}^{\mathrm{e}}$. From (4.4.17) we obtain

$$\sum_{n=0}^{\infty} \sum_{k=0}^{n} C_{nk}^{\mathrm{e}} \frac{t^n}{n!} x^k = (1 - t^2)^{-x/2}. \tag{4.4.19}$$

Hence it follows that

$$\sum_{k=0}^{n} C_{nk}^{\mathrm{e}} x^k = \begin{cases} n! \binom{x/2 + n/2 - 1}{n/2}, & n \text{ is even}, \\ 0, & n \text{ is odd}. \end{cases} \tag{4.4.20}$$

Setting $x = 1$ in equalities (4.4.19) and (4.4.20), we obtain

$$\sum_{n=0}^{\infty} C_n^{\mathrm{e}} \frac{t^n}{n!} = (1 - t^2)^{-1/2},$$

$$C_n^{\mathrm{e}} = \begin{cases} n! \binom{n/2 - 1/2}{n/2}, & n \text{ is even}, \\ 0, & n \text{ is odd}. \end{cases}$$

Now let A be the sequence of odd numbers $A = \{1, 3, 5, \ldots\}$. In this case we put $C_n^A = C_n^{\mathrm{o}}$, $C_{nk}^A = C_{nk}^{\mathrm{o}}$, $C_{nkl}^A = C_{nkl}^{\mathrm{o}}$. From (4.4.17) we obtain

$$\sum_{n=0}^{\infty} \sum_{k=0}^{n} C_{nk}^{\mathrm{o}} \frac{t^n}{n!} x^k = \left(\frac{1+t}{1-t} \right)^{x/2}. \tag{4.4.21}$$

Hence it follows that

$$\sum_{k=0}^{n} C^{\mathrm{o}}_{nk} x^k = n! \sum_{k=0}^{n} \binom{x/2}{n-k}\binom{x/2+k-1}{k}. \tag{4.4.22}$$

Setting $x = 1$ in equalities (4.4.21) and (4.4.22), we obtain

$$\sum_{n=0}^{\infty} C^{\mathrm{o}}_n \frac{t^n}{n!} = \left(\frac{1+t}{1-t}\right)^{1/2}$$

$$C^{\mathrm{o}}_n = \begin{cases} n!\binom{(n+1)/2-1}{n/2}, & n \text{ is even}, \\ n!\binom{n/2-1}{(n-1)/2}, & n \text{ is odd}. \end{cases}$$

4.4.3 The number of solutions of the equation $s^d = e$

Let s be a substitution satisfying the equation

$$s^d = e, \tag{4.4.23}$$

where d is a fixed natural number and e is the identity substitution. It is clear that the only substitutions which satisfy equation (4.4.23) are those whose cycle lengths are divisors of the number d. Therefore, the total number of solutions of equation (4.4.23) is equal to $Q_n(d) = C_n^A$, where $A = \{j : j \mid d\}$ and $j \mid d$ means that j divides d. Hence it follows that

$$\sum_{n=0}^{\infty} Q_n(d) \frac{t^n}{n!} = \exp\left\{\sum_{j|d} \frac{t^j}{j}\right\}. \tag{4.4.24}$$

Differentiating both sides of this equality and then equating the coefficients of $t^n/n!$, we obtain the recurrence relation

$$Q_{n+1}(d) = \sum_{j|d} (n)_{j-1} Q_{n-j+1}(d), \qquad (n)_0 = 1. \tag{4.4.25}$$

In particular, if $d = p$, where p is a prime number, then from equalities (4.4.24) and (4.4.25) we obtain

$$\sum_{n=0}^{\infty} Q_n(p) \frac{t^n}{n!} = e^{t+t^p/p}, \qquad Q_{n+1}(p) = Q_n(p) + (n)_{p-1} Q_{n-p+1}(p). \tag{4.4.26}$$

Therefore,

$$Q_n(p) = \sum_{k=0}^{[n/p]} \frac{n!}{p^k k!(n-pk)!}.$$

The following asymptotic formula for $p > 2$ and $n \to \infty$ can be proved by the saddle point method (see (Moser and Wyman, 1955b)):

$$Q_n(p) = \frac{n^{(1-1/p)n}}{\sqrt{p}} \exp\left\{-\left(1-\frac{1}{p}\right)n + n^{1/p}\right\}(1+o(1)).$$

4.4.4 Even and odd substitutions

The decrement of a cycle of length l is the number $l-1$. We say that the decrement of a substitution is equal to the sum of decrements of the cycles of this substitution. If a substitution s of degree n has k cycles, then its decrement is equal to $n-k$. A substitution s is called even if its decrement is an even number, and is called odd otherwise. We denote by $C_n^{\mathrm{e}}(x_1, x_2, \ldots, x_n)$ and $C_n^{\mathrm{o}}(x_1, x_2, \ldots, x_n)$ the cycle indicators of the sets of even and odd substitutions respectively. The following formulae are valid:

$$\begin{aligned} 2C_n^{\mathrm{e}}(x_1, x_2, \ldots, x_n) &= C_n(x_1, x_2, \ldots, x_n) + C_n(x_1, -x_2, x_3, -x_4, \ldots), \\ 2C_n^{\mathrm{o}}(x_1, x_2, \ldots, x_n) &= C_n(x_1, x_2, \ldots, x_n) - C_n(x_1, -x_2, x_3, -x_4, \ldots). \end{aligned} \tag{4.4.27}$$

The decrements of cycles of odd lengths are even; therefore the parity of a substitution is determined by the sum of decrements of the cycles of even lengths entering into its partition. If a substitution s belongs to a cycle class $[1^{\alpha_1}2^{\alpha_2}\ldots n^{\alpha_n}]$, then it is even if and only if $(-1)^{\alpha_2+\alpha_4+\cdots} = 1$. This fact means that

$$\begin{aligned} 2C_n^{\mathrm{e}}(x_1, x_2, \ldots, x_n) &= \sum_{1\alpha_1+\cdots+n\alpha_n=n} (1+(-1)^{\alpha_2+\alpha_4+\cdots}) \\ &\qquad \times C(\alpha_1, \alpha_2, \ldots, \alpha_n) x_1^{\alpha_1} x_2^{\alpha_2} \cdots x_n^{\alpha_n}, \\ 2C_n^{\mathrm{o}}(x_1, x_2, \ldots, x_n) &= \sum_{1\alpha_1+\cdots+n\alpha_n=n} (1-(-1)^{\alpha_2+\alpha_4+\cdots}) \\ &\qquad \times C(\alpha_1, \alpha_2, \ldots, \alpha_n) x_1^{\alpha_1} x_2^{\alpha_2} \cdots x_n^{\alpha_n}; \end{aligned}$$

hence formulae (4.4.27) follow.

Denote by $C_{nk}(\mathrm{e})$ and $C_{nk}(\mathrm{o})$ the number of even and odd substitutions with k cycles respectively. From equalities (4.4.27) it follows that

$$2\sum_{k=0}^{n} C_{nk}(\mathrm{e})x^k = x(x+1)\cdots(x+n-1) + x(x-1)\cdots(x-n+1), \tag{4.4.28}$$

$$2\sum_{k=0}^{n} C_{nk}(\mathrm{o})x^k = x(x+1)\cdots(x+n-1) - x(x-1)\ldots(x-n+1). \tag{4.4.29}$$

These equalities yield the formulae

$$C_{nk}(\mathrm{e}) = \frac{1}{2}(|s(n,k)| + s(n,k)),$$
$$C_{nk}(\mathrm{o}) = \frac{1}{2}(|s(n,k)| - s(n,k)),$$

where $s(n,k)$ are Stirling numbers of the first kind.

Denote by $C_{nk}^{(l)}(\mathrm{e})$ and $C_{nk}^{(l)}(\mathrm{o})$ the number of even and odd substitutions with k cycles of length l respectively. From formulae (4.4.27) it follows that

$$2\sum_{n=0}^{\infty}\sum_{k=0}^{[n/l]} C_{nk}^{(l)}(\mathrm{e})\frac{t^n}{n!}x^k = \frac{1}{1-t}e^{(x-1)t^l/l} + (1+t)e^{(-1)^{l-1}(x-1)t^l/l},$$

$$2\sum_{n=0}^{\infty}\sum_{k=0}^{[n/l]} C_{nk}^{(l)}(\mathrm{o})\frac{t^n}{n!}x^k = \frac{1}{1-t}e^{(x-1)t^l/l} - (1+t)e^{(-1)^{l-1}(x-1)t^l/l}.$$

Hence we obtain

$$2\sum_{k=0}^{[n/l]} C_{nk}^{(l)}(\mathrm{e})x^k = n!\left(\sum_{k=1}^{[n/l]}\frac{(x-1)^k}{l^k k!} + \frac{(-1)^{(l-1)n/l}(x-1)^{n/l}}{l^{n/l}(n/l)!}\delta_{n,l} + \frac{(-1)^{(l-1)(n-1)/l}(x-1)^{(n-1)/l}}{l^{(n-1)/l}((n-1)/l)!}\delta_{n-1,l}\right), \tag{4.4.30}$$

$$2\sum_{k=0}^{[n/l]} C_{nk}^{(l)}(\mathrm{o})x^k = n!\left(\sum_{k=1}^{[n/l]}\frac{(x-1)^k}{l^k k!} - \frac{(-1)^{(l-1)n/l}(x-1)^{n/l}}{l^{n/l}(n/l)!}\delta_{n,l} - \frac{(-1)^{(l-1)(n-1)/l}(x-1)^{(n-1)/l}}{l^{(n-1)/l}((n-1)/l)!}\delta_{n-1,l}\right), \tag{4.4.31}$$

where $\delta_{ij} = 1$ if $j \mid i$, and $\delta_{ij} = 0$ otherwise.

4.4.5 Cycles of even and odd lengths

We denote by C_{nk}^{odd} and C_{nk}^{even} the numbers of substitutions of degree n with k cycles of odd and even lengths respectively. Note that

$$\sum_{k=0}^{n} C_{nk}^{\text{odd}} x^k = C_n(x, 1, x, 1, \dots),$$

$$\sum_{k=0}^{n} C_{nk}^{\text{even}} x^k = C_n(1, x, 1, x, \dots).$$

Hence it follows that

$$\sum_{n=0}^{\infty}\sum_{k=0}^{n} C_{nk}^{\text{odd}} \frac{t^n}{n!} x^k = (1-t^2)^{-1/2}\left(\frac{1+t}{1-t}\right)^{x/2},$$

$$\sum_{n=0}^{\infty}\sum_{k=0}^{n} C_{nk}^{\text{even}} \frac{t^n}{n!} x^k = (1+t)(1-t^2)^{-(x+1)/2}.$$

From these formulae we derive the expressions for the generating functions:

$$\sum_{k=0}^{n} C_{nk}^{\text{odd}} x^k = n! \sum_{j=0}^{n} \binom{(x-1)/2}{n-j}\binom{(x+1)/2+j-1}{j}, \tag{4.4.32}$$

$$\sum_{k=0}^{n} C_{nk}^{\text{even}} x^k = n! \binom{(x+1)/2+[n/2]-1}{[n/2]}, \tag{4.4.33}$$

where $[a]$ means the integer part of a.

4.5 Mappings with constraints

4.5.1 Basic definitions

Consider the symmetric semigroup $\mathfrak{S}_n$ of mappings of a set X consisting of n elements into itself. We recall that the height of an element $x \in X$ with respect to a mapping $\sigma \in \mathfrak{S}_n$ is the least integer $h \geqslant 0$ such that $\sigma^{h+p}(x) = \sigma^h(x')$ for some natural number p. An element $x \in X$ of zero height is a cyclic element with respect to $\sigma \in \mathfrak{S}_n$. Let us introduce the equivalence relation where $x \sim x'$ means that there exist natural p and q such that $\sigma^p(x) = \sigma^q(x)$; this equivalence relation determines the equivalence classes on X which are called the components of σ. The restriction of σ on the set of cyclic elements is a substitution called the spanning set of σ. Every component contains exactly one cycle of the spanning set.

If $\Gamma(X,\sigma)$ is the directed graph of a mapping σ, then every component of σ is associated with exactly one connected component of the graph $\Gamma(X,\sigma)$ which contains a unique simple cycle joining the cyclic elements and rooted trees whose roots are the vertices of the cycle. The arcs of the trees are directed from elements of greater height to elements of lesser height.

Let $A=\{a_1,a_2,\dots\}$ be a subsequence of the natural series $\mathbf{N}$, and let $\mathfrak{S}_n(A)$ be the set of all mappings $\sigma\in\mathfrak{S}_n$ such that the sizes of the cycles of the graphs $\Gamma(X,\sigma)$ are elements of A. The elements of $\mathfrak{S}_n(A)$ are called A-mappings. A one-to-one A-mapping is called an A-substitution. It is clear that $\mathfrak{S}_n=\mathfrak{S}_n(\mathbf{N})$. If $A=\{1\}$, then $\mathfrak{S}_n(A)$ consists of the mappings $\sigma\in\mathfrak{S}_n$ such that the graphs $\Gamma(X,\sigma)$ are forests of rooted trees with distinct vertices.

4.5.2 Exact formulae

We denote by $U(k,j;n_1,n_2,\dots,n_k;A)$ the number of A-mappings $\sigma\in\mathfrak{S}_n(A)$, with j components and k cyclic elements, that are simultaneously the roots of k trees of the graph $\Gamma(X,\sigma)$ which have, under some fixed ordering, $n_1,n_2,\dots,n_k$ vertices respectively.

Lemma 4.5.1 *If $C(k,j;A)$ is the number of A-substitutions of degree k with j cycles, then*

$$U(k,j;n_1,\dots,n_k;A)=C(k,j;A)\frac{n!}{k!}\frac{n_1^{n_1-1}\cdots n_k^{n_k-1}}{n_1!\cdots n_k!},\qquad n=n_1+\cdots+n_k. \tag{4.5.1}$$

Proof The number of ways of forming the spanning set is equal to $\binom{n}{k}C(k,j;A)$; the number of rooted trees with a fixed label at the root and m vertices is equal to m^{m-2}; the number of ways of allocating the non-root vertices into groups for construction of the trees is $(n-k)!\times((n_1-1)!(n_2-1)!\cdots(n_k-1)!)^{-1}$. Hence formula (4.5.1) follows. □

We denote by $U_n(k,j;A)$, $U_n(k;A)$, $U_{nj}(A)$ the number of A-mappings with k cyclic elements and j components, with k cyclic elements, and with j components, respectively, and by $U_n(A)$, the total number of A-mappings. Let $C(k;A)$, $C(k,j;A)$ be the number of A-substitutions and of A-substitutions with j cycles of degree k.

Lemma 4.5.2 *The following formulae are true:*

$$U_n(k,j;A) = \binom{n-1}{k-1} n^{n-k} C(k,j;A), \qquad k = 0,1,\dots, \quad j = 0,1,\dots, \tag{4.5.2}$$

$$U_n(k;A) = \binom{n-1}{k-1} n^{n-k} C(k;A), \qquad k = 0,1,\dots, \tag{4.5.3}$$

$$U_{nj}(A) = \sum_{k=j}^{n} \binom{n-1}{k-1} n^{n-k} C(k,j;A), \qquad j = 0,1,\dots, \tag{4.5.4}$$

$$U_n(A) = \sum_{k=1}^{n} \binom{n-1}{k-1} n^{n-k} C(k;A), \qquad n = 0,1,\dots, \tag{4.5.5}$$

where

$$U_0(0,0;A) = U_0(0;A) = U_{00}(A) = U_0(A) = 1.$$

Proof Formulae (4.5.2)–(4.5.5) immediately follow from equality (4.5.1) after applying identity (4.2.12). For example,

$$\begin{aligned} U_n(k,j;A) &= \sum_{n_1+\cdots+n_k=n} C(k,j;A) \frac{n!}{k!} \frac{n_1^{n_1-1} n_2^{n_2-1} \cdots n_k^{n_k-1}}{n_1!\, n_2! \cdots n_k!} \\ &= C(k,j;A) \frac{n!\, k n^{n-k-1}}{k!\, (n-k)!} \\ &= \binom{n-1}{k-1} n^{n-k} C(k,j;A). \end{aligned}$$

□

Let us consider some special cases of the formulae given in Lemma 4.5.2.

Set $A = \mathbf{N} = \{1, 2, \dots\}$. From (4.5.3) and (4.5.4) we obtain expressions for the number of mappings $\sigma \in \mathfrak{S}_n$ with k cyclic elements, and for the number of mappings $\sigma \in \mathfrak{S}_n$ with j components, respectively (Rabin and Sitgreaves, 1954; Riordan, 1962):

$$U_n(k) = \frac{(n-1)!\, n^{n-k} k}{(n-k)!}, \qquad k = 0,1,\dots, \tag{4.5.6}$$

$$U_{nj} = \sum_{k=j}^{n} \binom{n-1}{k-1} n^{n-k} |s(k,j)|, \qquad j = 0,1,\dots. \tag{4.5.7}$$

Here $s(k, j)$ are Stirling numbers of the first kind. From formula (4.5.5) with $A = \mathbf{N}$ we obtain the identity

$$n^n = \sum_{k=1}^{n} \binom{n-1}{k-1} n^{n-k} k!,$$

which can also be proved directly.

Assuming that $A = \mathbf{N}$, from equality (4.5.4) with $j = 1$ we obtain the formula for the number of mappings $\sigma \in \mathfrak{S}_n$ with one component (Katz, 1955):

$$U_{n1} = (n-1)! \sum_{k=0}^{n-1} \frac{n^k}{k!}. \tag{4.5.8}$$

If $A = \{1\}$, then from equality (4.5.3) we obtain formula (4.2.11) for the number of forests of k rooted trees which was proved earlier.

4.5.3 Generating functions for mappings with constraints on cycles

Let us consider the generating function of an infinite number of variables:

$$Q(t, x, z, y_1, y_2, \dots; A)$$
$$= \sum_{k=0}^{\infty} \sum_{j=0}^{k} \sum_{n_1=0}^{\infty} \cdots \sum_{n_k=0}^{\infty} U(k, j, n_1, \dots, n_k; A) x^k z^j \frac{(y_1 t)^{n_1} \cdots (y_k t)^{n_k}}{(n_1 + \cdots + n_k)!}, \tag{4.5.9}$$

where $U(0, 0, 0, 0, \dots, 0; A) = 1$ and $U(k, j, n_1, n_2, \dots, n_k; A) = 0$ if there exists a variable with non-zero value. Multiplying both sides of equality (4.5.1) in Lemma 4.5.1 by $x^k z^j (y_1 t)^{n_1} \cdots (y_k t)^{n_k} / (n_1 + \cdots + n_k)!$ and carrying out the summation, we see that

$$Q(t, x, z, y_1, y_2, \dots; A) = \sum_{k=0}^{\infty} \sum_{j=0}^{k} C(k, j; A) \frac{x^k}{k!} z^j r(y_1 t) \cdots r(y_k t), \tag{4.5.10}$$

where $r(x)$ is the generating function for rooted trees (4.2.9).

Theorem 4.5.1 *If*

$$A(w) = \sum_{j \in A} \frac{w^j}{j},$$

then the generating function for A-mappings,

$$Q(t, x, z; A) = \sum_{n=0}^{\infty} \sum_{k=0}^{\infty} \sum_{j=0}^{k} U_n(k, j; A) \frac{t^n}{n!} x^k z^j, \tag{4.5.11}$$

satisfies the relation

$$Q(t,x,z;A) = \exp\{zA(xr(t))\}. \tag{4.5.12}$$

Proof Setting $y_1 = y_2 = \cdots = y_k = 1$, from formula (4.5.10) we obtain

$$Q(t,x,z;A) = \sum_{k=0}^{\infty}\sum_{j=0}^{k} C(k,j;A) r^k(t) \frac{x^k}{k!} z^j. \tag{4.5.13}$$

By virtue of the results of Section 4.4, the generating function for A-substitutions,

$$\Phi(w,z) = \sum_{k=0}^{\infty}\sum_{j=0}^{k} C(k,j;A) \frac{w^k}{k!} z^j, \qquad C(0,0;A) = 1, \tag{4.5.14}$$

is of the form

$$\Phi(w,z) = e^{zA(w)}. \tag{4.5.15}$$

Since $Q(t,x,z;A) = \Phi(xr(t),z)$, equality (4.5.12) holds. □

As a corollary to Theorem 4.5.1, we obtain the generating functions for the number of cyclic points, the number of components, and the total number of A-mappings, respectively:

$$\exp\{A(xr(t))\} = \sum_{n=0}^{\infty}\sum_{k=0}^{n} U_n(k;A) \frac{t^n}{n!} x^k, \tag{4.5.16}$$

$$\exp\{zA(r(t))\} = \sum_{n=0}^{\infty}\sum_{j=0}^{n} U_{nj}(A) \frac{t^n}{n!} z^j, \tag{4.5.17}$$

$$\exp\{A(r(t))\} = \sum_{n=0}^{\infty} U_n(A) \frac{t^n}{n!}. \tag{4.5.18}$$

Let us consider some special cases of the sequence A.

We set $A = \{1\}$ and note that in this case the generating functions (4.5.16) and (4.5.17) coincide and yield

$$e^{xr(t)} = \sum_{n=0}^{\infty}\sum_{k=0}^{n} r_n^{(k)} \frac{t^n}{n!} x^k, \tag{4.5.19}$$

where $r_n^{(k)}$ is the number of forests of k rooted trees. From formula (4.5.19) it follows that

$$\frac{1}{k!} r^k(t) = \sum_{n=k}^{\infty} r_n^{(k)} \frac{t^n}{n!}, \tag{4.5.20}$$

that is, the forests of rooted trees satisfy a formula similar to (4.2.20) for the forests of unrooted trees. Formula (4.5.18) with $A = \{1\}$ yields the expression for the generating function for the number of forests of rooted trees

$$e^{r(t)} = \sum_{n=0}^{\infty} L_n \frac{t^n}{n!}.$$

From relation (4.2.8) we obtain

$$r(t) = t \sum_{n=0}^{\infty} L_n \frac{t^n}{n!}. \tag{4.5.21}$$

Therefore,

$$L_{n-1} = n^{n-2}, \qquad n = 2, 3, \dots .$$

We now consider the case where $A = \mathbf{N} = \{1, 2, \dots\}$. We introduce the notation $U_n(k) = U_n(k; \mathbf{N})$, $U_{nj} = U_{nj}(\mathbf{N})$, $U_n = U_n(\mathbf{N})$, $U_n(k, j) = U_n(k, j; \mathbf{N})$, $C(k, j) = C(k, j; \mathbf{N})$. From Theorem 4.5.1 it follows that

$$(1 - xr(t))^{-z} = \sum_{n=0}^{\infty} \sum_{k=0}^{n} \sum_{j=0}^{n} U_n(k, j) \frac{t^n}{n!} x^k z^j. \tag{4.5.22}$$

Setting $z = 1$ and $x = 1$ successively, we obtain the special cases of formulae (4.5.16) and (4.5.17):

$$F(t, x) = (1 - xr(t))^{-1} = \sum_{n=0}^{\infty} \sum_{k=0}^{n} U_n(k) \frac{t^n}{n!} x^k, \tag{4.5.23}$$

$$G(t, z) = (1 - r(t))^{-z} = \sum_{n=0}^{\infty} \sum_{j=0}^{n} U_{nj} \frac{t^n}{n!} z^j. \tag{4.5.24}$$

Both these formulae are given in (Riordan, 1962). From formulae (4.5.23) and (4.5.24) we obtain

$$r^k(t) = \sum_{n=k}^{\infty} U_n(k) \frac{t^n}{n!}, \tag{4.5.25}$$

$$\sum_{n=j}^{\infty} U_{nj} \frac{t^n}{n!} = \sum_{k=j} C(k, j) \frac{r^k(t)}{k!}. \tag{4.5.26}$$

Using these generating functions and formula (4.5.20), we obtain

$$U_n(k) = r_n^{(k)} k!, \qquad U_{nj} = \sum_{k=j}^{n} C(k, j) r_n^{(k)}.$$

These formulae also follow from equalities (4.5.3) and (4.5.4). Putting $D_n = U_{n1}$ and taking into account the fact that $C(k, 1) = (k-1)!$, from equality (4.5.26) we obtain

$$D(t) = \sum_{n=1}^{\infty} D_n \frac{t^n}{n!} = -\ln(1 - r(t)).$$

Using this equality and (4.5.24), we see that

$$G(t, z) = e^{zD(t)}. \tag{4.5.27}$$

Equality (4.5.27) is given in (Riordan, 1962) and is a special case of relation (4.1.4). We set

$$U_n(z) = \sum_{j=0}^{n} U_{nj} z^j. \tag{4.5.28}$$

From (4.5.27) it follows that

$$\frac{\partial}{\partial t} G(t, z) = zG(t, z) \frac{\partial}{\partial t} D(t).$$

Evaluating the coefficient of $t^n/n!$ on both sides of the last equality, we obtain

$$U_{n+1}(z) = z \sum_{k=0}^{n} \binom{n}{k} D_{k+1} U_{n-k}(z). \tag{4.5.29}$$

Equating the coefficients of z^j on the left- and right-hand sides of this equality, we arrive at the relation

$$U_{n+1,j} = \sum_{k=0}^{n} \binom{n}{k} D_{k+1} U_{n-k,j-1}. \tag{4.5.30}$$

This can also be proved directly if we assume that the vertex of the graph $\Gamma(X, \sigma)$, $|X| = n+1$, labeled by $x_{n+1} \in X$, belongs to a component with $k+1$ vertices, and the remaining $n-k$ vertices belong to a graph with $j-1$ components.

4.5.4 Generating functions for mappings of bounded height

We say that an A-mapping has a height not exceeding h if the height of any element $x \in X$ with respect to this mapping does not exceed h. We denote the set of such mappings by $\mathfrak{S}_n^h(A)$. If for an A-mapping σ of height not exceeding h there exists an element $x \in X$ of height h, then this A-mapping is of height h. We denote the set of A-mappings of height h by $\mathfrak{S}_n^{(h)}(A)$. It is clear that

$$\mathfrak{S}_n^{(h)}(A) = \mathfrak{S}_n^h(A) \setminus \mathfrak{S}_n^{h-1}(A). \tag{4.5.31}$$

If we put $\mathfrak{S}_n^{(h)}(A) = \mathfrak{S}_n^h(A) = \varnothing$ for $h < 0$, then this equality holds for all integers $h \leqslant n$. In particular, $\mathfrak{S}_n^{(0)}(A) = \mathfrak{S}_n^0(A)$ is the set of A-substitutions of degree n. From equality (4.5.31) it follows that

$$\mathfrak{S}_n^h(A) = \mathfrak{S}_n^{(h)}(A) \cup \mathfrak{S}_n^{(h-1)}(A) \cup \cdots \cup \mathfrak{S}_n^{(0)}(A),$$
$$\mathfrak{S}_n^{(i)}(A) \cap \mathfrak{S}_n^{(j)}(A) = \varnothing, \qquad i \neq j.$$

We denote by $T_n(k, j; k_1, k_2, \ldots, k_h; A)$ the number of A-mappings $\sigma \in \mathfrak{S}_n^h(A)$ with j components, k cyclic elements and k_i elements $x \in X$ of height i, $i = 1, 2, \ldots, h$, $k + k_1 + k_2 + \cdots + k_h = n$.

Lemma 4.5.3 *If $C(k, j; A)$ is the number of A-substitutions of degree k with j cycles, then*

$$T_n(k, j; k_1, k_2, \ldots, k_h; A) = C(k, j; A)n!\frac{k^{k_1}k_1^{k_2}\cdots k_{h-1}^{k_h}}{k!k_1!\cdots k_h!}, \tag{4.5.32}$$

where $k_i = 0$ implies $k_{i+1} = \cdots = k_k = 0$.

Proof The number of partitions of the elements of X with respect to their height is equal to $n!(k!k_1!\cdots k_h!)^{-1}$, the number of mappings of the elements of the ith height into the elements of the $(i-1)$th height is equal to $k_{i-1}^{k_i}$, and the number of spanning sets of an A-mapping with k cyclic elements and j components is equal to $C(k, j; A)$. □

Theorem 4.5.2 *Let the function $a_h(t, y_1, \ldots, y_h)$ be defined recursively:*

$$a_0(t) = t, \qquad a_1(t, y) = te^{ty}, \tag{4.5.33}$$

$$a_h(t, y_1, y_2, \ldots, y_h) = t\exp\{y_1 a_{h-1}(t, y_2, \ldots, y_h)\}, \tag{4.5.34}$$

and let

$$F(t,x,z,y_1,\dots,y_h;A)$$
$$=\sum_{k=0}^{\infty}\sum_{j=0}^{k}\sum_{k_1=0}^{\infty}\cdots\sum_{k_h=0}^{\infty}T_n(k,j;k_1,\dots,k_h;A)x^k t^k z^j\frac{(y_1t)^{k_1}\cdots(y_ht)^{k_h}}{(k+k_1+\cdots+k_h)!}. \tag{4.5.35}$$

Then

$$F(t,x,z,y_1,\dots,y_h;A)=\exp\{zA(xa_h(t,y_1,\dots,y_h))\}, \tag{4.5.36}$$

where, as above,

$$A(w)=\sum_{j\in A}\frac{w^j}{j}.$$

The theorem can be proved by direct multiplication of the series and by the use of Lemma 4.5.3 and equality (4.5.15).

Let us introduce the following notation for the A-mappings $\sigma\in\mathfrak{S}_n^h(A)$: $T_n^{(h)}(k,j;A)$ is the number of A-mappings with k cyclic elements and j components; $T_n^{(h)}(k;A)$ is the number of mappings with k cyclic elements; $T_{nj}^{(h)}(A)$ is the number of mappings with j components; $T_n^{(h)}(A)$ is the total number of mappings in $\mathfrak{S}_n^h(A)$. From Lemma 4.5.3 it follows that, for $k=0,1,\dots,n$, $j=1,2,\dots,k$,

$$T_n^{(h)}=C(k,j;A)\frac{n!}{k!}\sum_{k_1+\cdots+k_h=n-k}\frac{k^{k_1}k_1^{k_2}\cdots k_{h-1}^{k_h}}{k_1!\,k_2!\cdots k_h!}. \tag{4.5.37}$$

In addition, the formulae

$$T_n^{(h)}(k;A)=\sum_{j=0}^{n}T_n^{(h)}(k,j;A),$$
$$T_{nj}^{(h)}(A)=\sum_{k=j}^{n}T_n^{(h)}(k,j;A),$$
$$T_n^{(h)}(A)=\sum_{j=0}^{n}T_{nj}^{(h)}(A)$$

hold. For the sake of convenience we assume that

$$T_0^{(h)}(0,0;A)=T_0^{(h)}(0;A)=T_{00}^{(h)}(A)=T_0^{(h)}(A)=1,$$
$$a_h(t)=a_h(t,1,\dots,1);$$

then from Theorem 4.5.2 we obtain

$$\exp\{zA(xa_h(t))\} = \sum_{n=0}^{\infty}\sum_{k=0}^{n}\sum_{j=0}^{k} T_n^{(h)}(k,j;A)\frac{t^n}{n!}x^k z^j, \qquad (4.5.38)$$

$$\exp\{A(xa_h(t))\} = \sum_{n=0}^{\infty}\sum_{k=0}^{n} T_n^{(h)}(k;A)\frac{t^n}{n!}x^k, \qquad (4.5.39)$$

$$\exp\{zA(a_h(t))\} = \sum_{n=0}^{\infty}\sum_{j=0}^{n} T_{nj}^{(h)}(A)\frac{t^n}{n!}z^j, \qquad (4.5.40)$$

$$\exp\{A(a_h(t))\} = \sum_{n=0}^{\infty} T_n^{(h)}(A)\frac{t^n}{n!}. \qquad (4.5.41)$$

Let us consider some special cases.

Setting $A = \mathbf{N} = \{1,2,\ldots\}$, from equalities (4.5.38)–(4.5.41) we derive the generating functions for mappings of bounded height:

$$(1-xa_h(t))^{-z} = \sum_{n=0}^{\infty}\sum_{k=0}^{n}\sum_{j=0}^{k} T_n^{(h)}(k,j)\frac{t^n}{n!}x^k z^j, \qquad (4.5.42)$$

$$(1-xa_h(t))^{-1} = \sum_{n=0}^{\infty}\sum_{k=0}^{n} T_n^{(h)}(k)\frac{t^n}{n!}x^k,$$

$$(1-a_h(t))^{-z} = \sum_{n=0}^{\infty}\sum_{j=0}^{n} T_{nj}^{(h)}\frac{t^n}{n!}z^j, \qquad (4.5.43)$$

$$(1-a_h(t))^{-1} = \sum_{n=0}^{\infty} T_n^{(h)}\frac{t^n}{n!},$$

where $T_n^{(h)}(k,j) = T_n^{(h)}(k,j;\mathbf{N})$, $T_n^{(h)}(k) = T_n^{(h)}(k;\mathbf{N})$, $T_{nj}^{(h)} = T_{nj}^{(h)}(\mathbf{N})$, and $T_n^{(h)} = T_n^{(h)}(\mathbf{N})$.

Now let us consider the case $A = \{1\}$. Under this assumption, $T_{nj}^{(h)}(\{1\})$ is the number of forests of j rooted trees, and $T_n^{(h)}(\{1\})$ is the total number of forests of rooted trees, provided that in both the cases the height of the trees does not exceed h. We have

$$\sum_{n=0}^{\infty}\sum_{j=0}^{n} T_{nj}^{(h)}(\{1\})\frac{t^n}{n!}z^j = e^{za_h(t)},$$

$$\sum_{n=0}^{\infty} T_n^{(h)}(\{1\})\frac{t^n}{n!} = e^{a_h(t)}. \qquad (4.5.44)$$

It can easily be seen that $T_n^{(h)}(\{1\})$ is equal to the number of mappings $\sigma \in \mathfrak{S}_n$ satisfying the condition $\sigma^{h+1} = \sigma^h$, and $T_{nj}^{(h)}(\{1\})$ is the number of these mappings with j fixed points.

If I_n is the number of idempotents of the symmetric semigroup $\mathfrak{S}_n$, that is, those elements satisfying the equation $\sigma^2 = \sigma$, and I_{nj} is the number of idempotents with j fixed points, then

$$I_{nj} = \binom{n}{j} j^{n-j}, \qquad I_n = \sum_{j=0}^{n} I_{nj}.$$

In addition, (4.5.44) yields (Harris and Schoenfeld, 1967)

$$\sum_{n=0}^{\infty}\sum_{j=0}^{n} I_{nj}\frac{t^n}{n!}z^j = e^{zte^t},$$

$$\sum_{n=0}^{\infty} I_n\frac{t^n}{n!} = e^{te^t}.$$

Now let us denote by $D_n^{(h)}$ and $d_n^{(h)}$ the number of rooted trees of height not exceeding h and of height h, respectively, provided that n vertices are labeled. The following formulae are valid:

$$D_n^{(h)} = n! \sum_{k_1+\cdots+k_h=n-1} \frac{k_1^{k_2}k_2^{k_3}\cdots k_{h-1}^{k_h}}{k_1!\,k_2!\cdots k_h!}, \tag{4.5.45}$$

$$d_n^{(h)} = D_n^{(h)} - D_n^{(h-1)}. \tag{4.5.46}$$

The derivation of (4.5.45) is analogous to the proof of Lemma 4.5.3. From (4.5.45) and (4.5.46) it follows that (Riordan, 1960)

$$a_h(t) = \sum_{n=0}^{\infty} D_n^{(h)}\frac{t^n}{n!}, \qquad a_h(t) - a_{h-1}(t) = \sum_{n=0}^{\infty} d_n^{(h)}\frac{t^n}{n!}. \tag{4.5.47}$$

We finally note that A-mappings of zero height are A-substitutions; therefore

$$C(n, j; A) = T_{nj}^{(0)}(A), \qquad j = 1, 2, \ldots, n.$$

Setting $a_0(t) = t$, from equalities (4.5.40) and (4.5.41) we obtain

$$e^{zA(t)} = \sum_{n=0}^{\infty}\sum_{j=0}^{n} C(n, j; A)\frac{t^n}{n!}z^j, \tag{4.5.48}$$

$$e^{A(t)} = \sum_{n=0}^{\infty} C(n; A)\frac{t^n}{n!}. \tag{4.5.49}$$

Hence, a simple rule for obtaining generating functions of the form (4.5.38)–(4.5.41) for A-substitutions of bounded height follows. To this end, one should substitute $xa_h(t)$ or $a_h(t)$, respectively, for t in the generating functions (4.5.48) and (4.5.49), where $a_h(t)$ is the generating function for rooted trees of height not exceeding h. This approach to the construction of the generating functions for mappings with constraints can be extended to the general case where the properties of the mappings are expressed in terms of constraints on the structure of the trees of the corresponding graphs (Stepanov, 1969).

5
The general combinatorial scheme

Beginning with Leibniz, a number of mathematicians have pointed out the fundamental importance of the distinguishability of objects in combinatorial problems. A rigorous definition of this notion requires certain formal notions which allow us to express the indistinguishability of objects in mathematical terms.

Consider as an example the formalization of the notion of indistinguishability in the well-known combinatorial scheme of allocation of particles to cells, usually referred as the urn model. There are m particles and n cells of unlimited capacity. The first problem consists of finding the number of possible allocations of the particles provided the allocations satisfy some restrictions. Obviously, in combinatorial problems of such a kind it is necessary to know which allocations are distinct and which ones are indistinguishable. First of all, the answer depends on whether the cells and the particles are distinguishable.

In Chapter 1 for such a formalization of combinatorial models we used the notion of a mapping of a finite set X into a set Y and introduced the notion of a configuration. With that approach the allocations of m particles into n cells were put into a one-to-one correspondence with configurations $\varphi : X \to Y$, $|X| = m$, $|Y| = n$, and the distinguishability of allocations was defined in terms of the corresponding configurations. In this chapter we use another approach to the formalization of combinatorial schemes of the type considered, which is also based on the notion of a mapping. Let us illustrate the essence of the approach using the urn model again.

As the basic model we consider the scheme of allocation of m distinct particles to n different cells. In this case each allocation is determined by a mapping $\varphi : X \to Y$, $|X| = m$, $|Y| = n$, without any additional restrictions on the mapping. To determine the scheme with different cells and indistinguishable particles, we can introduce an equivalence relation on the set of all allocations corresponding to the basic model. Under this equivalence relation two allocations are equivalent if they differ only in the numbering of the cells. This equivalence relation induces the corresponding equivalence relation on the set of mappings. Each of the distinct allocations is in correspondence with some equivalence class and, consequently, the number of such allocations coincides with the number of elements in the factor set of the set of allocations. A similar approach can be used for the construction of models with indistinguishable cells and distinct particles, or with indistinguishable particles and cells. As a result we have four types of urn model, each of which finds use in applied combinatorial problems. To introduce the equivalence relations, we can use, as in Pólya enumeration theory, groups of substitutions G and H acting on the sets X and Y respectively. The construction of the four above-mentioned models of the urn scheme can be realized if we take as G and H either the identity or symmetric groups of appropriate degree.

This approach is applicable not only to the formalization of urn models but also to a number of other combinatorial schemes. In particular, it can be used in the classification of schemes with respect to enumeration problems arising in the schemes. It is clear that the process of formalization itself involves two steps: the choice of the sets X and Y and the determination of the permutation groups G and H corresponding to the required notion of distinguishability in a given problem. Two combinatorial schemes are considered identical if they can be formalized with the same collection of X, Y, G and H.

The next step in the formalization of the combinatorial problem consists of constructing the generating function for the enumeration of distinguishable objects with respect to some characteristic that is often called a weight. The general method of such a construction is provided by the Pólya enumeration theory presented in Chapter 6. In this method it is supposed that the corresponding generating function F for the basic model is known. The required generating function Φ for the enumeration of the distinguishable objects corresponding to the elements of the factor set is represented in terms of the function F as a polynomial of several variables which depends on the groups G and H. The application of this approach leads to cumbersome expressions, even in some typical situations. Therefore, it is natural to separate several simple cases in terms of the groups G and H, and to present the method of finding the generating functions for the enumeration of objects possessing certain characteristics that are common in combinatorial problems. Such typical characteristics are the so-called primary and secondary specifications of the corresponding mappings. These specifications can be thought of as determining in some sense the composition of those elements that are the images of the mapping, taking into account repetitions. Let each of the groups G and H be either the identity group or the symmetric group of appropriate degree. Using the primary and secondary specifications as the characteristics, we can give a method for constructing the generating functions for the enumeration of non-equivalent mappings in each of the four possible cases under various restrictions on these specifications. The described formalization of certain classes of combinatorial problems, together with the method for finding the corresponding generating functions, is called the general combinatorial scheme.

This chapter is devoted to investigations of the general combinatorial scheme which lets us unify the study of various well-known combinatorial models. Some of the methods for constructing generating functions in the general combinatorial scheme have been previously applied in particular combinatorial models and can be found in Riordan's book (Riordan, 1958). In this chapter, considerable attention is given to asymptotic formulae for those problems which represent particular cases of the general combinatorial scheme. The most important asymptotic results concern the Hardy–Ramanujan formula for the number of decompositions of a natural number into summands, the asymptotics of the Bell numbers which are the numbers of partitions of a finite set into blocks, and the asymptotic formula for the number of m-samples in a commutative asymmetric n-basis with restrictions on the primary specifications.

5.1 Definition of the general combinatorial scheme

5.1.1 Equivalence of configurations

In Chapter 1 we defined a configuration as a single-valued mapping $\varphi : X \to Y$, satisfying some conditions Λ, where $X = \{1, \ldots, m\}$ and Y is a finite set. In what follows we consider configurations σ without any restrictions, that is, configurations with $\Lambda = \varnothing$.

Let σ be a configuration which is a mapping of the set $X = \{1, \ldots, m\}$ into a set $A = \{a_1, \ldots, a_n\}$. The configuration σ can be represented in the form

$$\sigma = \begin{pmatrix} 1 & 2 & \ldots & m \\ \sigma(1) & \sigma(2) & \ldots & \sigma(m) \end{pmatrix} = \begin{pmatrix} 1 & 2 & \ldots & m \\ a_{i_1} & a_{i_2} & \ldots & a_{i_m} \end{pmatrix},$$

where $\sigma(j) = a_{i_j}$, $j = 1, \ldots, m$. It is sometimes convenient to represent σ as the m-dimensional vector $\sigma = (\sigma(1), \ldots, \sigma(m))$ or $\sigma = (a_{i_1}, \ldots, a_{i_m})$.

We now define two interrelated characteristics of the configuration σ which are called the primary and secondary specifications.

The expression

$$[\sigma] = [a_1^{\alpha_1} a_2^{\alpha_2} \ldots a_n^{\alpha_n}], \qquad \alpha_1 + \alpha_2 + \cdots + \alpha_n = m,$$

is called the primary specification of the configuration σ if the vector $(a_{i_1}, \ldots, a_{i_m})$ has exactly α_j elements a_j from the set A, $j = 1, \ldots, n$. The number α_j is the a_j-index of the primary specification σ.

If the set A is transformed in such a way that the element a_j produces α_j identical elements, where $j = 1, \ldots, n$ and $\alpha_1 + \cdots + \alpha_n = m$, then we obtain a multiset A' of the images of the mapping $\sigma : X \to A$. The composition of the multiset A' is uniquely determined by the vector $(\alpha_1 \ldots, \alpha_n)$, $\alpha_1 + \cdots + \alpha_n = m$. We say that the multiset A' has the primary specification $[a_1^{\alpha_1} \ldots a_n^{\alpha_n}]$. Since the vector $(a_{i_1}, \ldots, a_{i_m})$ corresponding to the configuration σ can be considered as the ordered multiset of the images of the mapping σ, the primary specifications of σ and A' coincide. Therefore, in what follows we do not differentiate the primary specifications of the mapping itself and of the multiset of the images of the mapping.

If, among the numbers $\alpha_1, \ldots, \alpha_n$ which determine the primary specification of a configuration σ, the number 0 occurs β_0 times, the number 1 occurs β_1 times, the number 2 occurs β_2 times, and so on, then the expression

$$[[\sigma]] = [[0^{\beta_0} 1^{\beta_1} 2^{\beta_2} \ldots m^{\beta_m}]],$$

where

$$\beta_0 + \beta_1 + \cdots + \beta_m = n, \qquad \beta_1 + 2\beta_2 + \cdots + m\beta_m = m,$$

is called the secondary specification of the configuration σ. The number β_j is the j-index of the secondary specification, $j = 0, 1, \ldots, m$. We sometimes omit the 0-index of the secondary specification and write

$$[[1^{\beta_1} 2^{\beta_2} \ldots m^{\beta_m}]],$$

where

$$\beta_0 + \beta_1 + \cdots + \beta_m = n, \qquad \beta_1 + 2\beta_2 + \cdots + m\beta_m = m.$$

Similarly, we say that the the multiset A' of the images of the mapping σ has the secondary specification

$$[[0^{\beta_0} 1^{\beta_1} 2^{\beta_2} \ldots m^{\beta_m}]],$$

with $\beta_0 + \cdots + \beta_m = n$ and $\beta_1 + 2\beta_2 + \cdots + m\beta_m = m$ if the number of elements of A appearing in A' exactly i times is equal to β_i, $i = 0, 1, \ldots, m$.

The set of all mappings $\sigma : X \to A$ will be denoted by A^X. It is clear that there exists a bijective correspondence between A^X and the mth Cartesian power $A^{(m)}$.

Let G be a group of substitutions of degree m acting on the set X. Let us introduce an operation, called the action from the left of a substitution $g \in G$ on a configuration $\sigma \in A^X$, whose result is some configuration $\sigma' \in A^X$. We denote this operation by $\circ$ and assume that

$$g \circ \sigma = \sigma', \qquad g \in G, \tag{5.1.1}$$

if the equality

$$(\sigma(g(1)), \ldots, \sigma(g(m))) = (\sigma'(1), \ldots, \sigma(m)), \tag{5.1.2}$$

where $g(i)$ is the image of i under the substitution g, $i = 1, \ldots, m$, holds for the vectors corresponding to the configurations σ and σ'. It is not difficult to see that if configurations σ and σ' are related by the equality $g \circ \sigma = \sigma'$, then the corresponding vectors differ only by a permutation of the coordinates determined by the substitution g. Hence the operation, introduced above, of action from the left of a substitution on a configuration preserves the primary specification of the configuration.

A group of substitutions induces an equivalence relation on the set of configurations A^X which will be referred to as the G-equivalence. Indeed, put $\sigma \overset{G}{\sim} \sigma'$ if there exists a substitution $g \in G$ such that $g \circ \sigma = \sigma'$. The binary relation $\overset{G}{\sim}$ is reflexive, symmetric and transitive, and is consequently an equivalence relation. Now find the number of elements in the G-equivalence class containing a given configuration. The set of substitutions g, $g \in G$, satisfying the condition $g \circ \sigma = \sigma'$ is called the G-stabilizer of σ. It is clear that the G-stabilizer of σ forms a subgroup G_σ of the group G. We suppose here that

$$(gg') \circ \sigma = g \circ (g' \circ \sigma),$$

where $g, g' \in G$, $\sigma \in A^X$. Consider the decomposition of G into residue classes with respect to the subgroup G_σ

$$G = G_\sigma \cup g_1 G_\sigma \cup \cdots \cup g_{l-1} G_\sigma,$$

where the number of residue classes l is equal to the index of the G-stabilizer G_σ in the group G. There exists a one-to-one correspondence between the residue classes $g_i G_\sigma$, $i = 1, \ldots, l-1$, and those configurations which are the images of the configuration σ under the action of the permutations from the residue classes. Hence it follows that the number of distinct elements in the G-equivalence class $K_\sigma(G)$ which contains the configuration σ is equal to

$$|K\sigma(G)| = \frac{|G|}{|G_\sigma|}. \tag{5.1.3}$$

For example, if $G = S_m$, where S_m is the group of all substitutions of degree m acting on the set X, and

$$[\sigma] = [a_1^{\alpha_1} a_2^{\alpha_2} \ldots a_n^{\alpha_n}], \qquad \alpha_1 + \cdots + \alpha_n = m,$$

then $|G_\sigma| = \alpha_1! \cdots \alpha_n!$ and

$$|K_\sigma(S_m)| = \frac{m!}{\alpha_1! \cdots \alpha_n!}. \tag{5.1.4}$$

This corresponds to the fact that two configurations belong to the same equivalence class if and only if their primary specifications coincide.

Now consider a group of substitutions H acting on the set A. Define an operation called action from the right of a substitution $h \in H$ on a configuration $\sigma \in A^X$. We denote this operation by $*$ and assume that configurations $\sigma, \sigma' \in A^X$ are related by the equality

$$\sigma * h = \sigma', \qquad h \in H, \tag{5.1.5}$$

if the equality

$$(h(\sigma(1)), \ldots, h(\sigma(m))) = (\sigma'(1), \ldots, \sigma'(m)) \tag{5.1.6}$$

holds. Here we assume that $h(a_i)$ is the image of $a_i \in A$, $i = 1, \ldots, m$, under the substitution h. The action of a substitution $h \in H$ on a configuration σ preserves the secondary specification of σ, that is, if $\sigma' = \sigma * h$, then $[[\sigma']] = [[\sigma]]$.

A group H of substitutions induces an equivalence relation on the set of configurations A^X. Indeed, define on the set A^X a binary relation putting $\sigma \overset{H}{\sim} \sigma'$ if there exists a substitution $h \in H$ such that $\sigma * h = \sigma'$. This binary relation is reflexive, symmetric and transitive. Thus it is an equivalence relation, which will be called the H-equivalence. If two configurations belong to the same H-equivalence class, then the corresponding vectors can be transformed one to another by some substitution of the set A. Hence, the configurations contained in the same H-equivalence class have the same secondary specification.

Now find a formula for the number $K_\sigma(H)$ of elements in the H-equivalence class containing σ. The H-stabilizer of a configuration $\sigma \in A^X$ is the set H_σ of substitutions $h \in H$ such that $\sigma * h = \sigma'$. It is clear that the H-stabilizer H_σ is a subgroup of H and therefore, as in the case of G-equivalence, we obtain

$$|K_\sigma(H)| = \frac{|H|}{|H_\sigma|}. \tag{5.1.7}$$

If $H = S_n$, where S_n is the symmetric group of degree n acting on the set A and

$$[[\sigma]] = [[0^{\beta_0}1^{\beta_1}\ldots m^{\beta}]], \qquad \beta_1 + 2\beta_2 + \cdots + m\beta_m = m,$$

then $|H_\sigma| = \beta_0!$. Indeed, none of the $\beta_0!$ substitutions of the elements of A which are not contained in σ change the configuration σ. On the other hand, any substitution of elements of A which appear in σ changes σ. Thus in this case

$$|K_\sigma(S_n)| = \frac{n!}{\beta_0!}. \tag{5.1.8}$$

Let us now introduce the operation of joint action of a substitution $g \in G$ from the left and a substitution $h \in H$ from the right on a configuration $\sigma \in A^X$. Obviously, the order of the action is irrelevant and leads to the same result in both the cases. We assume that

$$g \circ \sigma * h = \sigma' \tag{5.1.9}$$

if the vector corresponding to the configuration σ' is related to the vector corresponding to the configuration σ by the equality

$$(\sigma'(1), \ldots, \sigma'(m)) = (h(\sigma(g(1))), \ldots, h(\sigma(g(m)))). \tag{5.1.10}$$

Now consider the equivalence relation induced on the set of configurations A^X by the groups of substitutions G and H acting on the sets X and A respectively.

Define a binary relation on the set A^X assuming that $\sigma \sim \sigma'$ if there exist substitutions $g \in G$ and $h \in H$ such that $g \circ \sigma * h = \sigma'$. This binary relation is an equivalence relation called GH-equivalence. It is clear that if H is the identity group, then GH-equivalence coincides with G-equivalence, and if G is the identity group, then GH-equivalence coincides with H-equivalence.

Now consider the connections between G-equivalence, H-equivalence and GH-equivalence. To this end we introduce the notion of composition of binary relations. Let R and R' be binary relations on a set Z. The composition of the binary relations R and R' is the binary relation $R \otimes R'$ defined as follows: $(z, z') \in R \otimes R'$ if there exists an element $z'' \in Z$ such that $(z, z'') \in R$ and $(z'', z') \in R'$. Note that the operation $\otimes$ can be considered as a composition law on the set of binary relations on the set Z. This law is associative, so that the set of binary relations with the operation $\otimes$ forms a subgroup. Now put $Z = A^X$ and denote the G-equivalence, H-equivalence and GH-equivalence relations on the set A^X by R_G, R_H and R_{GH} respectively. Let $(\sigma, \sigma') \in R_{GH}$, where $\sigma, \sigma' \in A^X$. This means that there exists a configuration $\sigma'' \in A^X$ such that $(\sigma, \sigma'') \in R_G$ and $(\sigma'', \sigma') \in R_H$. Here σ'' may coincide with σ or σ'. Thus,

$$R_{GH} = R_G \otimes R_H, \tag{5.1.11}$$

that is, the GH-equivalence relation is the composition of the G-equivalence and H-equivalence relations.

By the same reasoning we obtain

$$R_{GH} = R_H \otimes R_G, \tag{5.1.12}$$

that is, the operation $\otimes$ is commutative. Note that this property is a consequence of the fact that the order of fulfilment of the operations $\circ$ and $*$ is irrelevant for each configuration $\sigma \in A^X$. It should be noted that commutativity is a corollary of the following general theorem (Kurosh, 1973).

Theorem 5.1.1 *The composition $R \otimes R'$ of equivalence relations R and R' is an equivalence relation if and only if these relations are permutable, that is, if and only if $R \otimes R' = R' \otimes R$.*

We now describe the GH-equivalence class $K_\sigma(G, H)$ containing $\sigma \in A^X$. Let

$$K_\sigma(G) = \{\sigma, \sigma_1, \ldots, \sigma_{l-1}\}, \qquad K_\sigma(H) = \{\sigma, \sigma'_1, \ldots, \sigma'_{s-1}\}$$

be the G-equivalence and H-equivalence classes, respectively, containing the element $\sigma \in A^X$. Then

$$\begin{aligned} K_\sigma(G, H) &= K_\sigma(G) \cup K_{\sigma'_1}(G) \cup \cdots \cup K_{\sigma'_{s-1}} \\ &= K_\sigma(H) \cup K_{\sigma_1}(H) \cup \cdots \cup K_{\sigma_{l-1}}. \end{aligned} \tag{5.1.13}$$

This equality means that the class $K_\sigma(G, H)$ is a union of some G-equivalence or H-equivalence classes. It is not difficult to show that, generally speaking, these unions are not partitions.

For given groups of substitutions G and H acting on the sets X and

A, respectively, we construct a group of substitutions H^G acting on the set A^X. The action of an element $(g;h) \in H^G$ on $\sigma \in A^X$ is defined as follows: $(g;h)(\sigma) = \sigma'$ if and only if $g \circ \sigma * h = \sigma'$. The group H^G is usually called the power group. The degree of the power group is equal to $|A|^{|X|}$ and the order is $|G||H|$.

Denote the stabilizer of an element $\sigma \in A^X$ in the group H^G by

$$H_\sigma^G = \{(g;h) : (g;h)(\sigma) = \sigma, \ g \in G, \ h \in H\}.$$

In the same way that equality (5.1.3) was obtained, we find the number of distinct elements in the GH-equivalence class $K_\sigma(G,H)$ containing the element σ is equal to

$$|K_\sigma(G,H)| = \frac{|G||H|}{|H_\sigma^G|}. \tag{5.1.14}$$

5.1.2 The number of equivalence classes

The problem of finding the number $N(G)$ of the G-equivalence classes generated by a group of substitutions G acting on a set X can be solved by Burnside's lemma. Below we prove a theorem which lets us find the number $N_k(G)$ of G-equivalence classes consisting of k elements. Burnside's lemma will be deduced from this theorem.

Let P be a partially ordered set (poset) with the order relation $\preccurlyeq$. An element t' is called the lower bound of P if $t' \preccurlyeq x$ for any $x \in P$. The lower bound is also called the zero element. Similarly, we can introduce the notion of the upper bound of P; the upper bound is usually called the unity element. The lower and upper bounds are unique if they exist. An interval $[x,y]$ is the set of elements z such that $x \preccurlyeq z \preccurlyeq y$. A poset P is locally finite if the number of elements in any interval $[x,y]$ is finite. Consider the class of real-valued functions $f(x,y)$ defined for all $x, y \in P$, where P is a locally finite poset. Let $f(x,y) = 0$ if the relation $x \preccurlyeq y$ is not valid. Define addition of the functions and multiplication by a scalar in the usual way. The product of functions is defined by the relation

$$h(x,y) = \sum_{x \preccurlyeq z \preccurlyeq y} f(x,z)g(z,y).$$

Multiplication of functions is associative and distributive. The unity element with respect to multiplication is the usual Kronecker's function $\delta(x,y)$. The set of the functions with the two composition laws, addition and multiplication, forms the algebra $I(P)$, which is called the incidence algebra. It can be shown that a function $f(x,y) \in I(P)$ has both left and right inverse elements with respect to multiplication if and only if $f(x, x) \neq 0$ for any $x \in P$. These inverse elements coincide.

In $I(P)$ we define the zeta function $\zeta(x, y)$ putting $\zeta(x, y) = 1$ if $x \preccurlyeq y$ and $\zeta(x, y) = 0$ otherwise. The function $\zeta(x, y)$ has the inverse function with respect to multiplication. The inverse function is usually denoted by $\mu(x, y)$ and is called the Möbius function in the algebra $I(P)$. It is not difficult to see that $\mu(x, x) = 1$ and $\mu(x, y) = 0$ if the relation $x \preccurlyeq y$ is not valid. In addition,

$$\sum_{x \preccurlyeq z \preccurlyeq y} \mu(x, z) = 0, \qquad x \prec y.$$

Using the function $\mu(x, y)$ we can prove the following useful inversion formula.

Theorem 5.1.2 *Let P be a locally finite poset with zero element; let a function $F(x)$ be defined for all $x \in P$; and let $g(x)$ be defined by the formula*

$$g(x) = \sum_{y \preccurlyeq x} f(y), \qquad x \in P.$$

Then

$$f(x) = \sum_{y \preccurlyeq x} g(x)\mu(y, x), \qquad x \in P,$$

where $\mu(y, z)$ is the Möbius function of the poset P.

Proof Consider the sum

$$W = \sum_{y \preccurlyeq x} g(y)\mu(y, x) = \sum_{y \preccurlyeq x} \sum_{z \preccurlyeq y} f(z)\mu(y, x).$$

Changing the order of summation, we obtain

$$W = \sum_{y \preccurlyeq x} \sum_{z} f(z)\zeta(z, y)\mu(y, x)$$
$$= \sum_{z} f(z) \sum_{z \preccurlyeq y \preccurlyeq x} \zeta(z, y)\mu(y, x) = \sum_{z} f(z)\delta(z, x).$$

Hence it follows that $W = f(x)$, and the theorem is proved. □

Let G be a finite group of permutations acting on a non-empty set X. The set of all subgroups of the finite group G is a locally finite poset if $H \preccurlyeq L$, where H and L are subgroups of G, means that $H \subseteq L$. For $L \subseteq G$ consider the number F_L defined by the equality

$$F_L = |\{x : g(x) = x, \ x \in X \text{ for all } g \in L\}|.$$

Theorem 5.1.3 (Klass (1976)) *The number of G-equivalence classes of X consisting of k elements is equal to*

$$N_k(G) = \frac{1}{k} \sum_{H \preccurlyeq G:\ [G:H]=k} \sum_{L \preccurlyeq G} \mu(H, L) f_L,$$

where $[G : H]$ *is the index of the subgroup H in the group G.*

Proof Define functions $\delta_{L,x}$ and $\tilde{\delta}_{L,H}$ putting

$$\delta_{L,x} = \begin{cases} 1, & \text{if } g(x) = x \text{ for all } g \in L, \\ 0 & \text{otherwise,} \end{cases}$$

$$\tilde{\delta}_{L,H} = \begin{cases} 1, & \text{if } H = L, \\ 0 & \text{otherwise.} \end{cases}$$

Put

$$G(x) = \{g : g(x) = x,\ g \in G\}.$$

From the definition of $N_k(G)$ it follows that

$$kN_k(G) = |\{x : [G : G(x)] = k,\ x \in X\}|.$$

On the other hand,

$$\begin{aligned} A &= \sum_{H \preccurlyeq G:\ [G:H]=k} \sum_{L \preccurlyeq G} \mu(H, L) F_L \\ &= \sum_{H \preccurlyeq G:\ [G:H]=k} \sum_{H \preccurlyeq L \preccurlyeq G} \mu(H, L) \sum_{x \in X} \delta_{L,x}. \end{aligned}$$

Changing the order of summation, we obtain

$$A = \sum_{x \in X} \sum_{H \preccurlyeq G:\ [G:H]=k} \sum_{H \preccurlyeq L \preccurlyeq G} \mu(H, L) \delta_{L,x}.$$

Note that

$$\sum_{H \preccurlyeq L \preccurlyeq G} \mu(H, L) \delta_{L,x} = \sum_{H \preccurlyeq L \preccurlyeq G(x)} \mu(H, L) = \tilde{\delta}_{H,G(x)}.$$

Thus,

$$A = \sum_{x \in X} \sum_{H \preccurlyeq G:\ [G:H]=k} \tilde{\delta}_{H,G(x)} = |\{x : [G : G(x)] = k,\ x \in X\}|.$$

The theorem is thus proved. □

Now define a function $\tilde{\varphi}$ on the subgroups of a finite group G as follows: $\tilde{\varphi}(H) = \varphi(|H|)$ if H is a cyclic subgroup of G and $\tilde{\varphi}(H) = 0$ otherwise, where $\varphi(k)$ is the ordinary Euler totient function equal to the number of j such that j and k are relatively prime and $1 \leqslant j \leqslant k$.

Lemma 5.1.1 *For any subgroup L of a finite group G,*

$$|L| = \sum_{H \preccurlyeq L} \tilde{\varphi}(H).$$

Proof Any cyclic subgroup H has $\varphi(|H|)$ generators and the systems of generators of different subgroups have no common elements. Thus

$$|L| \geqslant \sum_{H \preccurlyeq L} \tilde{\varphi}(H).$$

On the other hand, each element $g \in L$ generates a cyclic subgroup of L. This implies the inverse inequality. The lemma is thus proved. □

The following assertion is a direct corollary of Lemma 5.1.1 and Theorem 5.1.2.

Lemma 5.1.2 *Let L be a subgroup of a finite group G. Then*

$$\tilde{\varphi}(L) = \sum_{H \preccurlyeq L} |H| \mu(H, L).$$

Lemma 5.1.3 (Burnside's lemma) *Let G be a finite group of permutations acting on a non-empty set X. The number $N(G)$ of G-equivalence classes generated by G on X is equal to*

$$N(G) = \frac{1}{|G|} \sum_{g \in G} F_g,$$

where

$$F_g = |\{x : g(x) = x, \ x \in X\}|.$$

Proof By Theorem 5.1.3

$$N(G) = \sum_{k=0}^{\infty} N_k(G) = \sum_{k=1}^{\infty} \sum_{H \preccurlyeq G \,:\, [G:H]=k} \frac{|H|}{|G|} \sum_{L \preccurlyeq G} \mu(H, L) F_L.$$

Hence it follows that

$$N(G) = \frac{1}{|G|} \sum_{H \preccurlyeq G} |H| \sum_{L \preccurlyeq G} \mu(H,L) F_L$$
$$= \frac{1}{|G|} \sum_{L \preccurlyeq G} F_L \sum_{H \preccurlyeq G} |H| \mu(H,L).$$

It follows from Lemma 5.1.2 that

$$N(G) = \frac{1}{|G|} \sum_{L \preccurlyeq G} F_L \tilde{\varphi}(L).$$

By definition of the function $\tilde{\varphi}(L)$,

$$N(G) = \frac{1}{|G|} \sum_{L_C \preccurlyeq G} F_{L_C} \varphi(|L_C|),$$

where the summation is taken over the cyclic subgroups L_C of the group G. Any cyclic subgroup L_C has $\varphi(|L_C|)$ elements; therefore

$$N(G) = \frac{1}{|G|} \sum_{g \in G} F_{L_C(g)},$$

where $L_C(g)$ is the cyclic subgroup with the generator $g \in G$. The lemma is proved since $F_g = F_{L_C(g)}$. □

Burnside's lemma plays an important role in the representation of Pólya enumeration theory. Therefore, in Chapter 6 we give a simple direct proof of this lemma.

Now we use Burnside's lemma for the proof of a theorem which gives a formula for the number of GH-equivalence classes on the set of configurations A^X.

Theorem 5.1.4 *Let $j_k(g)$ and $j_k(h)$ be the numbers of cycles of length k in permutations $g \in G$ and $h \in H$ respectively. Then the number of GH-equivalence classes on the set A^X is equal to*

$$N(G,H) = \frac{1}{|G||H|} \sum_{g \in G} \sum_{h \in H} \prod_{k=1}^{|X|} \left(\sum_{s|k} s j_s(h) \right)^{j_k(g)},$$

where the last summation is over all divisors of k.

Proof According to Burnside's lemma

$$N(G,H) = \frac{1}{|G||H|} \sum_{g \in G} \sum_{h \in H} \sum_{\sigma \,:\, g \circ \sigma * h = \sigma} 1,$$

where $\sigma \in A^X$. Let the transitivity domain of a substitution $g \in G$ containing an element $x \in X$ consist of the elements $x, g(x), \ldots, g^{k-1}(x)$. If the equality $g \circ \sigma * h = \sigma$ holds, then these elements are transformed by $\sigma \in A^X$ into $\sigma(x), h^{-1}(\sigma(x)), \ldots, h^{-k+1}(\sigma(x))$. Thus the size of the transitivity domain of the substitution $h \in H$ containing the element $\sigma(x)$ is a divisor of k. The number of ways of choosing the image of a fixed element $x \in X$ under a mapping $\sigma \in A^X$ is equal to

$$\sum_{s|k} s j_s(h),$$

where we assume that the equality $g \circ \sigma * h = \sigma$ holds and that x belongs to the transitivity domain of the substitution $g \in G$ which consists of k elements. The choice of images of different elements $x \in X$ is carried out independently; therefore

$$\sum_{\sigma\,:\, g\circ\sigma*h=\sigma} 1 = \prod_{k=1}^{|X|} \left(\sum_{s|k} s j_s(h) \right)^{j_k(g)} .$$

The theorem is proved. □

5.1.3 The general combinatorial scheme and its particular cases

Choose an arbitrary representative from each GH-equivalence class. A set of representatives of the GH-equivalence classes is called a system of representatives. In solving the considered class of combinatorial problems, the construction of a system of representatives corresponds to the first step, called the step of formalization.

The next step concerns the construction of generating functions for enumerating the system of representatives according to some characteristics. For arbitrary groups G and H the methods for constructing the generating functions are not simple and often lead to unobservable results. These methods will be presented in the next chapter. In this chapter we restrict ourselves to four cases only. These cases correspond to two possibilities for the group G acting on the set X and two possibilities for the group H acting on the set A. Namely, we assume that either $G = \{e\}$, where e is the identity substitution of degree m, or $G = S_m$, where S_m is the symmetric group of degree m. Similarly, we assume that either $H = \{e'\}$, where e' is the identity substitution of degree n, or $H = S_n$, where $n = |A|$. For these cases there are rather simple methods for constructing the generating function if the primary or secondary specifications are taken as the characteristics of the representatives. The collection of these four cases of formalization and the methods of finding the generating functions will be called the general combinatorial scheme. A choice of the groups G and H determines a particular case of the general combinatorial scheme, which allows us to consider from a unified point of view a number of combinatorial problems which have previously been treated independently.

To describe the particular cases of the general combinatorial scheme, it is convenient to introduce special terminology. The set A with n elements will be called an n-basis. The elements of a system of representatives will be referred to as an m-sample. If $G = \{e\}$ and $H = \{e'\}$, then an m-sample is said to be an m-sample in the non-commutative asymmetric n-basis. If $G = S_m$ and $H = \{e'\}$, then we have an m-sample in the commutative asymmetric n-basis. The case where $G = \{e\}$ and $H = S_n$ gives an m-sample in the non-commutative symmetric n-basis. Finally, the case where $G = S_m$ and $H = S_n$ corresponds to an m-sample in the commutative symmetric n-basis. Thus the combinations of two of these four attributes, namely, combinations of commutative or non-commutative with symmetric or asymmetric, added to an n-basis, completely characterize m-samples and all the particular cases of the general combinatorial scheme considered.

We now turn to consideration of the four particular cases of the general combinatorial scheme.

5.1.4 Non-commutative asymmetric n-basis

This particular case is determined by the choice $G = \{e\}$ and $H = \{e'\}$, where e and e' are the identity substitutions of degree m and n respectively. In this case, each GH-equivalence class consists of one element and the system of representatives coincides with the set A^X. Each element of the system of representatives will be called an m-sample in the non-commutative asymmetric n-basis.

Consider combinatorial schemes which are covered by this particular case of the general combinatorial scheme.

(a) Allocations Each allocation of size m of n distinct objects with unlimited repetitions can be put into a one-to-one correspondence with an element of A^X and therefore with an m-sample in the non-commutative asymmetric n-basis. An allocation of size m of n distinct objects without repetitions corresponds to one, and only one, element from A^X with the condition of injectivity.

(b) Allocations of distinct particles to different cells Let m distinct particles and n different cells be given. Label the m particles with elements of the set X. Each allocation of the particles to cells of unlimited capacity can be put in a one-to-one correspondence with an element of A^X and therefore with an m-sample in the non-commutative asymmetric n-basis. Each allocation without empty cells corresponds to an m-sample in the same basis with the condition of surjectivity. Many other properties of allocations can be transferred to properties of m-samples in a similar way. The scheme of allocation of distinct particles to different cells is directly related to Maxwell–Boltzmann statistics in physics which describes the behavior of particles in the corresponding phase space.

5.1.5 Commutative asymmetric n-basis

This particular case of the general combinatorial scheme is distinguished by the conditions $G = S_m$, $H = \{e'\}$, where S_m is the symmetric group of degree m. In this case those configurations from A^X which differ by permutations of images constitute a GH-equivalence class. In other words, two configurations from A^X belong to the same equivalence class if and only if their primary specifications coincide. This means that the primary specification $[a_1^{\alpha_1} \ldots a_n^{\alpha_n}]$, $\alpha_1 + \cdots + \alpha_n = m$, of a representative of an equivalence class uniquely determines this class. In this case an element of a system of representatives is called an m-sample in the commutative asymmetric n-basis. Let us list some combinatorial schemes which can be described by this particular case of the general combinatorial scheme.

(a) Combinations Each combination of size m of n distinct objects with unlimited repetitions can be put in a one-to-one correspondence with an equivalence class of elements from A^X with the corresponding primary specification and, therefore, an m-sample in the commutative asymmetric n-basis. A combination of size m from n distinct objects without repetitions corresponds to an m-sample in the commutative asymmetric n-basis with the condition of injectivity.

(b) Allocation of identical particles to different cells Divide the set of all allocations of m distinct particles to n different cells into equivalence classes. Put into the same class those allocations which differ only by their labels but not by the number of particles in each of the cells. The equivalence classes obtained thereby and the S_m-equivalence classes of A^X can be put in a one-to-one correspondence. Therefore there exists a one-to-one correspondence between the allocations of m identical particles to n different cells and the m-samples in the commutative asymmetric n-basis. Each allocation without empty cells corresponds to one and only one m-sample in this n-basis with the condition of surjectivity.

It should be noted that Bose–Einstein statistics considered in physics is determined by allocations of indistinguishable particles into different cells and is consequently included in the combinatorial scheme of the allocation of identical particles to different cells. If each cell can accept only one particle, then the scheme leads to Fermi–Dirac statistics.

(c) Composition of an integer into summands Consider the problem of solving the equation

$$\alpha_1 + \cdots + \alpha_n = m, \tag{5.1.15}$$

for non-negative integers α_j, $j = 1, \ldots, n$. Any primary specification $[a_1^{\alpha_1} \ldots a_n^{\alpha_n}]$, $\alpha_1 + \cdots + \alpha_n = m$, uniquely determines an equivalence class; therefore each m-sample in the commutative asymmetric n-basis can be put in a one-to-one correspondence with a solution of equation (5.1.15). Each m-sample in the commutative asymmetric n-basis with the additional condition of surjectivity can be put in a one-to-one correspondence with a solution of the equation

$$\alpha_1 + \cdots + \alpha_n = m \tag{5.1.16}$$

for positive integers α_j, $j = 1, \ldots, n$. A solution of equation (5.1.16) is called a composition of m into n summands.

5.1.6 Non-commutative symmetric n-basis

Let $G = \{e\}$ be the identity group and $H = S_n$, where S_n is the symmetric group of degree n. In this case two configurations from A^X belong to the same GH-equivalence class if one of them is transformed into the other by the action of some substitution $h \in H$ on the elements of the n-basis. It is clear that the secondary specifications of the configurations from the same equivalence class coincide. In this case an element of a system of representatives will be called an m-sample in the non-commutative symmetric n-basis. Let us give examples of the combinatorial models which can be included in this particular case of the general combinatorial scheme.

(a) Allocation of distinct particles into identical cells Suppose we are given m distinct particles and n identical cells of unlimited capacity. Each allocation of the particles to the cells can be put in a one-to-one correspondence with an m-sample in the non-commutative symmetric n-basis. Indeed, each m-sample in the non-commutative symmetric n-basis is associated with the GH-equivalence class containing the m-samples from A^X which are transformed one into the other by the action of a substitution $h \in H$ on the elements of the n-basis. On the other hand, each allocation of m distinct particles to n identical cells can be put in correspondence with a class of equivalent allocations of m distinct particles to n different cells such that the allocations in the class differ only in the labels of the cells. There exists an obvious one-to-one correspondence between the GH-equivalence classes of m-samples and the classes of equivalent allocations. Hence it follows that there exists a one-to-one correspondence between the allocations of m distinct particles to n identical cells and the m-samples in the non-commutative symmetric n-basis.

(b) Partitions of finite sets Consider a partition of a set X consisting of m elements into n non-empty disjoint subsets

$$X = X_1 \cup \cdots \cup X_n, \quad X_i \cap X_j = \varnothing, \quad i \neq j.$$

Without loss of generality, we assume that $X = \{1, \ldots, n\}$. Under these conditions each allocation of m distinct particles to n identical cells, with the additional condition that there are no empty cells, can be put in a one-to-one correspondence with a partition of the set X into n blocks. If empty blocks are allowed, then the condition that the cells have to be non-empty can be omitted.

(c) Rhymed sequences Denote by N_k the number of the position among the images of $\sigma \in A^X$ where the element $a_k \in A$ appears for the first time. If all the elements of A are among the images of $\sigma \in A^X$ and $N_k < N_{k'}$ for $k < k'$, then σ is called a rhymed sequence. This means that those symbols with lower numbers appear earlier than the symbols with greater numbers.

Each rhymed sequence of length m constructed from n symbols can be put in a one-to-one correspondence with an m-sample in the non-commutative symmetric n-basis with the condition of surjectivity. To this end we can take the elements of A^X with the properties of the rhymed sequence as the representatives of the GH-equivalence classes.

5.1.7 Commutative symmetric n-basis

In this case $G = S_m$ and $H = S_n$. Under these conditions two configurations from A^X belong to the same GH-equivalence class if one of them is transformed into the other either by a permutation of the elements of the m-sample or by a permutation of the elements of the n-basis. Let us show that configurations σ and σ', $\sigma, \sigma' \in A^X$, belong to the same equivalence class if and only if their secondary specifications coincide. Indeed, if $g \circ \sigma * h = \sigma'$, then the action of g preserves the primary, and, consequently, the secondary specification of σ. The action of h preserves the secondary specification. Thus $[[\sigma']] = [[\sigma]]$. Suppose now that

$$[[\sigma]] = [[\sigma']] = [[0^{\beta_0} 1^{\beta_1} \dots m^{\beta_m}]], \qquad \beta_1 + 2\beta_2 + \dots + n\beta_m = m,$$

and elements

$$a_1^{(j)}, \dots, a_{\beta_j}^{(j)}, \qquad \bar{a}_1^{(j)}, \dots, \bar{a}_{\beta_j}^{(j)}$$

appear j times as the images of σ and σ' respectively. Consider the permutation

$$h = \begin{pmatrix} a_1^{(1)} & \dots & a_{\beta_1}^{(1)} & \dots & a_1^{(j)} & \dots & a_{\beta_j}^{(j)} \\ \bar{a}_1^{(1)} & \dots & \bar{a}_{\beta_1}^{(1)} & \dots & \bar{a}_1^{(j)} & \dots & \bar{a}_{\beta_j}^{(j)} \end{pmatrix},$$

such that the mapping of those elements that do not appear as the images of σ, is an arbitrary bijection. It is clear that the configuration $\bar{\sigma} = \sigma * h$ has the same primary specification as σ', that is, $[\bar{\sigma}] = [\sigma']$. Therefore, $\sigma' = g \circ \bar{\sigma}$, that is, $\sigma' = g \circ \sigma * h$. In this case an element of a system of representatives will be called an m-sample in the commutative symmetric n-basis.

Consider examples of combinatorial models covered by this particular case of the general combinatorial scheme.

(a) Allocation of identical particles to identical cells Let m identical particles be allocated to n identical cells of unlimited capacity. Each allocation of the particles to the cells can be put in a one-to-one correspondence with an m-sample in the commutative symmetric n-basis. Indeed, each

m-sample in the commutative symmetric n-basis can be put in correspondence with a GH-equivalence class consisting of the samples from A^X which can be transformed one into the other either by a permutation of their elements or by a permutation of the elements of the basis. On the other hand, any allocation of m identical particles to n identical cells corresponds to a class of equivalent allocations of m distinct particles to different cells such that the allocations within the class differ only by permutations of labels of particles and cells. There exists a one-to-one correspondence between the GH-equivalence classes and the classes of equivalent allocations. This means that there exists a one-to-one correspondence between the m-samples in the commutative symmetric n-basis and the allocations of m identical particles to identical cells.

(b) Partitions of natural numbers A representation of a natural number m as a sum of natural numbers arranged in non-increasing order is called a partition of m, which is uniquely determined by the numbers $\beta_1, \ldots, \beta_m$ such that $\beta_1 + 2\beta_2 + \cdots + m\beta_m = m$, where β_j is the number of summands in the partition equal to j. Obviously, $\beta_1 + \cdots + \beta_m$ is the total number of summands in the partition. Roughly speaking, a partition of m is a representation of m as a sum of natural numbers whose order is irrelevant. This feature makes the partition of a number different from its composition, which depends on the order of summands.

Each partition of m into no more than n summands can be put in a one-to-one correspondence with an m-sample in the commutative symmetric n-basis. This assertion follows from the fact that such an m-sample can be put in a one-to-one correspondence with a secondary specification

$$[[0^{\beta_0} 1^{\beta_1} \ldots m^{\beta_m}]],$$

where

$$\beta_1 + 2\beta_2 + \cdots + \beta_m = m, \qquad \beta_0 + \beta_1 + \cdots + \beta_m = n,$$

which determines a partition of m into no more that n summands.

5.2 Commutative asymmetric n-basis

In this section we give a method for constructing the generating functions for the enumeration of m-samples in the commutative asymmetric n-basis with restrictions on the primary specifications. The method is most transparent in comparison with corresponding methods in the other particular cases of the general combinatorial scheme. This clearity stems from the possibility of putting an m-sample in the commutative asymmetric n-basis in a one-to-one correspondence with exactly one term of a formal power series of several variables called an enumerator.

Let $\Lambda = (\Lambda_1, \ldots, \Lambda_n)$ be an ordered set of sequences $\Lambda_j \subseteq \mathbf{N}_0 = \{0, 1, \ldots\}$, $j = 1, \ldots, n$. Introduce the generating function of $n+1$ variables:

$$\Phi(t; x_1, \ldots, x_n; \Lambda) = \prod_{j=1}^{n} \sum_{\alpha_j \in \Lambda_j} t^{\alpha_j} x_j^{\alpha_j}, \tag{5.2.1}$$

where the summation is over α_j, taking sequential values from Λ_j, $j = 1, \ldots, n$. This generating function is called an enumerator in the commutative asymmetric basis.

Represent the enumerator in the form

$$\Phi(t; x_1, \ldots, x_n; \Lambda) = \sum_{m=0}^{\infty} t^m \sum_{\alpha_1 + \cdots + \alpha_n = m} x_1^{\alpha_1} \cdots x_n^{\alpha_n}, \tag{5.2.2}$$

where the second summation is over all solutions of the equation $\alpha_1 + \cdots + \alpha_n = m$ for non-negative integers such that $\alpha_j \in \Lambda_j$, $j = 1, \ldots, n$, and the second sum is zero for $m = 0$.

It follows from expression (5.2.2) that each term of the enumerator $\Phi(t; x_1, \ldots, x_n; \Lambda)$ of the form $t^m x_1^{\alpha_1} \cdots x_n^{\alpha_n}$ with $\alpha_1 + \cdots + \alpha_n = m$ can be put in a one-to-one correspondence with an m-sample in the commutative asymmetric n-basis with the primary specification $[a_1^{\alpha_1} \ldots a_n^{\alpha_n}]$, provided that the index α_j of the primary specification is an element of the sequence Λ_j, $j = 1, \ldots, n$. Therefore it can be said that $\Phi(t; x_1, \ldots, x_n; \Lambda)$ enumerates the m-samples in the commutative asymmetric n-basis with the indices of the primary specification determined by Λ.

Denote by $C_{nm}(\Lambda)$ the number of m-samples in the commutative asymmetric n-basis with primary specifications of the form $[a_1^{\alpha_1} \ldots a_n^{\alpha_n}]$, where $\alpha_1 + \cdots + \alpha_n = m$, $\alpha_j \in \Lambda_j$, $j = 1 \ldots, n$, and $\Lambda = (\Lambda_1, \ldots, \Lambda_n)$. We can write the equality formally as

$$C_{nm}(\Lambda) = \sum_{\alpha_1 + \cdots + \alpha_n = m} 1,$$

where $\alpha_j \in \Lambda_j$, $j = 1, \ldots, n$. Using this equality and the expression for the enumerator, we see that if

$$\psi_n(t; \Lambda) = \sum_{m=0}^{\infty} C_{nm}(\Lambda) t^m,$$

then

$$\psi_n(t;\Lambda) = \Phi(t;1,\dots,1;\Lambda) = \prod_{j=1}^{n} \sum_{\alpha_j \in \Lambda_j} t^{\alpha_j}. \qquad (5.2.3)$$

This expression for the generating function of $C_{nm}(\Lambda)$ gives an easy method of constructing it for a given $\Lambda = (\Lambda_1,\dots,\Lambda_n)$. We simply write the product of n identical factors of the form

$$(1+t+t^2+\cdots)(1+t+t^2+\cdots)\cdots(1+t+t^2+\cdots),$$

put $t^0 = 1$, and in the jth factor delete the terms t^ν with $\nu \notin \Lambda_j$, $j = 1,\dots,n$. As a result we obtain the required generating function.

Let us consider some particular cases, assuming that t takes real values and $|t| < 1$.

5.2.1 The number of m-samples without restrictions

Let $\Lambda_1 = \cdots = \Lambda_n = \mathbf{N}_0$, that is, there are no restrictions on the indices of the primary specification. Putting $C_{nm} = C_{nm}(\mathbf{N}_0)$, we have

$$\psi_n(t;\mathbf{N}_0) = \sum_{m=0}^{\infty} C_{nm}t^m = (1-t)^{-n}. \qquad (5.2.4)$$

Hence it follows that

$$C_{nm} = \binom{n+m-1}{m}. \qquad (5.2.5)$$

It is clear that C_{nm} is the number of m-samples in the commutative asymmetric n-basis or, what amounts to the same, the number of combinations of size m from n distinct elements with unlimited repetitions. In addition, C_{nm} is equal to the number of allocations of m identical particles into n different cells, and consequently it is equal to the number of solutions of the equation $\alpha_1 + \cdots + \alpha_n = m$ for non-negative integers. All these numbers are determined by formula (5.2.5).

5.2.2 The number of m-samples with restrictions from below

Let $q_1,\dots,q_n$ be non-negative integers and $\Lambda_j = \{q_j, q_j+1,\dots\}$, $j = 1,\dots,n$. Put $q = q_1 + \cdots + q_n$ and find the generating function of the numbers $C_{nm}(\Lambda)$ in this case. It turns out that the numbers $C_{nm}(\Lambda)$

depend only on q and it is natural to denote them by $C_{nm}^{(q)}$. Using the enumerator, we obtain

$$\psi_n(t;q) = \sum_{m=q}^{\infty} C_{nm}^{(q)} t^m = t^q(1-t)^{-n}.$$

Hence

$$C_{nm}^{(q)} = \binom{n+m-q-1}{n-1}, \qquad q = q_1 + \cdots + q_n.$$

If K_{nm} is the number of compositions of m into n summands and W_{nm} is the number of allocations of m identical particles to n different cells without empty cells, then

$$K_{nm} = W_{nm} = C_{nm}^{(n)} = \binom{m-1}{n-1}. \tag{5.2.6}$$

Obviously, this value is equal to the number of m-samples in the commutative asymmetric n-basis with the condition of surjectivity.

5.2.3 The number of m-samples with homogeneous restrictions

Denote by $C_{nm}^{(k)}(s)$ the number of m-samples in the commutative asymmetric n-basis such that k fixed elements of the n-basis, and only these elements, appear exactly s times. In other words, $\alpha_1 = \cdots = \alpha_k = s$ for some fixed $i_1, \ldots, i_k$.

Using the enumerator, we obtain

$$\sum_{m=ks}^{\infty} C_{nm}^{(k)}(s) t^m = t^{ks}\left(\frac{1}{1-t} - t^s\right)^{n-k}. \tag{5.2.7}$$

Hence it follows that

$$C_{nm}^{(k)}(s) = \sum_{j=0}^{n-k} (-1)^j \binom{n-k}{j} \binom{n+m-(s+1)(k+j)-1}{m-s(k+j)}. \tag{5.2.8}$$

In particular, this formula determines the number of allocations of m identical particles to n different cells such that k fixed cells contain exactly s particles.

Formula (5.2.7) implies that for $s = 0$

$$\sum_{m=0}^{\infty} C_{nm}^{(k)}(0) t^m = t^{n-k}(1-t)^{-(n-k)}. \tag{5.2.9}$$

Hence it follows that

$$C_{nm}^{(k)}(0) = \binom{m-1}{n-k-1}. \tag{5.2.10}$$

This formula determines the number of m-samples in the commutative asymmetric n-basis which do not contain exactly k fixed elements of the basis. It also gives the number of allocations of m identical particles to n different cells such that exactly k fixed cells are empty.

5.2.4 The number of m-samples with non-homogeneous restrictions

Denote by $Q_{nm}(k_1, \dots, k_s)$ the number of m-samples in the commutative asymmetric n-basis such that elements $a_{i_1}, \dots, a_{i_s}$ of the basis appear $k_1, \dots, k_s$ times, respectively; that is, for the primary specification the equalities $\alpha_{i_1} = k_1, \dots, \alpha_{i_s} = k_s$ hold. Obviously, $Q_{nm}(k_1, \dots, k_s)$ is also equal to the number of allocations of m identical particles to N different cells such that the cells with labels $i_1, \dots, i_s$ contain $k_1, \dots, k_s$ particles respectively. Using the enumerator, we obtain

$$\sum_{m=k}^{\infty} Q_{nm}(k) t^m = t^k (1-t)^{-(n-s)}, \tag{5.2.11}$$

where $k = k_1 + \cdots + k_s$ and $Q_{nm}(k) = Q_{nm}(k_1, \dots, k_s)$. It follows from (5.2.11) that

$$Q_{nm}(k) = \binom{n+m-s-k-1}{m-k}. \tag{5.2.12}$$

Denote by $\bar{Q}_{nm}(k)$ the number of m-samples in the commutative asymmetric n-basis such that all symbols of the basis appear and s fixed symbols appear exactly $k_1, \dots, k_s$ times, $k = k_1 + \cdots + k_s$. In this case the generating function is of the form

$$\sum_{m=n-s-k}^{\infty} \bar{Q}_{nm}(k) t^m = t^{n-s+k} (1-t)^{-(n-s)}.$$

Hence it follows that

$$\bar{Q}_{nm}(k) = \binom{m-k-1}{n-s-1}. \tag{5.2.13}$$

5.2.5 Runs

Consider the configurations obtained as a result of mappings of the set $X = \{1, \ldots, m\}$ into the set $A = \{0, 1\}$ provided the images contain m_0 zeros and m_1 units and $m_0 + m_1 = m$. Each configuration of such a form corresponds to a sequence of length m containing m_0 zeros and m_1 units. The number of such sequences is equal to $\binom{m}{m_0}$.

The maximal number of consecutive zeros (units) among the images of a configuration is called a run of zeros (units). Denote by $N_m^0(k)$ and $N_m^1(k)$ the numbers of configurations with k runs of zeros and units respectively. Divide all configurations with k runs of zeros into three classes. The first class consists of the configurations which begin with a zero run and terminate in a zero run; the second class consists of the configurations which begin with and terminate in unit runs; and finally the third class is formed by the configurations which begin with and terminate in runs of different symbols. The number of configurations in the first class is equal to the number of compositions of m_0 into k summands multiplied by the number of compositions of m_1 into $k-1$ summands, that is,

$$\binom{m_0 - 1}{k - 1}\binom{m_1 - 1}{k - 2}.$$

Similarly, the numbers of configurations in the second and third classes are equal to

$$\binom{m_0 - 1}{k - 1}\binom{m_1 - 1}{k}, \qquad \binom{m_0 - 1}{k - 1}\binom{m_1 - 1}{k - 1}$$

respectively. Hence it follows that

$$N_m^0(k) = \binom{m_0 - 1}{k - 1}\binom{m_1 + 1}{k}. \tag{5.2.14}$$

For reasons of symmetry,

$$N_m^1(k) = \binom{m_0 + 1}{k - 1}\binom{m_1 - 1}{k - 1}.$$

If $N_m(k)$ is the number of configurations with k runs of both kinds, then similarly we obtain

$$N_m(k) = \begin{cases} 2\binom{m_0-1}{s-1}\binom{m_1-1}{s-1}, & k = 2s, \\ \binom{m_0-1}{s-1}\binom{m_1-1}{s} + \binom{m_0-1}{s}\binom{m_1-1}{s-1}, & k = 2s+1. \end{cases} \tag{5.2.15}$$

Assume now that the images of configurations of the type considered are placed on the m equidistant points of a circle. Note that the number of runs in configurations on a circle is always even. Denote by $M_m(2k)$ the number of different configurations on a circle with $2k$ runs. Then

$$M_m(2k) = \frac{m}{k}\binom{m_0 - 1}{k-1}\binom{m_1 - 1}{k-1}. \qquad (5.2.16)$$

Formula (5.2.16) follows from the relation

$$M_m(2k) = N_m(2k) + N_m(2k+1),$$

which can be derived by an arbitrary cutting of the circle and placing the configuration on the line obtained. Substituting the values of $N_m(2k)$ and $N_m(2k+1)$, we get the formula.

5.2.6 Inversions in permutations

In Section 3.3 we obtained the generating function for the numbers $B(n,r)$ which determine the number of permutations of degree n with r inversions. Let us derive this generating function by using the technique of constructing the generating functions for m-samples in the commutative asymmetric n-basis.

We consider the set of vectors

$$M_n = \{(\delta_1, \delta_2, \dots, \delta_n) : 0 \leqslant \delta_i \leqslant n - i,\ 1 \leqslant i \leqslant n\}.$$

If S_n is the set of all permutations of the elements $1, 2, \dots, n$, then there exists a bijection of S_n into the set M_n such that the vector $(\delta_1, \delta_2, \dots, \delta_n)$ is put in one-to-one correspondence with a permutation $a_1, a_2, \dots, a_n$, where $\delta_1 + \delta_2 + \cdots + \delta_n = J(a_1, a_2, \dots, a_n)$. Here $J(a_1, a_2, \dots, a_n)$ is the number of inversions in the permutation $a_1, a_2, \dots, a_n$. Let us give the algorithm which establishes this bijection (Stanley, 1986). At the zeroth step, we have the permutation $n, n-1, \dots, 2, 1$. At the ith step, we move the element $n - i$ in such a manner that it has δ_{n-i} elements of the permutation on its left, $0 \leqslant i \leqslant n-1$. The permutation obtained at the last step of the algorithm has $\delta_1 + \delta_2 + \cdots + \delta_n$ inversions, and $\delta_n = 0$. It is clear how to construct, beginning with a permutation $a_1, a_2, \dots, a_n$, the vector $(\delta_1, \delta_2, \dots, \delta_n)$ that determines the number of inversions of the elements of the permutation.

From the bijection thus established it follows that

$$B(n,r) = |\{(\delta_1, \delta_2, \dots, \delta_{n-1}) : \delta_1 + \delta_2 + \cdots + \delta_{n-1} = r,\ 0 \leqslant \delta_i \leqslant n - i,\ 1 \leqslant i \leqslant n-1\}|.$$

Therefore, formula (5.2.3) implies that

$$b_n(x) = \sum_{r=0}^{\binom{n}{2}} B(n,r)x^r = (1+x+\cdots+x^{n-1})(1+x+\cdots+x^{n-2})\cdots(1+x).$$

Hence formula (3.3.3) derived above follows.

5.3 The asymptotics of the number of m-samples

5.3.1 The main results

Let $\bar{s} = (s_1, \ldots, s_n)$ and let $C_{nm}(\bar{s})$ be the number of m-samples in the commutative asymmetric n-basis with primary specifications $[a_1^{\alpha_1} \ldots a_n^{\alpha_n}]$, $\alpha_1 + \cdots + \alpha_n = m$, satisfying the restrictions $\alpha_i \leqslant s_i$, $i = 1, \ldots, n$. It is clear that $C_{nm}(\bar{s})$ is the number of allocations of m identical particles to n different cells such that the capacity of the ith cell is limited by s_i, $i = 1, \ldots, n$. On the other hand, $C_{nm}(\bar{s})$ is equal to the number of solutions of the equation $x_1 + \cdots + x_n = m$ for non-negative integers with the conditions $0 \leqslant x_i \leqslant s_i$, $i = 1, \ldots, n$.

Taking $\Lambda_j = \{0, 1, \ldots, s_j\}$, $j = 1, \ldots, n$, and putting $x_1 = \cdots = x_n = 1$ in the enumerator for the commutative asymmetric n-basis, we obtain

$$\sum_{m=0}^{2N} C_{nm}(\bar{s})t^m = \prod_{j=1}^{n}(1 + t + t^2 + \cdots + t^{s_j}), \tag{5.3.1}$$

where

$$2N = s_1 + \cdots + s_n.$$

In particular, if $s_1 = \cdots = s_n = s$, then, denoting $C_{nm}(\bar{s}) = C_{nm}(s)$, we obtain the equality

$$\sum_{m=0}^{ns} C_{nm}(s) = (1-t)^{-n}(1-t^{s+1})^n.$$

Hence it follows that

$$C_{nm}(s) = \sum_{j=0}^{n} (-1)^j \binom{n}{j} \binom{n+m-(s+1)j-1}{m-(s+1)j}. \tag{5.3.2}$$

The complexity of the formula for large n and m raises the problem of finding an asymptotic approximation for $C_{nm}(\bar{s})$ as $m \to \infty$.

The primary object of this section is to prove the following theorem (Anderson, 1967; Deshouillers, 1974).

Theorem 5.3.1 *Let*

$$A = \frac{1}{3}\sum_{j=1}^{n} s_j(s_j+2), \quad b = \max_{1\leqslant j\leqslant n}(s_j+1), \quad 2N = s_1 + \cdots + s_n,$$

and let $A \to \infty$, $b/\sqrt{A} \to 0$ *and* $\left|(N-m)/\sqrt{S}\right|$ *be bounded. Then*

$$C_{nm}(\bar{s}) = \sqrt{\frac{2}{\pi A}}e^{-2(m-N)^2/A}\prod_{j=1}^{n}(s_j+1)(1+o(1)). \tag{5.3.3}$$

If we put $s_1 = \cdots = s_n = s$ in the theorem, then we obtain the following assertion.

Corollary 5.3.1 *If* $n \to \infty$ *and* $\left|(ns/2-m)/\sqrt{ns(s+2)/3}\right|$ *is bounded, then*

$$C_{nm}(s) = \sqrt{\frac{6}{\pi s(s+2)n}}(s+1)^n \exp\left\{-\frac{6(m-ns/2)^2}{s(s+2)n}\right\}(1+o(1)). \tag{5.3.4}$$

5.3.2 Proof of the main theorem

As a preliminary we prove several lemmas.

Lemma 5.3.1 *The representation*

$$C_{nm}(\bar{s}) = \frac{1}{\pi}\prod_{j=1}^{n}(s_j+1)I \tag{5.3.5}$$

holds, where

$$I = \int_{-\pi/2}^{\pi/2} e^{2i(N-m)\theta}\prod_{j=1}^{n}\frac{\sin(s_j+1)\theta}{(s_j+1)\sin\theta}\,d\theta.$$

Proof By the Cauchy formula

$$C_{nm}(\bar{s}) = \frac{1}{2\pi i}\oint_C \prod_{j=1}^{n}(1+z+z^2+\cdots+z^{s_j})\frac{dz}{z^{m+1}}, \tag{5.3.6}$$

where C is a closed contour crossing over the origin. If we take the unit circle with the center at the origin as the contour C, then putting $z = e^{i\theta}$, $-\pi \leqslant \theta \leqslant \pi$, we find that

$$C_{nm}(\bar{s}) = \frac{1}{2\pi}\int_{\pi}^{\pi} e^{i(N-m)\theta}\prod_{j=1}^{n}\frac{\sin((s_j+1)\theta/2)}{\sin(\theta/2)}\,d\theta. \tag{5.3.7}$$

Substituting θ' for $\theta/2$, we obtain the formula

$$C_{nm}(\bar{s}) = \frac{1}{\pi}\int_{-\pi/2}^{\pi/2} e^{2i(N-m)\theta'} \prod_{j=1}^{n} \frac{\sin(s_j+1)\theta'}{\sin\theta'}\, d\theta', \qquad (5.3.8)$$

which implies (5.3.5). □

Choose $\Delta = b^{-1/2}a^{-1/4}$ and divide the integral I into three integrals so that

$$I = I_1 + I_2 + I_3, \qquad (5.3.9)$$

where the domains of integration in the integrals I_1, I_2 and I_3 are $-\pi/2 \leqslant \theta \leqslant -\Delta$, $-\Delta \leqslant \theta \leqslant \Delta$ and $\Delta \leqslant \theta \leqslant \pi/2$ respectively.

Lemma 5.3.2 *If $A \to \infty$, $b/\sqrt{A} \to 0$ and $\left|(N-m)/\sqrt{A}\right|$ is bounded, then*

$$I_2 = \sqrt{\frac{2\pi}{A}} e^{2(m-N)^2/A}(1+o(1)). \qquad (5.3.10)$$

Proof It is well known that, for $|x| < \pi$,

$$\ln\frac{\sin x}{x} = \sum_{k=1}^{\infty}(-1)^k B_{2k} 2^{2k} \frac{x^{2k}}{2k(2k)!}, \qquad (5.3.11)$$

where B_k, $k = 1, 2, \ldots$, are the Bernoulli numbers.

It follows from (5.3.11) that, for $-\Delta \leqslant \theta \leqslant \Delta$,

$$\ln \prod_{j=1}^{n} \frac{\sin(s_j+1)\theta}{(s_j+1)\sin\theta} = -\frac{1}{2}A\theta^2 + R, \qquad (5.3.12)$$

where

$$R = \sum_{k=2}^{\infty}(-1)^k B_{2k} \frac{2^{2k}\theta^{2k}}{2k(2k)!} \sum_{j=1}^{n}((s_j+1)^{2k} - 1).$$

Using the inequality

$$(-1)^{k-1}B_{2k} > 0, \qquad k = 1, 2, \ldots,$$

we obtain

$$\left|\frac{2R}{A\theta^2}\right| \leqslant \sum_{k=2}^{\infty}(-1)^{k-1}B_{2k}\frac{2^{2k}}{2k(2k)!}\left(\frac{b}{\sqrt{A}}\right)^{k-1} = o\left(\frac{b}{\sqrt{A}}\right)$$

uniformly in $-\Delta \leqslant \theta \leqslant \Delta$. From this estimate and equality (5.3.12) it follows that, as $A \to \infty$,

$$I_2 = \int_{-\Delta}^{\Delta} \exp\left\{-\frac{A\theta^2}{2}(1+o(1)) + 2i(N-m)\theta\right\} d\theta,$$

uniformly with respect to $|\theta| \leqslant \Delta$. This formula implies that

$$I_2 = \frac{1}{\sqrt{A}} \int_{-\delta}^{\delta} \exp\left\{-\frac{\varphi^2}{2} + \frac{2i(N-m)}{\sqrt{A}}\varphi\right\} d\varphi(1+o(1)),$$

where $\delta = \left(\sqrt{A}/b\right)^{1/2}$. If we take the integral with infinite limits instead of one above, thereby making an error of exponential order, and carry out the change of variables $\psi = \varphi - 2i(N-m)/\sqrt{A}$, we obtain

$$I_2 = \frac{1}{\sqrt{A}} e^{-2(N-m)^2/A} \int_{-\infty-ih}^{\infty-ih} e^{-\psi^2/2}\, d\psi(1+o(1)),$$

where $h = 2(N-m)/\sqrt{A}$. Using the well-known formula

$$\int_{-\infty-ih}^{\infty-ih} e^{-u^2/2}\, du = \sqrt{2\pi},$$

we obtain the assertion of the lemma. □

Lemma 5.3.3 *For any natural $r \geqslant 2$ there exists Δ such that for any θ, $\Delta < \theta \leqslant \pi/2$,*

$$\frac{\sin r\theta}{r \sin\theta} < \frac{\sin r\Delta}{r \sin\Delta}. \tag{5.3.13}$$

Proof For $0 < \theta \leqslant \pi/(2r)$ we prove the inequality by induction on r. For $r = 2$ the inequality is true, since it is equivalent to

$$\cos\theta < \cos\Delta, \qquad \Delta < \theta \leqslant \pi/2.$$

If the inequality (5.3.13) holds, then

$$\frac{\sin(r+1)\theta}{(r+1)\sin\theta} = \frac{\sin r\theta}{r\sin\theta}\frac{r\cos\theta}{r+1} + \frac{\cos r\theta}{r+1} < \frac{\sin(r+1)\Delta}{(r+1)\sin\Delta}.$$

Now consider the domain $\pi/(3r) < \theta \leqslant \pi/2$. In this domain, for $r = 3, 4, \ldots$,

$$\left|\frac{\sin r\theta}{r\sin\theta}\right| \leqslant \frac{1}{r\sin(\pi/(3r))} \leqslant \frac{1}{3\sin(\pi/9)} \leqslant 0.99. \tag{5.3.14}$$

On the other hand, for sufficiently small Δ

$$\left|\frac{\sin r\Delta}{r\sin\Delta}\right| > 0.99, \tag{5.3.15}$$

since $\sin x/x \to 1$ as $x \to 0$. From inequalities (5.3.14) and (5.3.15) it follows that (5.3.13) is valid for θ such that $\pi/(3r) < \theta \leqslant \pi/2$. The lemma is proved since the intervals $(0, \pi/(2r)]$ and $(\pi/(3r), \pi/2]$ overlap. □

Lemma 5.3.4 *For all natural $r \geqslant 2$ and $|r\Delta| < \pi$,*

$$\left|\frac{\sin r\Delta}{r\sin\Delta}\right| \leqslant (\cos\Delta)^{(r^2-1)/3}. \tag{5.3.16}$$

Proof We prove the equivalent inequality

$$\ln\left|\frac{\sin r\Delta}{r\sin\Delta}\right| \leqslant \frac{r^2-1}{3}\ln\cos\Delta. \tag{5.3.17}$$

Using expansion (5.3.11) for $0 \leqslant x \leqslant \pi/2$ and the following expansion:

$$\ln\cos x = \sum_{k=1}^{\infty}(-1)^k B_{2k}2^{2k}(2^{2k}-1)\frac{x^{2k}}{2k(2k)!}, \tag{5.3.18}$$

we can write the following inequality, equivalent to inequality (5.3.17):

$$\frac{3}{r^2-1}\sum_{k=1}^{\infty}(-1)^{k-1}B_{2k}2^{2k}(r^{2k}-1)\frac{\Delta^{2k}}{2k(2k)!}$$
$$\geqslant \sum_{k=1}^{\infty}(-1)^{k-1}B_{2k}2^{2k}(2^{2k}-1)\frac{\Delta^{2k}}{2k(2k)!}.$$

For series with positive coefficients, this inequality is valid for all $r \geqslant 2$. □

Lemma 5.3.5 *If $A \to \infty$, then*

$$|I_1| = O\left(e^{-\sqrt{A}/(2b)}\right), \qquad |I_3| = O\left(e^{-\sqrt{A}/(2b)}\right).$$

Proof Both the integrals are estimated similarly; therefore we prove the estimate for I_3 only. It follows from Lemmas 5.3.3 and 5.3.4 that, for $\Delta < \theta \leqslant \pi/2$,

$$\prod_{j=1}^{n}\frac{\sin(s_j+1)\theta}{(s_j+1)\sin\theta} < \prod_{j=1}^{n}\frac{\sin(s_j+1)\Delta}{(s_j+1)\sin\Delta} \leqslant (\cos\Delta)^A.$$

It can easily be deduced from expansion (5.3.18) that

$$A \ln \cos \Delta < -\sqrt{A}/(2b).$$

Consequently,

$$|I_3| \leqslant \int_{\Delta}^{\pi/2} \left| \prod_{j=1}^{n} \frac{\sin(s_j+1)\theta}{(s_j+1)\sin\theta} \right| d\theta < Ce^{-\sqrt{A}/(2b)},$$

where C is an absolute constant. The lemma is proved. □

We now turn to the proof of the main theorem. Taking into account the exponential order of decrease of $|I_1|$ and $|I_3|$, we find from (5.3.7)–(5.3.9) that

$$C_{nm}(\bar{s}) = \frac{1}{\pi} \prod_{j=1}^{n} (s_j+1) I_2 (1+o(1)).$$

Substituting the asymptotic estimate of I_2 into this relation, we obtain (5.3.3).

5.4 Non-commutative asymmetric *n*-basis

The particular case of the general combinatorial scheme characterized by the non-commutative asymmetric n-basis includes the scheme of allocation of distinct particles to different cells. This model is a combinatorial version of the classical occupancy problem that is well known in probability theory. Many investigations are devoted to the occupancy problem; a number of results, essentially of an asymptotic character, can be found in (Kolchin, Sevastyanov and Chistyakov, 1978).

In this section attention is concentrated on the method for constructing the generating functions for the enumeration of m-samples in the non-commutative asymmetric n-basis with restrictions of a general character on the primary specifications.

At the end of the section we consider a method for obtaining asymptotic formulae and apply it to a problem on the number of arrangements.

Let Λ_j be a subsequence of $\mathbf{N}_0 = \{0, 1, 2, \dots\}$, that is, $\Lambda_j \subseteq \mathbf{N}_0$, $j = 1, \dots, n$, and let $\Lambda = (\Lambda_1, \dots, \Lambda_n)$ be an ordered set of such sequences. The generating function of $n+1$ variables:

$$\tilde{\Phi}(t; x_1, \dots, x_n; \Lambda) = \prod_{j=1}^{n} \sum_{\alpha_j \in \Lambda_j} \frac{x_j^{\alpha_j} t^{\alpha_j}}{\alpha_j!}, \tag{5.4.1}$$

is called the enumerator in the non-commutative asymmetric n-basis.

Since $\tilde{\Phi}$ is an exponential generating function, (5.4.1) will be referred to as the enumerator of exponential type.

Represent the enumerator (5.4.1) in the form

$$\tilde{\Phi}(t; x_1, \ldots, x_n; \Lambda) = \sum_{m=0}^{\infty} \frac{t^m}{m!} \sum_{\alpha_1+\cdots+\alpha_n=m} \frac{m!}{\alpha_1 \cdots \alpha_n!} x_1^{\alpha_1} \cdots x_n^{\alpha_n}, \quad (5.4.2)$$

where the summation in the second sum is carried out over all solutions of the equation $\alpha_1 + \cdots + \alpha_n = m$ for non-negative integers such that $\alpha_j \in \Lambda_j$, $j = 1, \ldots, n$.

The number of configurations in A^X with primary specification $[a_1^{\alpha_1} \ldots a_n^{\alpha_n}]$, $\alpha_1 + \cdots + \alpha_n = m$, is equal to $m!/(\alpha_1! \cdots \alpha_n!)$. This means that each m-sample in the commutative asymmetric n-basis with a given primary specification $[a_1^{\alpha_1} \ldots a_n^{\alpha}]$ corresponds to $m!/(\alpha_1! \cdots \alpha_n!)$ m-samples in the non-commutative asymmetric n-basis with this primary specification. Thus the summand

$$\frac{t^m x_1^{\alpha_1} \cdots x_n^{\alpha_n}}{\alpha_1! \cdots \alpha_n!}, \qquad \alpha_1 + \cdots + \alpha_n = m,$$

in the enumerator $\tilde{\Phi}$ is put in a one-to-one correspondence with a class of m-samples in the non-commutative asymmetric n-basis with primary specification $[a_1^{\alpha_1} \ldots a_n^{\alpha_n}]$. In this sense we can say that $\tilde{\Phi}$ enumerates the m-samples in the non-commutative asymmetric n-basis with restrictions on the primary specifications determined by Λ.

Denote by $D_{nm}(\Lambda)$ the number of m-samples in the non-commutative asymmetric n-basis with primary specifications determined by Λ. This means that $\alpha_j \in \Lambda_j$, $j = 1, \ldots, n$, for the primary specification $[a_1^{\alpha_1} \ldots a_n^{\alpha_n}]$ of any m-sample. It is clear that

$$D_{nm}(\Lambda) = \sum_{\alpha_1+\cdots+\alpha_n=m} \frac{m!}{\alpha_1! \cdots \alpha_n!},$$

where $\alpha_j \in \Lambda_j$, $j = 1, \ldots, n$. Putting $x_1 = \cdots = x_n = 1$ in the enumerator $\tilde{\Phi}$, we find that

$$\tilde{\Phi}(t; 1, \ldots, 1; \Lambda) = \sum_{m=0}^{\infty} d_{nm}(\Lambda) \frac{t^m}{m!} = \prod_{j=1}^{n} \sum_{\alpha_j \in \Lambda_j} \frac{t^{\alpha_j}}{\alpha_j!}. \quad (5.4.3)$$

Formula (5.4.3) gives a general method for constructing the generating functions for the number of m-samples in the non-commutative asymmetric n-basis for given sequences $\Lambda_1, \ldots, \Lambda_n$ determining the primary specifications. Take the product of n identical factors:

$$\left(1+\frac{t}{1!}+\frac{t^2}{2!}+\cdots\right)\left(1+\frac{t}{1!}+\frac{t^2}{2!}+\cdots\right)\cdots\left(1+\frac{t}{1!}+\frac{t^2}{2!}+\cdots\right).$$

Delete the terms $T^k/k!$ in the ith factor such that $k \notin \Lambda_i$, $i = 1,\dots,n$. The product of the factors obtained by the deletion determines the required generating function.

Consider some examples which illustrate the described method for finding the generating functions.

(a) Let $\Lambda_j = \mathbf{N}_0 = \{0,1,2,\dots\}$, $j = 1,2,\dots,n$. In this case there are no restrictions on the primary specifications and $D_{nm} = D_{nm}(\Lambda)$ is equal to the total number of m-samples in the non-commutative asymmetric n-basis. From (5.4.3) it follows that

$$\sum_{m=0}^{\infty} D_{nm}\frac{t^m}{m!} = \prod_{j=1}^{n}\sum_{\alpha_j\in\Lambda_j}\frac{t^{\alpha_j}}{\alpha_j!} = e^{nt}.$$

Hence $D_{nm} = n^m$. Obviously, this value coincides with the number of allocations of m distinct particles to n different cells if there is no restriction on the capacity of the cells.

(b) Let $\Lambda_j = \mathbf{N} = \{1,2,\dots\}$, $j = 1,\dots,n$. In this case $D^0_{nm} = D_{nm}(\mathbf{N})$ is equal to the number of m-samples such that each element of the non-commutative asymmetric n-basis appears in each sample at least once. From (5.4.3) we obtain

$$\sum_{m=n}^{\infty} D^0_{nm}\frac{t^m}{m!} = \prod_{j=1}^{n}\sum_{\alpha_j=1}^{\infty}\frac{t^{\alpha_j}}{\alpha_j!} = (e^t-1)^n. \tag{5.4.4}$$

Hence it follows that

$$D^0_{nm} = \Delta^n 0^m = \sum_{k=0}^{n}(-1)^k\binom{n}{k}(n-k)^m. \tag{5.4.5}$$

(c) Let $\Lambda_j = \mathbf{N}_1 = \{2,3,\dots\}$, $j = 1,\dots,n$, and put $D^1_{nm} = D_{nm}(\mathbf{N}_1)$. In this case,

$$\sum_{m=2n}^{\infty} D^1_{nm}\frac{t^m}{m!} = \prod_{j=1}^{n}\sum_{\alpha_j=2}^{\infty}\frac{t^{\alpha_j}}{\alpha_j!} = (e^t-1-t)^n.$$

Taking the coefficient of $t^m/m!$ in the expansion of $(e^t - 1 - t)^n$, we obtain the formula

$$D^1_{nm} = \sum_{k=0}^{n} (-1)^k \binom{n}{k} (m)_k \Delta^{n-k} 0^{m-k},$$

which represents D^1_{nm} in terms of the Morgan numbers.

(d) Denote by $D^{(k)}_{nm}(s)$ the number of m-samples in the non-commutative asymmetric n-basis such that each of the fixed elements $a_{i_1}, \ldots, a_{i_s}$ of the basis appears exactly s times and none of the remaining elements satisfy this requirement. Using (5.4.3), we obtain

$$\sum_{m=ks}^{\infty} D^{(k)}_{nm}(s) \frac{t^m}{m!} = \left(\frac{t^s}{s!}\right)^k \left(e^t - \frac{t^s}{s!}\right)^{n-k}. \tag{5.4.6}$$

Comparing the coefficients of $t^m/m!$ on both sides of equality (5.4.6), we obtain

$$D^{(k)}_{nm}(s) = \sum_{j=0}^{n-k} (-1)^j \binom{n-k}{j} \frac{1}{(s!)^{j+k}} (n-k-j)^{m-s(j+k)} (m)_{s(j+k)}. \tag{5.4.7}$$

It follows from (5.4.4)–(5.4.6) that

$$D^{(k)}_{nm}(0) = \Delta^{n-k} 0^m. \tag{5.4.8}$$

(e) Let $Q(n, m, k)$ be the number of m-samples in the non-commutative asymmetric n-basis such that a fixed element of the basis appears exactly k times. Then

$$\sum_{m=k}^{\infty} Q(n, m, k) \frac{t^m}{m!} = \frac{t^k}{k!} r^{(n-1)t}. \tag{5.4.9}$$

(f) A problem on the number of arrangements Suppose that there are n fires and m fire-engines that can be used to put them out. Suppose also that if $n \leqslant m$, then an attempt to extinguish each fire is made. Under these conditions an arrangement of m fire-engines to n fires is a surjective mapping $\varphi : X \to Y$, where $|X| = m$ and $|Y| = n$. Therefore the number of different arrangements is equal to $\Delta^n 0^m$.

If the number of fires is not fixed in advance but does not exceed m, then the number of arrangements of fire-engines is equal to

$$\omega_m = \sum_{n=0}^{m} \Delta^n 0^m.$$

It is easy to see that

$$\sum_{m=0}^{\infty} \omega_m \frac{x^m}{m!} = \frac{1}{2-e^x}.$$

Hence we obtain the recurrence relation

$$\omega_m = \delta_{0,m} + \sum_{j=0}^{m-1} \binom{m}{j} \omega_j,$$

where $\delta_{i,j}$ is Kronecker's symbol. From the generating function given above we can easily see that

$$\omega_m = \sum_{k=1}^{\infty} \frac{k^m}{2^{k+1}}. \tag{5.4.10}$$

Let us now find an asymptotic formula for ω_m as $m \to \infty$. Choose ε such that $0 < \varepsilon < 1/6$ and represent the sum in (5.4.10) in the form

$$\omega_m = \sum_{k=1}^{\infty} \frac{k^m}{2^{k+1}} = S_1 + S_2 + S_3, \tag{5.4.11}$$

where the summation in the sums S_1, S_2, S_3 is carried out over k in the intervals

$$(1, [m/\ln 2 - Am^{1/2+\varepsilon}/\ln 2] - 1),$$
$$([m/\ln 2 - Am^{1/2+\varepsilon}/\ln 2], [m/\ln 2 + Am^{1/2+\varepsilon}/\ln 2]),$$
$$([m/\ln 2 + Am^{1/2+\varepsilon}/\ln 2] + 1, \infty)$$

respectively; here A is a positive constant and $[a]$ is the integer part of a. Introducing the new variable $k' = k - [m/\ln 2]$ and obtaining an asymptotic formula for the terms of the sum S_2, we find that

$$S_2 = \frac{1}{2}\left(\frac{m}{2^{1/\ln 2}\ln 2}\right)^m \sum_{Am^{1/2+\varepsilon}/\ln 2 \leqslant k \leqslant -Am^{1/2+\varepsilon}} e^{-k^2(\ln 2)^2/(2m)}(1+o(1)).$$

Making an error of exponential order, we change the sum for an integral and obtain the formula

$$S_2 = \frac{1}{2}\left(\frac{m}{2^{1/\ln 2}\ln 2}\right)^m \frac{\sqrt{2\pi m}}{\ln 2}(1+o(1)). \tag{5.4.12}$$

Taking into account the well-known inequality

$$\int_u^{\infty} e^{-x^2/2}\,dx < \frac{1}{u} e^{-u^2/2}, \qquad u > 0,$$

and the fact that the terms of S_3 decrease, we see that

$$S_3 = O\left(\frac{S_2}{m^\varepsilon} e^{-A^2 m^{2\varepsilon}/6}\right). \tag{5.4.13}$$

Similarly,

$$S_1 = O\left(\frac{S_2}{m^\varepsilon} e^{-A^2 m^{2\varepsilon}/6}\right). \tag{5.4.14}$$

Thus, formulae (5.4.11)–(5.4.14) imply that, as $m \to \infty$,

$$\omega_m = \frac{1}{2}\left(\frac{m}{2^{1/\ln 2}\ln 2}\right)^m \frac{\sqrt{2\pi m}}{\ln 2}(1 + o(1)).$$

By obvious transformations the formula can be represented as follows:

$$\omega_m = \sqrt{\frac{\pi m}{2}} \log_2 e \left(\frac{m \log_2 e}{e}\right)^m (1 + o(1)).$$

Using Stirling's formula, we obtain

$$\omega_m = \frac{1}{2} m! (\log_2 e)^{m+1} (1 + o(1)).$$

(g) Trees In Section 4.2 we have obtained a number of basic formulae for enumerating trees with a given number of vertices. We also gave Prüfer's algorithm which establishes a bijection between the unrooted trees with n vertices and the $(n-2)$-samples in the non-commutative asymmetric n-basis. Let us consider this bijection in more detail.

Let $d(x_i)$ be the degree of a vertex x_i, and introduce the value $\varkappa(x_i) = d(x_i) - 1$. For a tree with vertices $x_1, x_2, \ldots, x_n$ the notation $[x_1^{\varkappa_1} x_2^{\varkappa_2} \ldots x_n^{\varkappa_n}]$ will be referred to as the primary specification of this tree. Prüfer's algorithm establishes a bijection between the trees of a primary specification $[x_1^{\varkappa_1} x_2^{\varkappa_2} \ldots x_n^{\varkappa_n}]$ and the $(n-2)$-samples in the non-commutative asymmetric basis $a_1, a_2, \ldots, a_n$ of the primary specification $[a_1^{\varkappa_1} a_2^{\varkappa_2} \ldots a_n^{\varkappa_n}]$. Hence it follows that the number of trees of the primary specification $[x_1^{\varkappa_1} x_2^{\varkappa_2} \ldots x_n^{\varkappa_n}]$ is equal to

$$\tilde{r}(\varkappa_1, \varkappa_2, \ldots, \varkappa_n) = \frac{(n-2)!}{\varkappa_1!\, \varkappa_2! \cdots \varkappa_n!}.$$

This equality leads to the formula for the number of unrooted trees with n vertices

$$\tilde{r}_n = \sum_{\substack{\varkappa_1 + \cdots + \varkappa_n = n-2, \\ \varkappa_1, \ldots, \varkappa_n \geqslant 0}} \frac{(n-2)!}{\varkappa_1!\, \varkappa_2! \cdots \varkappa_n!} = n^{n-2}, \qquad n \geqslant 2,$$

which was obtained earlier. Further, since the number $\tilde{r}_n(k)$ of unrooted trees with the degree of a fixed vertex is equal to k, it coincides with the number of $(n-2)$-samples in the non-commutative asymmetric n-basis where a fixed element occurs exactly k times. Therefore,

$$\tilde{r}_n(k) = \binom{n-2}{k-1}(n-1)^{n-k-1}.$$

5.5 Commutative symmetric n-basis

An m-sample in the commutative symmetric n-basis corresponds to a partition of the number m into no more than n summands. The problem of enumeration of partitions of natural numbers was considered by Euler, who developed the generating function method just for solving such combinatorial problems. In this section we give a method for the construction of generating functions enumerating the m-samples in the commutative symmetric n-basis with restrictions of a general character on the secondary specifications.

Consider sequences $\Lambda_j \subseteq \mathbf{N}_0 = \{0, 1, 2, \dots\}$, $j = 1, 2, \dots$, and put $\Lambda = (\Lambda_1, \Lambda_2, \dots)$. The enumerator in the commutative symmetric n-basis is defined as the generating function of an infinite number of variables

$$\psi(t; x_1, x_2, \dots; \Lambda) = \prod_{j=1}^{\infty} \sum_{\beta_j \in \Lambda_j} (x_j t^j)^{\beta_j}. \tag{5.5.1}$$

If we multiply the sums, then the enumerator ψ takes the form

$$\psi(t; x_1, x_2, \dots; \Lambda) = \sum_{m=0}^{\infty} t^m \sum_{\beta_1 + 2\beta_2 + \cdots + m\beta_m = m} x_1^{\beta_1} \cdots x_m^{\beta_m}, \tag{5.5.2}$$

where the inner summation is carried out over all partitions of m such that the term j appears β_j times and $\beta_j \in \Lambda_j$, $j = 1, 2, \dots$. It is clear that a summand of the enumerator of the form $t^m x_1^{\beta_1} \cdots x_m^{\beta_m}$ can be put in one-to-one correspondence with the partition of the number m, with the summands j appearing β_j times, $j = 1, 2, \dots$. Thus, the ψ enumerates the partitions with restrictions determined by Λ. The summands of the form $t^m x_1^{\beta_1} \cdots x_m^{\beta_m}$ such that $\beta_1 + \cdots + \beta_m \leqslant n$ correspond to the m-samples in the commutative symmetric n-basis.

We can introduce the enumerator

$$\psi(t; x_1, x_2, \dots; A, \Lambda) = \prod_{j \in A} \sum_{\beta_j \in \Lambda_j} (x_j t^j)^{\beta_j},$$

where A is a subsequence of the sequence of natural numbers. According to the equality

$$\psi(t;x_1,x_2,\ldots;A,\Lambda)=\sum_{m=0}^{\infty}t^m\sum_{\sum_{j\in A}j\beta_j=m,\ \beta_j\in\Lambda_j,\ j\in A}x_1^{\beta_1}\cdots x_m^{\beta_m},$$

this enumerator enumerates the partitions with two restrictions, namely, the summands must take values from the sequence A and each possible summand j appears exactly β_j times, $j\in A$. Obviously, this enumerator can be obtained from $\psi(t;x_1,x_2,\ldots;\Lambda)$ if we take $\Lambda_j=\{0\}$ for $j\notin A$.

Let $R_{mn}(\Lambda)$ be the number of partitions of m into n summands determined by Λ, and let $R_m(\Lambda)$ be the total number of such partitions. Formally we can write

$$R_{mn}(\Lambda)=\sum 1, \tag{5.5.3}$$

where the summation is over $\beta_1,\ldots,\beta_m$ such that

$$\beta_1+2\beta_2+\cdots+m\beta_m=m,\quad \beta_1+\cdots+\beta_m=n,\quad \beta_j\in\Lambda_j,\quad j=1,2,\ldots,$$

and

$$R_m(\Lambda)=\sum 1, \tag{5.5.4}$$

where the summation is over $\beta_1,\ldots,\beta_m$ such that

$$\beta_1+2\beta_2+\cdots+m\beta_m=m,\quad \beta_j\in\Lambda_j,\quad j=1,2,\ldots.$$

From representations (5.5.1) and (5.5.2) of the enumerators and formulae (5.5.3) and (5.5.4) it follows that

$$\psi(t;1,1,\ldots;\Lambda)=\sum_{m=0}^{\infty}R_m(\Lambda)t^m=\prod_{j=1}^{\infty}\sum_{\beta_j\in\Lambda_j}t^{j\beta_j}, \tag{5.5.5}$$

$$\psi(t;x,x,\ldots;\Lambda)=\sum_{m=0}^{\infty}\sum_{n=0}^{m}R_{mn}(\Lambda)t^mx^n=\prod_{j=1}^{\infty}\sum_{\beta_j\in\Lambda_j}(xt^j)^{\beta_j}. \tag{5.5.6}$$

The number of m-samples in the commutative symmetric n-basis with secondary specifications determined by Λ is equal to

$$Q_m(n;\Lambda)=\sum_{k=0}^{n}R_{mk}(\Lambda), \tag{5.5.7}$$

where, by definition, $R_{00}(\Lambda)=1$ and $R_{mk}(\Lambda)=0$ for $m<k$. Consider particular cases determined by Λ.

(a) Let $\Lambda_j = \mathbf{N}_0$, $j = 1,2,\ldots$, and let $R_m = R_m(\mathbf{N}_0)$, $R_{mn} = R_{mn}(\mathbf{N}_0)$ be the corresponding numbers of partitions without restrictions on the summands. It follows from (5.5.5) and (5.5.6) that

$$R(t) = \sum_{m=0}^{\infty} R_m t^m = \prod_{j=1}^{\infty}(1-t^j)^{-1},$$

$$R(t,x) = \sum_{m=0}^{\infty}\sum_{n=0}^{m} R_{mn} t^m x^n = \prod_{j=1}^{\infty}(1-xt^j)^{-1}.$$

Consider the generating function

$$R_n(t) = \sum_{m=n}^{\infty} R_{mn} t^m.$$

It follows from the obvious functional relation

$$(1-xt)R(t,x) = R(t,xt)$$

that

$$R_n(t) = \frac{t}{1-t^n} R_{n-1}(t), \quad R_0(t) = 1, \quad n = 1,2,\ldots.$$

By the sequential application of this relation we obtain

$$R_n(t) = \frac{t^n}{(1-t)(1-t^2)\cdots(1-t^n)}. \tag{5.5.8}$$

Put $\Lambda_j = \mathbf{N}_0$ in (5.5.7) and denote by $Q_m(n) = Q_m(n;\mathbf{N}_0)$ the number of partitions of m into no more than n summands. Note that

$$Q(t,n) = \sum_{m=0}^{\infty} Q_m(n) t^m = \sum_{k=0}^{n} R_k(t).$$

Hence, taking (5.5.8) into account, we obtain

$$Q(t,n) = \frac{1}{(1-t)(1-t^2)\cdots(1-t^n)}. \tag{5.5.9}$$

(b) Put

$$\Lambda_j = \begin{cases} \mathbf{N}_0, & j = 1,\ldots,n, \\ \{0\}, & j = n+1, n+2,\ldots \end{cases}$$

and denote by $R_m^{(n)}$ the number of partitions of m into summands which do not exceed n. It can easily be seen from the form of the generating function (5.5.5) that

$$R^{(n)}(t) = \sum_{m=0}^{\infty} R_m^{(n)} t^m = \frac{1}{(1-t)(1-t^2)\cdots(1-t^n)}. \qquad (5.5.10)$$

(c) Take $\Lambda_j = \{0, 1\}$, $j = 1, 2, \ldots$, and denote by V_m the number of partitions of m into different summands and by V_{mn} the number of partitions of m into n different summands. From (5.5.5) and (5.5.6) it follows that

$$V(t) = \sum_{m=0}^{\infty} V_m t^m = \prod_{j=1}^{\infty}(1 + t^j),$$

$$V(t, x) = \sum_{m=0}^{\infty}\sum_{n=0}^{m} V_{mn} t^m x^n = \prod_{j=1}^{\infty}(1 + xt^j).$$

Consider the generating function

$$V_n(t) = \sum_{m=n}^{\infty} V_{mn} t^m.$$

Using the functional relation

$$V(t, x) = (1 + xt)V(t, xt),$$

we find that

$$V_n(t) = \frac{t^{n(n+1)/2}}{(1-t)(1-t^2)\cdots(1-tn)}. \qquad (5.5.11)$$

(d) The results obtained can be summarized as follows. From (5.5.8)–(5.5.11) it follows that

$$R_n(t) = t^n Q(t, n), \quad R^{(n)}(t) = Q(t, n), \quad V_n(t) = t^{n(n+1)/2} Q(t, n).$$

Hence

$$R_{m+n,n} = Q_m(n) = R_m^{(n)} = V_{m+n(n+1)/2,n}.$$

Thus, the following theorem is valid.

Theorem 5.5.1 *The number of partitions of $m+n$ into n summands is equal to the number of partitions of m into no more than n summands; is equal to the number of partitions of m into the summands which do not exceed n; and is equal to the number of partitions of $m + n(n+1)/2$ into n different summands.*

For $m = 5$ and $n = 2$ the partitions of 7 into two summands are $1+6$, $2+5$, $3+4$, the partitions of 5 into one and two summands are 5, $1+4$, $2+3$, the partitions of 5 into summands not exceeding 2 are $1+1+1+1+1$, $1+1+1+2$, $1+2+2$, and the partitions of 8 into two different summands are $1+7$, $2+6$, $3+5$.

A number of results of such a type can be obtained with the help of Ferrers graphs of partitions. The Ferrers graph of a partition $m = \beta_1 + 2\beta_2 + \cdots + m\beta_m$ has β_i rows consisting of i equidistant points and beginning from the same vertical line. For example, the Ferrers graph of the partition $8 = 1 + 3 + 4$ is of the form

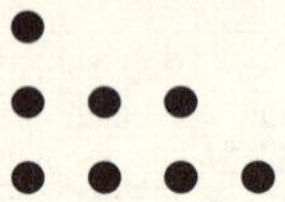

If we transform a Ferrers graph so that its rows become columns and vice versa, then the graph obtained is the Ferrers graph corresponding to the conjugate partition. For the partition $8 = 1 + 3 + 4$ the conjugate partition is $8 = 1 + 2 + 2 + 3$ and the corresponding Ferrers graph is of the form

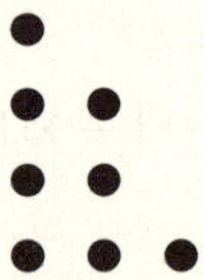

It is clear that the partition conjugate to a conjugate partition coincides with the initial partition.

In particular, it follows from the construction of the Ferrers graphs of an initial and the conjugate partitions that the number of partitions of m into summands which do not exceed n is equal to the number of partitions of m into no more than n summands, that is, the equality obtained above

$$R_m^{(n)} = Q_m(n)$$

can easily be deduced in such a way. Similarly, if $W_{m,n}$ is the number of partitions of m with the maximal summand equal to n, then this number coincides with the number of partitions of m into n summands, that is,

$$W_{m,n} = R_{m,n}.$$

(e) Partitions with different summands Let $R_{mn}^{(k)}(\Lambda)$ be the number of partitions of m into n summands such that they take exactly k different values and the number of appearances of the summand j is determined by a sequence Λ_j, $j = 1, 2, \ldots$. If each Λ_j contains zero and Λ'_j is the subsequence obtained from Λ by deleting zero, then

$$R(t, x, y; \Lambda) = \sum_{m=0}^{\infty}\sum_{n=0}^{\infty}\sum_{k=0}^{n} R_{mn}^{(k)}(\Lambda)t^m x^n y^k$$
$$= \prod_{j=1}^{\infty}\left(1 + y \sum_{\beta_j \in \Lambda'_j} (xt^j)^{\beta_j}\right).$$

In particular, if $R_{mn}^{(k)} = R_{mn}^{(k)}(\mathbf{N}_0)$, that is, $R_{mn}^{(k)}$ is the number of partitions of m into n summands of which exactly k are different, then

$$R(t, x, y) = \sum_{m=0}^{\infty}\sum_{n=0}^{m}\sum_{k=0} nR_{mn}^{(k)}t^m x^n u^k = \prod_{j=1}^{\infty} \frac{1 + xt^j(y-1)}{1 - xt^j}.$$

Consider the generating function

$$R_n(t, y) = \sum_{m=n}^{\infty}\sum_{k=0}^{n} R_{mn}^{(k)} t^m y^k.$$

From the functional relation

$$(1 - xt)R(t, x, y) = (1 + xt(y-1))R(t, xt, y)$$

we obtain

$$R_n(t, y) = \frac{t^n(1 + (y-1))(1 + t(y-1)) \cdots (1 + t^{n-1}(y-1))}{(1-t)(1-t^2)\cdots(1-t^n)}.$$

Using (5.5.8) we find that

$$R_n(t, y) = R_n(t)\prod_{j=1}^{n-1}(1 + t^j(y-1)), \qquad n = 1, 2, \ldots. \tag{5.5.12}$$

This formula expresses the multiplicative property of the generating functions, useful in calculations.

(f) Perfect partitions A partition of m is perfect if it contains one and only one partition of each number less than m provided that all equal summands are considered to be indistinguishable.

The perfect partitions of the number 7 are $1 + 1 + 1 + 1 + 1 + 1 + 1$, $1 + 1 + 1 + 4$, $1 + 2 + 4$, $1 + 2 + 2 + 2$.

Let a perfect partition contain $q_1 - 1$ summands equal to 1, $m \geqslant q_1 > 1$. Then all numbers less than q_1 have the unique partition with summands equal to 1, and q_1 has to be a summand of the perfect partition. If the number of summands equal to q_1 is $q_2 - 1$, $q_2 > 1$, then the next summand of the perfect partition is $q_1 q_2$, and so on. Therefore,

$$m = (q_1 - 1) + q_1(q_2 - 1) + \cdots + q_1 \cdots q_{k-1}(q_k - 1) = q_1 \cdots q_k - 1,$$

provided that $q_i > 1$, $i = 1, \ldots, k$. Thus, we obtain the following assertion.

Theorem 5.5.2 *Each perfect partition of a natural number m can be put in a one-to-one correspondence with an ordered decomposition of $m + 1$ into factors different from one.*

For example, the number 8 decomposes into ordered factors

$$8, \quad 4 \cdot 2, \quad 2 \cdot 2 \cdot 2, \quad 2 \cdot 4,$$

which correspond to the perfect partitions of the number 7 given above.

As an obvious illustration of the above, the problem of making an economical choice of a set of resistances such that it is possible to choose uniquely a subset of it with total resistance equal to any number $j = 1, \ldots, m$ is reduced to constructing a perfect partition of m. Naturally, partitions with the smallest number of summands are of the most interest.

5.6 The Hardy–Ramanujan formula

The Hardy–Ramanujan asymptotic formula for the number of unlimited partitions of m as $m \to \infty$:

$$R_m = \frac{1}{4\sqrt{3}m} e^{\pi\sqrt{2m/3}}(1 + O(m^{-1/4+\varepsilon})), \tag{5.6.1}$$

where $0 < \varepsilon < 1/4$, is one of the famous formulae of additive number theory. The aim of the present section is to derive it (Postnikov, 1971).

Using the Cauchy integral formula, we can write

$$R_m = \frac{1}{2\pi i} \oint_C \frac{F(z)}{z^{m+1}} dz,$$

where

$$F(z) = \prod_{k=1}^{\infty} (1 - z^k)^{-1}$$

and C is a circle with radius less than one and the center at the origin.

Making the change of variables $z = e^u$, $u = v + iw$, and putting

$$f(u) = \prod_{k=1}^{\infty} \frac{1}{1 - e^{-ku}}$$

for $v > 0$, we obtain

$$R_m = \frac{1}{2\pi} \int_{-\pi}^{\pi} f(v + iw) e^{m(v+iw)}\, dw. \tag{5.6.2}$$

Take $w_0 = m^{-3/4+\varepsilon/3}$, $\varepsilon > 0$, and divide the last integral in such a way that

$$R_m = I_1 + I_2 + I_3, \tag{5.6.3}$$

where

$$I_1 = \frac{1}{2\pi} \int_{-\pi}^{-w_0} f(v + iw) e^{m(v+iw)}\, dw,$$
$$I_2 = \frac{1}{2\pi} \int_{-w_0}^{w_0} f(v + iw) e^{m(v+iw)}\, dw,$$
$$I_3 = \frac{1}{2\pi} \int_{w_0}^{\pi} f(v + iw) e^{m(v+iw)}\, dw.$$

The idea of obtaining the asymptotic formula for R_m from the last equality is rather simple. We need to show that the integrals I_1 and I_3 are small in comparison with I_2 as $m \to \infty$. Then we obtain an asymptotic representation of the principal value of the logarithm $\ln f(u)$ for $-w_0 \leqslant w \leqslant w_0$, which lets us find an asymptotic formula for I_2 and, consequently, for R_m. The asymptotic representation of $\ln f(u)$ is given by the following lemma.

Lemma 5.6.1 *Let $\Re u > 0$ and $u \to 0$ within an angle in the right half-plane. Then*

$$\ln f(u) = \frac{\pi^2}{6u} + \frac{1}{2} \ln \frac{u}{2\pi} + O(|u|). \tag{5.6.4}$$

Here ln means the principal value of the logarithm.

We omit the proof of the lemma which can be found in (Postnikov, 1971), where the derivation of (5.6.1) is also given.

We now turn to the study of the integral I_2. Using the lemma for $|w| \leqslant w_0 = m^{-3/4+\varepsilon/3}$, we can easily see that

$$\ln f(v + iw) = \frac{\pi^2}{6(v + iw)} + \frac{1}{2} \ln(v + iw) - \frac{1}{2} \ln 2\pi + O(|v + iw|).$$

Choosing $v = \pi(6m)^{-1/2}$ and taking into account the fact that, for $|w| < w_0$,

$$\frac{1}{v+iw} = \frac{1}{v} - \frac{iw}{v^2} - \frac{w^2}{v^3} + O\left(\frac{w_0^3}{v^4}\right), \qquad |\ln(1+iw/v)| = O\left(\frac{w_0}{v}\right),$$

we obtain

$$\ln f(v+iw) = \frac{\pi\sqrt{m}}{\sqrt{6}} - iwm - \frac{m\sqrt{6m}w^2}{\pi} + \frac{1}{4}\ln\frac{1}{24m} + O(m^{-1/4+\varepsilon}).$$

Using this relation, we find that

$$I_2 = \frac{1}{2\pi}\exp\left\{2\pi\sqrt{\frac{m}{6}} + \frac{1}{4}\ln\frac{1}{24m} + O(m^{-1/4+\varepsilon})\right\}\int_{-w_0}^{w_0} e^{-m\sqrt{6m}w^2/\pi}\,dw.$$

Making the change of variables

$$\frac{1}{2}y^2 = \frac{m\sqrt{6m}}{\pi}w^2$$

in the integral and carrying out further transformations, we obtain

$$I_2 = \frac{e^{\pi\sqrt{2m/3}}}{4\sqrt{6\pi m}}\int_{-\delta_0}^{\delta_0} e^{-y^2/2}\,dy(1+O(m^{-1/4+\varepsilon})), \tag{5.6.5}$$

where

$$\delta_0 = \frac{24^{1/4}m^{\varepsilon/3}}{\sqrt{\pi}}.$$

Changing the integral in (5.6.5) for the corresponding integral with infinite limits, we introduce an error of exponential order. Substituting the value of the integral obtained into (5.6.5), we find the asymptotic representation

$$I_2 = \frac{1}{4\sqrt{3}m}e^{\pi\sqrt{2m/3}}(1+O(m^{-1/4+\varepsilon})). \tag{5.6.6}$$

We now turn to the estimation of I_1 and I_3. Both integrals are treated similarly; therefore we obtain an estimate for I_3 only. Take a sufficiently large constant C and put $w_1 = Cv$. Represent the integral I_3 as a sum of two integrals in such a way that

$$I_3 = I_3' + I_3'', \tag{5.6.7}$$

where

$$I_3' = \frac{1}{2\pi}\int_{w_0}^{w_1} f(v+iw)e^{m(v+iw)}\,dw,$$

$$I_3'' = \frac{1}{2\pi}\int_{w_1}^{\pi} f(v+iw)e^{m(v+iw)}\,dw.$$

Applying Lemma 5.6.1, we find that for $w_0 \leqslant w \leqslant w_1$

$$\ln f(u) = \frac{\pi^2}{6u} + \frac{1}{2}\ln u + O(1).$$

Hence it follows that, for some $C_0 > 0$,

$$|I_3'| \leqslant C_0 \int_{w_0}^{w_1} \sqrt{|u|} \exp\left\{\frac{\pi^2}{6}\Re\frac{1}{u} + mv\right\} dw.$$

Using the estimate

$$\Re\frac{1}{u} = \frac{v}{v^2+w^2} \leqslant \frac{v}{v^2+w_0^2}$$

and noting that $|u| \leqslant v\sqrt{1+C^2}$, we obtain the inequality

$$|I_3'| \leqslant C_1\sqrt{v}w_1 \exp\left\{\frac{\pi^2}{6(v^2+w_0^2)} + mv\right\},$$

where $C_1 > 0$. Hence it follows that, for sufficiently large m and some $C_2 > 0$,

$$|I_3'| \leqslant \frac{C_2}{m^{3/4}} e^{\pi\sqrt{2m/3}} \exp\left\{-mv(1-(1+w_0^2/v^2)^{-1})\right\}.$$

Since

$$1 - \left(1+\frac{w_0^2}{v^2}\right)^{-1} = \frac{w_0^2}{v^2}\left(1+\frac{w_0^2}{v^2}\right)^{-1} \geqslant \frac{w_0^2}{2v^2},$$

and

$$mv\frac{w_0^2}{2v^2} = \frac{\sqrt{6}}{2\pi}m^{2\varepsilon/3},$$

we obtain the estimate

$$|I_3'| = e^{\pi\sqrt{2m/3}} O\left(m^{-3/4}\exp\left\{-\sqrt{6}m^{2\varepsilon/3}/(2\pi)\right\}\right). \qquad (5.6.8)$$

We now estimate the integral I_3''. Note first that

$$\ln f(u) = \sum_{n=1}^{\infty} \frac{1}{n(e^{nu}-1)}. \qquad (5.6.9)$$

The obvious formula

$$|e^{nu}-1| = \sqrt{(e^{nv}-1)^2 + 4e^{nv}\sin^2\frac{nw}{2}}$$

implies that $|e^{nu}-1| \geqslant e^{nv}-1$. In addition, for $|w| \leqslant \pi/n$,

$$|e^{nu}-1| \geqslant 2\left|\sin\frac{nw}{2}\right| \geqslant \frac{2}{\pi}n|w|.$$

Thus,

$$|\ln f(u)| \leqslant \sum_{n=1}^{\infty} \frac{1}{n|e^{nu}-1|} \leqslant \frac{\pi}{2|w|} + \sum_{n=2}^{\infty} \frac{1}{n(e^{nv}-1)}.$$

Since $|w| \geqslant Cv$ and $nv \leqslant e^{nv} - 1$, we obtain the estimate

$$|\ln f(u)| \leqslant \frac{\pi}{2Cv} + \frac{1}{v}\sum_{n=2}^{\infty}\frac{1}{n^2} = \frac{C_3}{v}, \tag{5.6.10}$$

where $C_3 = \pi/(2C) + \pi^2/6 - 1$. Using estimate (5.6.10), we obtain

$$|I_3''| \leqslant \frac{1}{2\pi}\int_{w_1}^{\pi} e^{C_3/v+mv}\, dw = O(e^{C_3/v+mv}).$$

For sufficiently large C there exists $\delta > 0$ such that

$$\frac{C_3}{v+mv} < \sqrt{m}\left(\pi\sqrt{2/3} - \delta\right).$$

Consequently,

$$|I_3''| = O\left(e^{\pi\sqrt{2m/3}-\delta\sqrt{m}}\right). \tag{5.6.11}$$

The Hardy–Ramanujan formula follows now from (5.6.3), (5.6.6), (5.6.8) and (5.6.11).

It should be noted that, in fact, Hardy and Ramanujan obtained a more accurate asymptotic formula (see e.g. (Hall, 1958)):

$$R_m = \frac{1}{2\pi\sqrt{2}}\frac{d}{dm}\left(\frac{e^{A\lambda_m}}{\lambda_m}\right) + O\left(e^{H\sqrt{m}}\right), \tag{5.6.12}$$

where $\frac{d}{dm}$ means differentiation with respect to m and

$$A = \pi\sqrt{2/3}, \quad \lambda_m = \sqrt{m - 1/24}, \quad H < A.$$

Rademacher found a representation of R_m as a convergent series (Hall, 1958) and thus essentially improved formula (5.6.12); namely, he proved that

$$R_m = \sum_{k=1}^{\infty} P_k(m) Q_k(m), \tag{5.6.13}$$

where

$$Q_k(m) = \frac{1}{\pi}\sqrt{k/2}\frac{d}{dm}\left(\frac{\sinh(A\lambda_m/k)}{\lambda_m}\right),$$

$$P_k(m) = \sum_j w_{j,k} e^{-2\pi imj/k},$$

and $w_{j,k}$ is some 24th root of unity.

In conclusion, we point out one more asymptotic result concerning the number R_{mn} of partitions of m into n summands. It is known (Hall, 1958) that R_{mn} is a polynomial of m of degree $n-1$ with the leading term equal to $m^{m-1}/(n!(n-1)!)$ and whose coefficients depend on the residue classes of m modulo $n!$. In particular, this means that if $m \to \infty$ and n is fixed, then

$$R_{mn} = \frac{m^{n-1}}{n!\,(n-1)!}(1+o(1)). \tag{5.6.14}$$

On the other hand, under the same conditions the number of compositions of m into n summands satisfies the asymptotic relation

$$K_{mn} = \binom{m-1}{n-1} = \frac{m^{n-1}}{(n-1)!}(1+o(1)). \tag{5.6.15}$$

Formulae (5.6.14) and (5.6.15) lead to the conclusion that if $m \to \infty$ and n is fixed, then there are $n!$ compositions per partition, that is, an overwhelming majority of the partitions consists of different summands.

5.7 Non-commutative symmetric *n*-basis

The particular case of the general combinatorial scheme corresponding to the non-commutative symmetric n-basis, which includes the scheme of partitions of a finite set into blocks, until recently did not appear in investigations on combinatorics as often as other cases.

Below we give the general method for constructing generating functions enumerating m-samples in the non-commutative symmetric n-basis with various restrictions on the secondary specifications.

Let X be a finite set with m elements. Denote by $T(\beta_1,\dots,\beta_m)$ the number of partitions of the set X into non-empty subsets with secondary specification

$$[[1^{\beta_1}2^{\beta_2}\dots m^{\beta_m}]], \qquad \beta_1+2\beta_1+\cdots+m\beta_m = m,$$

that is, containing β_i blocks of size i, $i = 1,\dots,m$. It is easy to check that, for any non-negative integers $\beta_1,\dots,\beta_m$ such that $\beta_1+2\beta_2+\cdots+m\beta_m = m$,

$$T(\beta_1,\dots,\beta_m) = \frac{m!}{(1!)^{\beta_1}(2!)^{\beta_2}\cdots(m!)^{\beta_m}\beta_1!\,\beta_2!\cdots\beta_m!}. \tag{5.7.1}$$

Indeed, in accordance with the partition $m = \beta_1 + 2\beta_2 + \cdots + \beta_m$ we divide the elements of X into β_1 blocks of size 1, β_2 blocks of size 2, and so on, and place these blocks on a line in the natural order. We obtain all the partitions of X with secondary specification $[[1^{\beta_1}2^{\beta_2}\dots m^{\beta_m}]]$, rearranging the elements of X in all possible ways. The total number of partitions is equal to the quotient obtained by division of $m!$ by the number of equivalent permutations of elements of X which lead to the same partitions. The equivalent permutations appear if we transpose the elements within a block and transpose blocks of equal sizes. Therefore, the number of equivalent permutations is equal to

$$(1!)^{\beta_1}(2!)^{\beta_2}\cdots(m!)^{\beta_m}\beta_1!\,\beta_2!\cdots\beta_m!.$$

Hence formula (5.7.1) follows.

Let Λ_j be a subsequence of $\mathbf{N}_0 = \{0, 1, \ldots\}$, $j = 1, 2, \ldots$, and $\Lambda = (\Lambda_1, \Lambda_2, \ldots)$. The generating function of an infinite number of variables,

$$\tilde{\varphi}(t; x_1, x_2, \ldots; \Lambda) = \prod_{j=1}^{\infty} \sum_{\beta_j \in \Lambda_j} \frac{1}{\beta_j!}\left(\frac{x_j t^j}{j!}\right)^{\beta_j}, \tag{5.7.2}$$

is called the enumerator in the non-commutative symmetric n-basis. This enumerator, just like the enumerator in the non-commutative asymmetric n-basis, is of exponential type. Rewrite $\tilde{\varphi}$ in the form

$$\tilde{\varphi}(t; x_1, \ldots; \Lambda) = \sum_{m=0}^{\infty} \frac{t^m}{m!} \sum_{\beta_j \in \Lambda_j} \frac{m!}{(1!)^{\beta_1}\cdots(m!)^{\beta_m}\beta_1!\cdots\beta_m!} x_1^{\beta_1}\cdots x_m^{\beta_m}. \tag{5.7.3}$$

The second summation is over all $\beta_j \in \Lambda_j$, $j = 1, \ldots, m$, such that $\beta_1 + 2\beta_2 + \cdots + m\beta_m = m$.

Note that the coefficient of $t^m x_1^{\beta_1}\cdots x_m^{\beta_m}/m!$ in the enumerator is equal to the number $T(\beta_1, \ldots, \beta_m)$ of partitions of the set X with a given secondary specification $[[1^{\beta_1}2^{\beta_2}\ldots m^{\beta_m}]]$. Therefore, we can say that $\tilde{\varphi}$ enumerates the classes of partitions determined by secondary specifications. If $\beta_1 + \cdots + \beta_m \leqslant n$, then the corresponding class is associated with a set of m-samples in the non-commutative symmetric n-basis with secondary specification

$$[[1^{\beta_1}2^{\beta_2}\ldots m^{\beta_m}]], \qquad \beta_1 + 2\beta_2 + \cdots + m\beta_m = m.$$

Let $T_{mn}(\Lambda)$ be the number of partitions of the set X into n blocks with secondary specifications determined by Λ, and let $T_m(\Lambda)$ be the total number of such partitions. It is clear that

$$T_m(\Lambda) = \sum_{\beta_j \in \Lambda_j} T(\beta_1, \ldots, \beta_m), \tag{5.7.4}$$

where the summation is over $\beta_j \in \Lambda_j$, $j = 1, \ldots, m$, such that $\beta_1 + 2\beta_2 +$

$\cdots + m\beta_m = m$, and

$$T_{mn}(\Lambda) = \sum_{\beta_j \in \Lambda_j} T(\beta_1, \ldots, \beta_m), \tag{5.7.5}$$

where the summation is over $\beta_j \in \Lambda_j$, $j = 1, \ldots, m$, such that $\beta_1 + 2\beta_2 + \cdots + m\beta_m = m$ and $\beta_1 + \cdots + \beta_m = n$. The number of m-samples in the non-commutative symmetric n-basis with secondary specifications determined by Λ can be written in the form

$$U_{nm}(\Lambda) = \sum_{\beta_j \in \Lambda_j} T(\beta_1, \ldots, \beta_m), \tag{5.7.6}$$

where the summation is over $\beta_j \in \Lambda_j$, $j = 1, \ldots, m$, such that $\beta_1 + 2\beta_2 + \cdots + m\beta_m = m$ and $\beta_1 + \cdots + \beta_m \leqslant n$. In all the formulae given above, it is assumed that $T_0(\Lambda) = T_{00}(\Lambda) = U_{00}(\Lambda) = 1$.

It follows from (5.7.2)–(5.7.5) that

$$\sum_{m=0}^{\infty} T_m(\Lambda) \frac{t^m}{m!} = \prod_{j=1}^{\infty} \sum_{\beta_j \in \Lambda_j} \frac{1}{\beta_j!} \left(\frac{t^j}{j!} \right)^{\beta_j}, \tag{5.7.7}$$

$$\sum_{m=0}^{\infty} \sum_{n=0}^{m} T_m(\Lambda) \frac{t^m}{m!} = \prod_{j=1}^{\infty} \sum_{\beta_j \in \Lambda_j} \frac{1}{\beta_j!} \left(\frac{t^j}{j!} \right)^{\beta_j}. \tag{5.7.8}$$

Denote by $T_m(\Lambda; l, k)$ the number of partitions of the set X into blocks with secondary specifications determined by Λ and with exactly k blocks of size l. It is clear that

$$T_m(\Lambda; l, k) = \sum_{\beta_j \in \Lambda_j, \beta_l = k} T(\beta_1, \ldots, \beta_m),$$

where the summation is over $\beta_j \in \Lambda_j$, $j = 1, \ldots, m$, $j \neq l$, such that $\beta_l = k$ and $\beta_1 + 2\beta_2 + \cdots + m\beta_m = m$. Putting $x_j = 1$, $j \neq l$, and $x_l = x$ in the enumerator $\tilde{\psi}$, we obtain

$$\sum_{m=0}^{\infty} \sum_{k=0}^{[m/l]} T_m(\Lambda; l, k) \frac{t^m}{m!} x^k = \sum_{\beta_l \in \Lambda_l} \frac{1}{\beta_l!} \left(\frac{x t^l}{l!} \right)^{\beta_l} \prod_{j \neq l}^{\infty} \sum_{\beta_j \in \Lambda_j} \frac{1}{\beta_j!} \left(\frac{t^j}{j!} \right)^{\beta_j}. \tag{5.7.9}$$

We now turn to particular cases corresponding to various choices of Λ.

(a) Let $\Lambda_j = \mathbf{N}_0 = \{0, 1, 2, \ldots\}$, $j = 1, 2, \ldots$, $T_m = T_m(\mathbf{N}_0)$ and $T_{mn} = T_{mn}(\mathbf{N}_0)$. Obviously, T_{mn} and T_m are equal to the number of partitions of a set with m elements into n blocks and the total number of partitions of the set into blocks respectively. From (5.7.7) and (5.7.8) it follows that

$$\sum_{m=0}^{\infty} T_m \frac{t^m}{m!} = e^{e^t-1}, \tag{5.7.10}$$

$$\sum_{m=0}^{\infty} \sum_{n=0}^{m} T_{mn} \frac{t^m}{m!} x^n = x^{x(e^t-1)}. \tag{5.7.11}$$

The numbers T_m are called the Bell numbers and they satisfy the recurrence relation

$$T_{m+1} = \sum_{k=0}^{m} \binom{m}{k} t_{m-k}, \qquad T_0 = 1, \tag{5.7.12}$$

which follows from the form of the generating function (5.7.10). Formula (5.7.10) also implies Dobinski's representation of the Bell numbers

$$T_m = \frac{1}{e} \sum_{k=0}^{\infty} \frac{k^m}{k!}. \tag{5.7.13}$$

From (5.7.11) we find that

$$\sum_{m=n}^{\infty} T_{mn} \frac{t^m}{m!} = \frac{1}{n!}(e^t - 1)^n. \tag{5.7.14}$$

Hence it follows that

$$T_{mn} = \sigma(m, n), \tag{5.7.15}$$

where $\sigma(m, n)$ is a Stirling number of the second kind. Hence we find that the Bell numbers can be represented as

$$T_m = \sum_{k=0}^{m} \sigma(m, k). \tag{5.7.16}$$

Let $T_m(l, k)$ be the number of partitions of a set with m elements containing exactly k blocks of size l. From (5.7.9) we obtain

$$\sum_{m=0}^{\infty} \sum_{k=0}^{[m/l]} T_m(l, k) \frac{T^m}{m!} x^k = \exp\left\{(x-1)\frac{t^l}{l!} + e^t - 1\right\}. \tag{5.7.17}$$

(b) Let $\Lambda^0 = (\Lambda_1^0, \Lambda_2^0, \dots)$, where

$$\Lambda_j^0 = \begin{cases} \mathbf{N}_0, & j \text{ is odd}, \\ \{0\}, & j \text{ is even}, \end{cases}$$

and $\Lambda^1 = (\Lambda^1_1, \Lambda^1_2, \ldots)$, where

$$\Lambda^1_j = \begin{cases} \mathbf{N}_0, & j \text{ is even}, \\ \{0\}, & j \text{ is odd}. \end{cases}$$

It follows from (5.7.8) that

$$\sum_{m=0}^{\infty}\sum_{n=0}^{m} T_{mn}(\Lambda^0)\frac{t^m}{m!}x^n = e^{x\sinh t},$$

$$\sum_{m=0}^{\infty}\sum_{n=0}^{m} T_{mn}(\Lambda^1)\frac{t^m}{m!}x^n = e^{x(\cosh t-1)},$$

where $\sinh t$ and $\cosh t$ are the hyperbolic sine and cosine. From the expressions obtained for the corresponding generating functions it follows that

$$T_{mn}(\Lambda^0) = \frac{1}{2^n n!}\sum_{j=0}^{n}(-1)^j\binom{n}{j}(n-2j)^m,$$

$$T_{mn}(\Lambda^1) = \frac{(2n)!}{2^n n!}\sum_{j=0}^{n}(-1)^j\binom{m}{j}n^j\sigma(m-j, 2n).$$

(c) Denote by $T^{(s)}_{mn}$ the number of partitions of a set with m elements into n blocks of sizes no greater than s, and let $T^{(s)}_m$ be the total number of such partitions. To obtain the generating functions of these numbers, we take

$$\Lambda_j = \begin{cases} \mathbf{N}_0, & j = 1, \ldots, s, \\ \{0\}, & j = s+1, \ldots. \end{cases}$$

Then from (5.7.7) and (5.7.8) we obtain

$$\sum_{m=0}^{\infty} T^{(s)}_m \frac{t^m}{m!} = \exp\left\{\frac{t}{1!} + \frac{t^2}{2!} + \cdots + \frac{t^s}{s!}\right\}, \tag{5.7.18}$$

$$\sum_{m=0}^{\infty}\sum_{n=0}^{m} T^{(s)}_{mn}\frac{t^m}{m!}x^n = \exp\left\{x\left(\frac{t}{1!} + \frac{t^2}{2!} + \cdots + \frac{t^s}{s!}\right)\right\}. \tag{5.7.19}$$

If $T^{(s)}_m(l,k)$ is the number of partitions considered, with the additional condition that each partition has exactly k blocks of size l, $l \leqslant s$, then from (5.7.9) we obtain

$$\sum_{m=0}^{\infty}\sum_{n=0}^{[m/l]} T^{(s)}_{mn}(l,k)\frac{t^m}{m!}x^n = \exp\left\{(x-1)\frac{t^l}{l!} + \frac{t}{1!} + \cdots + \frac{t^s}{s!}\right\}. \tag{5.7.20}$$

(d) Let A be a subsequence of the natural numbers and let T_m^A be the number of partitions of a set with m elements into blocks whose sizes take values from the sequence A. Let T_{mn}^A be the number of such partitions with n blocks and let $T_m^A(l,k)$ be the number of partitions of the form considered with exactly k blocks of size l. Take

$$\Lambda_j = \begin{cases} \mathbf{N}_0, & j \in A, \\ \{0\}, & j \notin A, \end{cases}$$

and consider the generating function

$$a(t) = \sum_{l \in A} \frac{t^j}{j!}.$$

Then, using the generating functions defined by (5.7.7)–(5.7.9), we find that

$$\sum_{m=0}^{\infty} T_m^A \frac{t^m}{m!} = e^{a(t)}, \tag{5.7.21}$$

$$\sum_{m=0}^{\infty} \sum_{n=0}^{m} T_{mn}^A \frac{t^m}{m!} x^n = e^{xa(t)}, \tag{5.7.22}$$

$$\sum_{m=0}^{\infty} \sum_{k=0}^{[m/l]} T_m^A(l,k) \frac{t^m}{m!} x^k = e^{(x-1)t^l/l! + a(t)}, \qquad l \in A. \tag{5.7.23}$$

5.8 Asymptotics of the Bell numbers

To find the asymptotics, we use Dobinski's formula, which represents the Bell numbers as the series

$$T_m = \frac{1}{e} \sum_{k=0}^{\infty} \frac{k^m}{k!}. \tag{5.8.1}$$

Let us find the asymptotics of the numbers T_m as $m \to \infty$. Split the sum

$$S = \sum_{k=0}^{\infty} \frac{k^m}{k!}$$

into three parts so that

$$S = S_1 + S_2 + S_3, \tag{5.8.2}$$

where

$$S_1 = \sum_{k=0}^{\theta_1 - 1} \frac{k^m}{k!}, \quad S_2 = \sum_{k=\theta_1}^{\theta_2} \frac{k^m}{k!}, \quad S_3 = \sum_{k=\theta_2+1}^{\infty} \frac{k^m}{k!},$$

$$\theta_1 = [e^r - A\sqrt{m}], \qquad \theta_2 = [e^r + A\sqrt{m}];$$

r is the unique solution of the equation

$$re^r = m, \tag{5.8.3}$$

and A is a positive constant.

Let us find an asymptotic estimate of S_2 as $m \to \infty$. Changing the variables according to the formula $k' = k - [e^r]$, denoting the new variable by k and using Stirling's formula in estimating factorials, we obtain

$$S_2 = \frac{1}{\sqrt{2\pi}} \sum_{-A\sqrt{m} \leqslant k \leqslant A\sqrt{m}} ([e^r] + k)^{m-[e^r]-k-1/2} e^{[e^r]+k}(1 + o(1)).$$

Taking (5.8.3) into account and carrying out the necessary transformations, we obtain

$$S_2 = \frac{e^{m(r+1/r-1)}}{\sqrt{2\pi(r+1)}} \sum_{-A\sqrt{m} \leqslant k \leqslant A\sqrt{m}} e^{k^2 r(r+1)/(2m)} \left(\frac{r(r+1)}{m}\right)^{1/2} (1 + o(1)).$$

The last sum is an integral sum, therefore as $m \to \infty$

$$\sum_{-A\sqrt{m} \leqslant k \leqslant A\sqrt{m}} e^{k^2 r(r+1)/(2m)} \left(\frac{r(r+1)}{m}\right)^{1/2}$$

$$= \int_{-A\sqrt{r(r+1)}}^{A\sqrt{r(r+1)}} e^{-y^2/2}\, dy(1 + o(1)) = \sqrt{2\pi}(1 + o(1)),$$

where the error of approximation is of an exponential order. Thus,

$$S_2 = \frac{1}{\sqrt{r+1}} e^{m(r+1/r-1)}(1 + o(1)). \tag{5.8.4}$$

Now estimate the sums S_1 and S_3. For large m and some $c > 0$,

$$S_1 = \sum_{k=0}^{[e^r - A\sqrt{m}]-1} \frac{k^m}{m!}$$

$$< \frac{e^{m(r+1/r-1)}}{\sqrt{2\pi(r+1)}} \sum_{-m/r \leqslant k \leqslant -A\sqrt{m}} e^{-(ck)^2 r(r+1)/(2m)}$$

$$\times \left(\frac{r(r+1)}{m}\right)^{1/2} (1 + o(1)).$$

Replacing the sum by an integral, we obtain the bound

$$S_1 < \frac{e^{m(r+1/r-1)}}{\sqrt{2\pi(r+1)}} \int_{-\infty}^{-A\sqrt{r(r+1)}} e^{-(cy)^2/2}\, dy.$$

Hence, taking $A = 1/c$, we find that

$$S_1 < \frac{1}{c\sqrt{2\pi r(r+1)}} \exp\left\{ m\left(r + \frac{1}{r} - 1\right) - \frac{r(r+1)}{2} \right\}.$$

Finally, we obtain

$$\frac{S_1}{S_2} = o\left(\frac{1}{r} \left(\frac{r}{m}\right)^{(r+1)/2} \right).$$

Similarly, it can be shown that

$$\frac{S_3}{S_2} = o\left(\frac{1}{r} \left(\frac{r}{m}\right)^{(r+1)/2} \right).$$

Using the two last asymptotic estimates, relation (5.8.4) and representation (5.8.2), we obtain the final estimate

$$T_m = \frac{1}{\sqrt{r+1}} e^{m(r+1/r-1)-1}(1 + o(1)), \tag{5.8.5}$$

where r is the unique root of the equation $re^r = m$. For sufficiently large m, this solution can be represented as the convergent series

$$r = \ln m - \ln\ln m + \sum_{k=0}^{\infty}\sum_{l=0}^{\infty} C_{kl} (\ln\ln m)^{l+1} (\ln m)^{-k-l-1}, \tag{5.8.6}$$

where the C_{kl} do not depend on m.

5.9 Coverings of sets by subsets

5.9.1 Coverings

A set of distinct, non-empty, unordered subsets $X_1, X_2, \ldots, X_k$ of a set $X = \{x_1, x_2, \ldots, x_n\}$ is called a covering of this set if $X = X_1 \cup X_2 \cup \cdots \cup X_k$. The subsets $X_1, X_2, \ldots, X_k$ are called the blocks of the covering, and a covering consisting of k blocks will be referred to as a k-block covering. We say that a family $X_1, X_2, \ldots, X_k$ is of specification $[1^{\beta_1} 2^{\beta_2} \ldots n^{\beta_n}]$ if β_j subsets of this family are of size j, $1 \leqslant j \leqslant n$. If $D(\beta_1, \beta_2, \ldots, \beta_n)$ is the number of coverings of specification $[1^{\beta_1}2^{\beta_2}\dots n^{\beta_n}]$, then the following formula is valid (Goulden and Jackson, 1983):

$$D(\beta_1,\beta_2,\dots,\beta_n)=\sum_{\nu=0}^{n-1}(-1)^\nu\binom{n}{\nu}\prod_{j=1}^{n-\nu}\binom{\binom{n-\nu}{j}}{\beta_j}. \tag{5.9.1}$$

This formula can be proved by the inclusion–exclusion method. If, for a family $X_1,X_2,\dots,X_k$, A_i is the property that the element $x_i\in X$, $1\leqslant i\leqslant n$, is not covered, then the number of families of a given specification possessing the properties $A_{i_1},A_{i_2},\dots,A_{i_\nu}$, $1\leqslant i_1<\dots<i_\nu\leqslant n$, is equal to

$$M(A_{i_1}A_{i_2}\dots A_{i_\nu})=\prod_{j=1}^{n-\nu}\binom{\binom{n-\nu}{j}}{\beta_j}. \tag{5.9.2}$$

Applying formulae (2.6.1), we obtain equality (5.9.1). If the condition

$$\frac{1}{n}\sum_{j=1}^{n-1}jb_j-\ln n\to\infty \tag{5.9.3}$$

holds as $n\to\infty$, then (5.9.1) yields the asymptotic estimate

$$D(\beta_1,\beta_2,\dots,\beta_n)=\prod_{j=1}^{n-1}\binom{\binom{n}{j}}{\beta_j}(1+o(1)), \tag{5.9.4}$$

which can be obtained if we apply the Bonferroni inequalities to formula (5.9.1) and then use the inequalities

$$\prod_{j=1}^{n-1}\binom{\binom{n-1}{j}}{\beta_j}\prod_{i=1}^{n}\binom{\binom{n}{i}}{\beta_i}^{-1}\leqslant\prod_{j=1}^{n-1}\left(1-\frac{j}{n}\right)^{\beta_j}\leqslant\exp\left\{-\frac{1}{n}\sum_{j=1}^{n-1}j\beta_j\right\}.$$

The number $D_{n,k}$ of k-block coverings of a set containing n elements is equal to (Sachkov, 1982)

$$D_{n,k}=\sum_{j=0}^{n}(-1)^j\binom{n}{j}\binom{2^{n-j}-1}{k},\qquad k=1,2,\dots,2^n-1. \tag{5.9.5}$$

Indeed, if, for a family $X_1,X_2,\dots,X_k$, A_i is the property that $x_i\notin X_1\cup X_2\cup\dots\cup X_k$, then the number of families possessing the properties $A_{i_1},\dots,A_{i_j}$ simultaneously is equal to $\binom{2^{n-j}-1}{k}$. Using the inclusion–exclusion method, we obtain formula (5.9.5). Further, summing both sides of equality (5.9.5), we obtain the following formula for the total number D_n of coverings of a set containing n elements (Sachkov, 1982):

$$D_n = \sum_{j=0}^{n} (-1)^j \binom{n}{j} 2^{2^{n-j}-1}, \quad n = 1, 2, \ldots, \quad D_0 = 1. \tag{5.9.6}$$

If the condition

$$\frac{n}{2^k} \to 0 \tag{5.9.7}$$

holds as $n \to \infty$, then the asymptotic formulae

$$D_{n,k} = \binom{2^n - 1}{k}(1 + o(1)), \tag{5.9.8}$$

$$D_n = 2^{2^n - 1}(1 + o(1)), \tag{5.9.9}$$

are valid, which can be proved if we apply the Bonferroni inequalities to formulae (5.9.5) and (5.9.6) and then use the inequality

$$\binom{2^{n-1} - 1}{k} \leqslant \frac{1}{2^k}\binom{2^n - 1}{k}.$$

5.9.2 Minimal coverings

A covering of a set $X = \{x_1, x_2, \ldots, x_n\}$ by blocks $X_1, X_2, \ldots, X_k$ is called minimal if after removing any one block the remainder do not form a covering. It is clear that a covering of X by blocks $X_1, X_2, \ldots, X_k$ is minimal if and only if for any i, $1 \leqslant i \leqslant k$, there exists an element $x \in X$ such that $x \in X_i$ but $x \notin X_j$ for all $j \neq i$. In this case we say that x is a 1-covered element.

We associate each element $x_j \in X$, $1 \leqslant j \leqslant n$, with the indicator vector $\delta_j = (\delta_j^{(1)}, \delta_j^{(2)}, \ldots, \delta_j^{(k)})$, where

$$\delta_j^{(i)} = \begin{cases} 1, & x_j \in X_i, \\ 0, & x_j \notin X_i, \end{cases} \qquad i = 1, 2, \ldots, k. \tag{5.9.10}$$

For coverings, any k-dimensional binary vectors, except the zero vector, can be chosen as the $\delta_1, \delta_2, \ldots, \delta_n$. For minimal coverings, among the vectors $\delta_1, \delta_2, \ldots, \delta_n$ there are at least k from the set $\{(1, 0, \ldots, 0), (0, 1, 0, \ldots, 0), \ldots, (0, 0, \ldots, 0, 1)\}$ which determine the 1-covered elements in the blocks $X_1, X_2, \ldots, X_k$. We denote by $L_n(k)$ the number of minimal k-block coverings of a set with n elements. The following formula (Sachkov, 1982) is valid:

$$L_n(k) = \frac{1}{k!} \sum_{s=0}^{k} (-1)^s \binom{k}{s} (2^k - s - 1)^n. \tag{5.9.11}$$

There exist $(2^k-1)^n$ indicator vectors $\delta_1, \delta_2, \dots, \delta_n$ determining the k-block coverings of a set with n elements. These indicator vectors possess the property A_i if none of them coincide with the vector $(0,\dots,0,1,0,\dots,0)$ where the unit is placed at the ith position, $1 \leqslant i \leqslant k$. The number of indicator vectors which simultaneously possess s fixed properties from the set $A_1, A_2, \dots, A_k$ is equal to $(2^k-s-1)^n$. Applying the inclusion–exclusion method, we find that the number of non-zero vectors $\delta_1, \delta_2, \dots, \delta_n$ among which the vectors $(1,0,\dots,0),\dots,(0,0,\dots,0,1)$ appear at least once is equal to

$$\tilde{L}_n(k) = \sum_{s=0}^{k} (-1)^s \binom{k}{s} (2^k - s - 1)^n. \tag{5.9.12}$$

We can establish $k!$ bijections between the blocks of a minimal covering $X_1, X_2, \dots, X_k$ and k distinct vectors $(1,0,\dots,0),\dots,(0,0,\dots,0,1)$ determining the 1-covered elements. Hence it follows that

$$\tilde{L}_n(k) = k!\,L_n(k). \tag{5.9.13}$$

Now formula (5.9.11) follows from equalities (5.9.12) and (5.9.13).

If the condition

$$k \exp\left\{-\frac{n}{2^k-1}\right\} \to \lambda, \qquad 0 \leqslant \lambda < \infty \tag{5.9.14}$$

holds as $n, k \to \infty$, then the asymptotic formula (Sachkov, 1982)

$$L_n(k) = \frac{1}{k!}(2^k-1)^n e^{-\lambda}(1+o(1)) \tag{5.9.15}$$

is valid.

5.9.3 Coverings with repetition

A family of, not necessarily distinct, subsets $X_1, X_2, \dots, X_k$ of a set $X = \{x_1, x_2, \dots, x_n\}$ such that $X = X_1 \cup X_2 \cup \cdots \cup X_k$ is called a covering with repetition of the set X. The subsets $X_1, X_2, \dots, X_k$ are called blocks, and a covering containing k blocks is referred to as a k-block covering. In the general case, the empty sets are allowed to be blocks. The elements $x_1, x_2, \dots, x_k$ can either be distinguishable or indistinguishable, and the family $X_1, X_2, \dots, X_k$ can either be ordered or unordered. In order to give a formal description of the various possible cases let us consider the groups of substitutions G and H acting on the sets $\{x_1, x_2, \dots, x_n\}$ and $\{X_1, X_2, \dots, X_k\}$ respectively. The GH-equivalence relation corresponding to these groups allows us to describe these cases. As in the general combinatorial scheme, we choose either the identity group $\{e_n\}$ or the symmetric group S_n of degree n as the group G. Similarly, either the group $\{e_k\}$ or S_k can be chosen as the group H. In line with these assumptions, we consider four particular cases (Clarke, 1990), where the number of equivalence classes is denoted by $D(e_n, e_k)$, $D(S_n, e_k)$, $D(e_n, S_k)$, $D(S_n, S_k)$.

As the indicator vector determined by formulae (5.9.10), we take any of $2^n - 1$ non-zero vectors. Therefore, a covering with repetition is determined by a mapping $\sigma : X \to B$, where B is the set of non-zero k-dimensional binary vectors. Hence it follows that

$$D(e_n, e_k) = (2^k - 1)^n. \tag{5.9.16}$$

On the set of mappings σ we define the S_n-equivalence relation, where S_n acts on the set X. It is obvious that $D(S_n, e_k)$ determines the number of n-samples in a $(2^k - 1)$-basis, and therefore,

$$D(S_n, e_k) = \binom{2^k + n - 2}{n}. \tag{5.9.17}$$

We evaluate $D(e_n, S_k)$ by the inclusion–exclusion method. Denote by A_j the property that $x_j \notin X_1 \cup \cdots \cup X_k$, $1 \leqslant j \leqslant n$. The number of unordered families of subsets such that $x_{j_1}, x_{j_2}, \ldots, x_{j_\nu} \notin X_1 \cup \cdots \cup X_k$ is equal to the number of k-samples in a $2^{n-\nu}$-basis, that is,

$$M(A_{j_1} \ldots A_{j_\nu}) = \binom{2^{n-\nu} + k - 1}{k}.$$

Therefore,

$$D(e_n, S_k) = \sum_{\nu=0}^{n} (-1)^\nu \binom{n}{\nu} \binom{2^{n-\nu} + k - 1}{k}. \tag{5.9.18}$$

Let $n \to \infty$ and let condition (5.9.7) be satisfied. Then from formula (5.9.18) we obtain

$$D(e_n, S_k) = \binom{2^n + k - 1}{k}(1 + o(1)). \tag{5.9.19}$$

Indeed, using the Bonferroni inequalities and the bound

$$\frac{(2^{n-1} + k - 1)_k}{(2^n + k - 1)_k} \leqslant \begin{cases} 2^{-k} e^{k^2/2^n}, & k \leqslant n/2, \\ 2^{-n/2} e, & k > n/2, \end{cases}$$

we derive the asymptotic estimate (5.9.19) from formula (5.9.18).

A more complicated derivation of the formula for $D(S_n, S_k)$ will be carried out in Section 6.1.

5.9.4 Minimal coverings with repetition

A covering with repetition $X_1, X_2, \ldots, X_k$ of a set $X = \{x_1, x_2, \ldots, x_n\}$ is called minimal if the family $X_1, X_2, \ldots, X_k$ contains no subfamily that is a covering of the set X. As in the case of coverings without repetition, we consider four cases of minimal coverings according to which pair of the groups $\{e_n\}$, S_n and $\{e_k\}$, S_k determines the GH-equivalence relation. The number of equivalence classes is denoted by $M(e_n, e_k)$, $M(S_n, e_k)$, $M(e_n, S_k)$, $M(S_n, S_k)$ respectively.

The formula (Sachkov, 1982)

$$M(e_n, e_k) = \sum_{s=0}^{k} (-1)^s \binom{k}{s} (2^k - s - 1)^n \qquad (5.9.20)$$

is true. We say that a covering $X_1, X_2, \ldots, X_k$ possesses property A_i if after removing the block X_i the remaining blocks form a covering. The number of coverings simultaneously possessing the properties $A_{i_1}, \ldots, A_{i_s}$ is $(2^k - s - 1)^n$. Applying the inclusion–exclusion method, we obtain formula (5.9.20).

For $n \geqslant k$ the equality

$$M(S_n, e_k) = D(S_{n-k}, e_k) \qquad (5.9.21)$$

holds. Indeed, if $X_1, X_2, \ldots, X_n$ is a minimal covering of X, then for each X_i there exists an element $x \in X_i$ such that $x \notin X_j$ for all $j \neq i$, $1 \leqslant i \leqslant k$. Such an x is called a 1-covered element for the given block. Eliminating one 1-covered element from every block, we obtain a k-block covering of a set containing $n - k$ elements. Adding, in turn, one element of every block of a k-block covering of a set with $n - k$ elements, we obtain a minimal k-block covering of a set with n elements. Hence equality (5.9.21) follows. From equality (5.9.21) and formula (5.9.17) it follows (Clarke, 1990) that

$$M(S_n, e_k) = \binom{2^k + n - k - 2}{n - k}. \qquad (5.9.22)$$

In the case where the equivalence relation is determined by the groups $\{e_n\}$ and S_k, all the blocks of a minimal covering must be distinct; therefore

$$M(e_n, S_k) = \frac{1}{k!} M(e_n, e_k). \qquad (5.9.23)$$

From (5.9.17) and formula (5.9.12) we obtain the equality (Clarke, 1990)

$$M(e_n, S_k) = \frac{1}{k!}\sum_{\nu=0}^{n}(-1)^{\nu}\binom{n}{\nu}\binom{2^{n-\nu}+k-1}{k}. \tag{5.9.24}$$

By the same reasoning as in the derivation of (5.9.21) we can obtain the relation

$$M(S_n, S_k) = D(S_{n-k}, S_k). \tag{5.9.25}$$

5.9.5 Diagnostics and minimal coverings

A diagnostic system (Flajolet and Odlyzko, 1990) based on a knowledge base is modeled by a binary relation $E \subseteq H \times O$, where H is the set of hypotheses and O is the set of observations, $H \cap O = \varnothing$. The binary relation E satisfies the following conditions:

(1) For any $h \in H$ there exists at least one $o \in O$ such that $(h, o) \in E$.
(2) For any $o \in O$ there exists at least one $h \in H$ such that $(h, o) \in E$.

The set of observations related to a hypothesis $h \in H$ is called the cluster associated with h. The cluster is a set of the form

$$\mathcal{U}(h) = \{o : (h, o) \in E\}.$$

Now let $X \subseteq O$ be the set of feasible observations. Then each cluster is restricted to its intersection with X.

A diagnostic of the set X is determined by a family of hypotheses Z such that the union of the corresponding associated clusters is a minimal covering of X, that is,

$$\bigcup_h \{\mathcal{U}(h) : h \in Z\} = X,$$

where every hypothesis $h' \in Z$ satisfies the condition

$$\bigcup_h \{\mathcal{U}(h) : h \in (Z \setminus \{h'\})\} \neq X.$$

Thus, the description of all minimal coverings of a set corresponds to the description of all diagnostics of this set.

5.9.6 Algorithms

There is a very simple algorithm for verifying whether a family $X_1, X_2, \ldots, X_k$ of subsets of a set X is a covering. If $X = \{x_1, x_2, \ldots, x_n\}$, then the elements of the sets $X_1, X_2, \ldots, X_k$ are allocated to n memory cells. A family $X_1, X_2, \ldots, X_k$ will form a covering of X if all the cells are occupied. Now let a family $X_1, X_2, \ldots, X_k$ form a partition of the set X. The algorithm for obtaining a minimal covering from the given one consists of the following. By item-by-item examination we find X_i, $1 \leqslant i \leqslant k$, such that the inclusion

$$X_i \subseteq \bigcup_{j \neq i} X_j$$

holds. If there exists no such X_i, then the covering is minimal; otherwise X_i is eliminated from the covering and then the algorithm is applied to the covering consisting of the remaining blocks. At each step of the recursive application of the algorithm there may exist several blocks satisfying the inclusion. This allows us to construct a branching process whose final stages give several minimal coverings. There are recursive algorithms which give all minimal coverings without repetition (Flajolet and Odlyzko, 1990).

6

Pólya's theorem and its applications

We have pointed out the important role of formalizing the notion of the indistinguishability of elements in enumerative combinatorial problems. A fundamental contribution to the development of the methods of such formalization was made by Pólya in his famous paper (Pólya, 1937). The ideas in this paper were developed by de Bruijn (Beckenbach, 1964) and other mathematicians. It should be noted that, prior to the paper (Pólya, 1937), a similar method had been suggested by Redfield in (Redfield, 1927).

The essence of the method of solving enumerative problems, referred to as Pólya's enumeration theory, can be described as follows. There is a set Y of elements such that each element possesses a characteristic or weight which takes values from a ring. The distribution of the elements of the set over the weights is determined by a generating function F, usually of several variables. The set of configurations of the form $f : X \to Y$, where X is a linearly ordered set, is considered. A characteristic or weight is also assigned to each configuration of the set. A group A of permutations generates an equivalence relation on the set of all configurations which determines the notion of the distinguishability of configurations. The non-equivalent configurations with given characteristics or weights are enumerated by a generating function Φ. The main theorem of Pólya's enumeration theory gives an expression of the generating function Φ in terms of the generating function F using some polynomial Z of several variables called the cycle index. This polynomial is determined by the cycle classes of the group A, which generates the equivalence classes of configurations.

In combinatorial problems whose formalization fits the scheme described above, the enumerative problem consists of finding the generating function Φ. Distinguishability of the elements is determined by the partition of the set of more simple elements, the configurations, into equivalence classes with respect to the group A of substitutions. The generating function F of these configurations is assumed to be known. Note that the cycle index Z of some groups can easily be written, but, in a number of cases, finding the cycle index is an independent problem.

The central part of this chapter is given over to Pólya's theorem, which describes the relation between the generating functions F and Φ. Certain modifications and applications of the theorem are also given.

Pólya's enumeration theory was developed for solving enumerative problems of graph theory and at present most of its applications lie, as before, in this area of combinatorics. Therefore, this chapter contains some applications of Pólya's theorem to enumerative problems of graph theory. Modifications of this theorem permit us to obtain some formulae for the number of non-isomorphic abstract automata; this is discussed in the final section.

6.1 Burnside's lemma

6.1.1 Cycle index

Let A be a group of substitutions of degree n acting on the set $X = \{1, \ldots, n\}$. Denote by $j_k(\alpha)$ the number of cycles of length k in the decomposition of a substitution $\alpha \in A$ into a product of independent cycles. It is clear that

$$\sum_{k=1}^{n} k j_k(\alpha) = n.$$

The cycle index of the group A is the polynomial of variables $t_1, \ldots, t_n$

$$Z(A) = Z(t_1, \ldots, t_n; A) = \frac{1}{|A|} \sum_{\alpha \in A} \prod_{k=1}^{n} t_k^{j_k(\alpha)}, \tag{6.1.1}$$

where $|A|$ is the order of the group A.

For example, the cycle index of the symmetric group S_n of degree n is equal to

$$Z(S_n) = \frac{1}{n!} \sum_{j_1 + 2j_2 + \cdots + nj_n = n} C(j_1, \ldots, j_n) t_1^{j_1} \cdots t_n^{j_n}, \tag{6.1.2}$$

where $C(j_1, \ldots, j_n)$ is the number of elements in the cycle class $[1^{j_1} 2^{j_2} \ldots n^{j_n}]$. This number is equal to

$$C(j_1, \ldots, j_n) = \frac{n!}{1^{j_1} 2^{j_2} \cdots n^{j_n} j_1! \, j_2! \cdots j_n!}. \tag{6.1.3}$$

For $Z(S_n)$ the recurrence relation

$$Z(S_n) = \frac{1}{n}\sum_{k=1}^{n} Z(S_{n-k})t_k$$

is valid (Harary and Palmer, 1973).

The set of all even substitutions of degree n with respect to the ordinary operation of multiplication constitutes a subgroup of the symmetric group S_n. This subgroup is called the alternating group and is denoted by A_n. The cycle index of the alternating group A_n can be represented in terms of the cycle index of the symmetric group S_n in the following way:

$$Z(A_n) = Z(t_1, t_2, t_3, t_4, \dots; S_n) + Z(t_1, -t_2, t_3, -t_4, \dots; S_n). \quad (6.1.4)$$

The powers $C^0, C^1, \dots, C^{n-1}$ of the unicyclic substitution $C = (1, 2, \dots, n)$ constitute the cyclic group of substitutions C_n. The cycle index of C_n is equal to

$$Z(C_n) = \frac{1}{n}\sum_{k|n} \varphi(k) t_k^{n/k}, \quad (6.1.5)$$

where $\varphi(k)$ is the Euler totient function and the summation is carried out over all divisors of n.

The dihedral group D_n is generated by the cycle $C = (1, 2, \dots, n)$ and the substitution $(1, n)(2, n-1)(3, n-2)\cdots$. Its cycle index is equal to

$$Z(D_n) = \frac{1}{2}Z(C_n) + \begin{cases} \frac{1}{2}t_1 t_2^{(n-1)/2}, & n \text{ is odd}, \\ \frac{1}{4}(t_2^{n/2} + t_1^2 t_2^{(n-2)/2}), & n \text{ is even}. \end{cases} \quad (6.1.6)$$

Finally, the cycle index of the identity group E_n of substitutions which consists of one identity substitution is obviously equal to

$$Z(E_n) = t_1^n. \quad (6.1.7)$$

The cycle index does not determine the corresponding group of substitutions uniquely. Moreover, as Pólya showed, two groups of substitutions A and B with the same cycle index can even be non-isomorphic.

6.1.2 Burnside's lemma

We say that elements $x, y \in X = \{1, 2, \dots, n\}$ are equivalent with respect to a group of substitutions A acting on X, or A-equivalent, if there exists a substitution $\alpha \in A$ such that $\alpha(x) = y$. In such a case we write $x \equiv y$. The equivalence classes on the set X are called the orbits of the group A or the transitivity domains of the group.

The stabilizer $A(x)$ of an element $x \in X$ with respect to the group A is the set of substitutions

$$A(x) = \{\alpha : \alpha(x) = x, \ \alpha \in A\}.$$

It is clear that $A(x)$ is a subgroup of the group A. If $x \equiv y$, then $\alpha(x) = y$ for some $\alpha \in A$, and, consequently,

$$A(y) = \alpha^{-1} A(x) \alpha,$$

that is, the subgroups $A(x)$ and $A(y)$ are conjugated in the group A. Hence, $|A(x)| = |A(y)|$.

Decompose the group A into residue classes with respect to the subgroup $A(y)$; then

$$A = A(y) \cup A(y)\alpha_1 \cup \cdots \cup A(y)\alpha_{l-1}.$$

Put $y_i = \alpha_i(y)$, $i = 1, \ldots, l-1$, and show that the elements $y, y_1, \ldots, y_{l-1}$ exhaust the orbit Y containing y. Suppose that there exists $\bar{y} \in Y$, $\bar{y} \neq y_i$, $i = 0, 1, \ldots, l-1$, $y_0 = y$. It is clear that there exists $\bar{\alpha} \in A$ such that $\bar{y} = \bar{\alpha}(y)$. If $\bar{\alpha} \in A(y)\alpha_i$, then $\bar{y} = \bar{\alpha}(y) = y_i$ for some i, $0 \leqslant i \leqslant l-1$. We obtain a contradiction; therefore $|Y| = l$, and, consequently,

$$|Y||A(y)| = |A|. \tag{6.1.8}$$

Lemma 6.1.1 (Burnside's lemma) *The number of orbits on the set X with respect to the group of substitutions A is equal to*

$$N(A) = \frac{1}{|A|} \sum_{\alpha \in A} j_1(\alpha). \tag{6.1.9}$$

In other words, the number of orbits is equal to the mean value of the unit cycles in a random substitution sampled from the group A with equal probabilities.

Proof Let $X_1, \ldots, X_{N(A)}$ be the orbits on the set X with respect to the group A and $x_i \in X_i$, $i = 1, \ldots, N(A)$. It follows from (6.1.8) that

$$\sum_{x \in X} |A(x)| = \sum_{i=1}^{N(A)} \sum_{x \in X_i} |A(x)|$$

$$= \sum_{i=1}^{N(A)} |A(x_i)||X_i| = N(A)|A|. \tag{6.1.10}$$

On the other hand,

$$\sum_{x\in X}|A(x)| = \sum_{x\in X}\sum_{\alpha\in A(x)} 1$$

$$= \sum_{\alpha\in A}\sum_{x:\,\alpha(x)=x} 1 = \sum_{\alpha\in A} j_1(\alpha). \tag{6.1.11}$$

Comparing equalities (6.1.10) and (6.1.11), we obtain the assertion of the lemma. □

Corollary 6.1.1 *If $A \mid Y$ is the set of substitutions which are restrictions to the set Y of substitutions from the group A, where Y is the union of the orbits of A, and $j_1(\alpha \mid Y)$ is the number of fixed points of the restriction α to the set Y, then*

$$N(A \mid Y) = \frac{1}{|A|}\sum_{\alpha\in A} j_1(\alpha \mid Y). \tag{6.1.12}$$

Let R be a commutative ring containing the rational numbers, and let $w : X \to R$ be a weight function on the orbits of the group A acting on the set X. The value $w(x)$ will be referred to as the weight of an element x. The weight $w(X_i)$ of the orbit X_i is defined as the weight of any of its elements, that is, $w(X_i) = w(x)$ for $x \in X_i$.

Lemma 6.1.2 (Burnside–de Bruijn lemma) *The total weight of orbits of the group of substitutions A is equal to the mean value of the total weight of the fixed elements in a random substitution sampled from A with equal probabilities, that is,*

$$\sum_{i=1}^{N(A)} w(X_i) = \frac{1}{|A|}\sum_{\alpha\in A}\sum_{x:\,\alpha(x)=x} w(x). \tag{6.1.13}$$

The assertion of the lemma is a consequence of the following chain of equalities:

$$\sum_{\alpha\in A}\sum_{x:\,\alpha(x)=x} w(x) = \sum_{x\in X}\sum_{\alpha\in A(x)} w(x)$$

$$= \sum_{i=1}^{N(A)} w(X_i)|X||A(x_i)| = |A|\sum_{i=1}^{N(A)} w(X_i).$$

6.1.3 Equivalence classes of families of sets and coverings

Let us consider the group of substitutions $G = S_n \times S_k$, where S_n and S_k are the symmetric groups acting on the sets $X = \{1, 2, \ldots, n\}$ and $X' = \{1, 2, \ldots, k\}$. We define the action of the group G on the set of vectors $U = \{(X_1, X_2, \ldots, X_k) : X_1, X_2, \ldots, X_k \subseteq X\}$ as follows. If $g = (g_1, g_2) \in G$, $g_1 \in S_n$, $g_2 \in S_k$, then $g(X_1, X_2, \ldots, X_k) = (Z_1, Z_2, \ldots, Z_k)$, where

$$Z_{g_2(i)} = g_1(X_i) = \{g_1(x) : x \in X_i\}, \qquad 1 \leqslant i \leqslant k.$$

We denote by $U(n,k)$ the number of orbits of the group G on the set U, and by $D(S_n, S_k)$, the number of orbits on the set U which consist of the vectors $(X_1, X_2, \ldots, X_k)$ such that $\bigcup_{i=1}^{k} X_i = X$. It is clear that (Clarke, 1990)

$$D(S_n, S_k) = U(n,k) - U(n-1,k). \tag{6.1.14}$$

To evaluate $U(n,k)$, we use Burnside's lemma. To this end we should calculate the mean number of fixed elements for the group G. Let $g(X_1, X_2, \ldots, X_k) = (X_1, X_2, \ldots, X_k)$, $g \in G$, $(X_1, X_2, \ldots, X_k) \in U$. Then for all $i \in X$ and $j \in X'$ the condition that $i \in X_j$ if and only if $g_1(i) \in X_{g_2(j)}$, $g_1 \in S_n$, $g_2 \in S_k$, is satisfied. We denote by $\chi(g)$ the number of elements $(X_1, X_2, \ldots, X_k) \in U$ satisfying this condition.

Let $g_1 = \lambda_1' \lambda_2' \cdots$ and $g_2 = \lambda_1'' \lambda_2'' \cdots$ be expansions of the substitutions g_1 and g_2 into products of independent cycles, and $\lambda_i' = (s_1, s_2, \ldots, s_r)$, $\lambda_j'' = (l_1, l_2, \ldots, l_t)$. We set

$$Y_{ij} = (X_{ij}^{(l_1)}, X_{ij}^{(l_2)}, \ldots, X_{ij}^{(l_t)}),$$

where

$$X_{ij}^{(l_v)} = X_{l_v} \cap \{s_1, s_2, \ldots, s_r\}, \qquad 1 \leqslant v \leqslant t.$$

It is clear that Y_{ij} and $Y_{i'j'}$ are chosen independently for $(i,j) \neq (i',j')$. Let us find the number of ways of choosing Y_{ij} for a fixed pair of cycles. Without loss of generality we assume that $\lambda_i' = (1, 2, \ldots, u)$ and $\lambda_j'' = (1, 2, \ldots, v)$. An element (a, b) belongs to a cycle containing the element (6.1.1) if and only if $a \equiv \nu \pmod u$ and $b \equiv \nu \pmod v$ for some ν. Hence it follows that there exist integers γ and δ satisfying the condition $a - b = \gamma u - \delta v$. If $d = (u, v)$ is the greatest common divisor of u and v, then the last equality implies that $d \mid (a - b)$.

Let us choose a subset $\sigma = \sigma(Y_{ij}) \subseteq \{1, 2, \ldots, d\}$. We put

$$X_{ij}^{(1)} = \{i : i \equiv j \pmod d \text{ for some } j \in \sigma,\ 1 \leqslant i \leqslant u\}.$$

Other elements X_{ij} are chosen as follows:

$$X_{ij}^{(l)} = \{i + l - 1 \pmod u) : i \in X_{ij}^{(1)}\}, \qquad l = 1, 2, \ldots, v.$$

Hence it follows that the number of ways of choosing Y_{ij} is equal to $2^d = 2^{(u,v)}$.

If α_i and β_j are the cycle lengths of λ'_i and λ''_j, respectively, then by virtue of the independence of the choice of Y_{ij} for distinct pairs (i, j) we obtain

$$\chi(g) = \prod_{(i,j)} 2^{(\alpha_i, \beta_j)}.$$

Hence it follows (Clarke, 1990) that

$$U(n,k) = \sum_{\substack{\sum_{i=1}^n i\alpha_i = n, \\ \sum_{j=1}^k j\beta_j = k}} \prod_{i=1}^{n} \frac{1}{i^{\alpha_i}\alpha_i!} \prod_{j=1}^{k} \frac{1}{j^{\beta_j}\beta_j!} \prod_{(i,j)} 2^{(\alpha_i,\beta_j)}. \tag{6.1.15}$$

Thus, the number $D(S_n, S_k)$ of coverings of a set containing n indistinguishable elements by k blocks, whose ordering is irrelevant, is determined by equality (6.1.14), where the values of $U(n,k)$ are calculated by formula (6.1.15).

6.2 Pólya's theorem

6.2.1 The main theorem

Consider finite sets X and Y such that $|X| = m$ and $|Y| = n$, and denote as usual the set of all configurations $f : X \to Y$ by Y^X. Suppose that a group A of substitutions of degree m acts on the set X and determines an equivalence relation $\equiv$ on the set Y^X. Under this equivalence relation $f_1 \sim f$, where $f, f_1 \in Y^X$, if there exists a substitution $\alpha \in A$ such that $f_1(x) = f(\alpha x)$ for all $x \in X$, where $\alpha x = \alpha(x)$.

Let a k-dimensional vector $(s_1, \ldots, s_k)$with components from an Abelian group be assigned to each element $y \in Y$. We call this vector the characteristic of y and write $[y] = (s_1, \ldots, s_k)$. We assume that the operation of component-wise addition is defined on the set of characteristics. The characteristic of a configuration f is defined as the sum of the characteristics of the values of the function, that is,

$$[f] = \sum_{x \in X} [f(x)]. \tag{6.2.1}$$

Denote by $a(s_1, \ldots, s_k)$ the number of elements $y \in Y$ with characteristic

$(s_1,\ldots,s_k)$, and by $b_m(s_1,\ldots,s_k)$, the number of non-equivalent configurations $f \in Y^X$ with the values of the characteristics equal to $(s_1,\ldots,s_k)$. Consider the generating function

$$F(y_1,\ldots,y_k) = \sum_{(s_1,\ldots,s_k)} a(s_1,\ldots,s_k) y_1^{s_1} \cdots y_k^{s_k}, \tag{6.2.2}$$

where the summation is over all characteristics of the elements of the set Y, and the generating function

$$\Phi_m(y_1,\ldots,y_k) = \sum_{(s_1,\ldots,s_k)} b_m(s_1,\ldots,s_k) y_1^{s_1} \cdots y_k^{s_k}, \tag{6.2.3}$$

where the summation is over all values of the characteristics of the non-equivalent configurations. Denote by $C(j_1,\ldots,j_m;A)$ the number of elements in the cycle class $[1^{j_1}2^{j_2}\ldots m^{j_m}]$ of the group A and introduce the cycle index of the group

$$Z(t_1,\ldots,t_m;A) = \frac{1}{|A|} \sum_{j_1+2j_2+\cdots+mj_m=m} C(j_1,\ldots,j_m;A) t_1^{j_1} \cdots t_m^{j_m}. \tag{6.2.4}$$

Pólya's theorem (Pólya, 1937) can be formulated as follows.

Theorem 6.2.1 (Pólya's theorem) *Let $F(y_1,\ldots,y_k)$ be the generating function given by* (6.2.2) *which enumerates the elements of Y with respect to the values of their characteristics, and let the group of substitutions A with the cycle index $Z(t_1,\ldots,t_m;A)$ given by* (6.2.4) *generate an equivalence relation on the set of configurations $f : X \to Y$, $|X| = m$. Then the generating function given by* (6.2.3) *enumerating the non-equivalent configurations is equal to*

$$\Phi_m(y_1,\ldots,y_k) = Z(F(y_1,\ldots,y_k), F(y_1^2,\ldots,y_k^2),\ldots,F(y_1^m,\ldots,y_k^m);A). \tag{6.2.5}$$

Proof First define a group $\bar{A}$ of substitutions on the set Y^X which is a homomorphic image of the group A. We assume that $\alpha \to \bar{\alpha}$ under the homomorphism $A \to \bar{A}$ if $\bar{\alpha}(f) = f(\alpha x)$ for all $x \in X$, $\alpha \in A$, $\bar{\alpha} \in \bar{A}$. The group $\bar{A}$ defines on Y^X an equivalence relation $\simeq$ such that $f \simeq f_1$ if $\bar{\alpha}(f) = f_1$. It is clear that if $f \sim f_1$, then $f \simeq f_1$ and vice versa, that is, the equivalence relations defined by the groups A and $\bar{A}$ coincide.

Denote by $b_m(s_1,\ldots,s_k;\bar{\alpha})$ the number of configurations with characteristic $(s_1,\ldots,s_k)$ which are fixed points of the substitution $\bar{\alpha}$. By Burnside's lemma

$$b_m(s_1, \ldots, s_k) = \frac{1}{|A|} \sum_{\bar{\alpha} \in \bar{A}} b_m(s_1, \ldots, s_k; \bar{\alpha}). \tag{6.2.6}$$

It is clear that if H is the kernel of the homomorphism $A \to \bar{A}$ under which $\alpha \to \bar{\alpha}$, then it follows from the condition $\bar{\alpha}(f) = f$ that $f(h\alpha x) = f(x)$, where $h \in H$. Thus, if Δ is an arbitrary system of representatives of the right residue classes of A with respect to the subgroup H, then

$$\frac{1}{|A|} \sum_{\alpha \in A} b_m(s_1, \ldots, s_k; \alpha) = \frac{1}{|A|} \sum_{\alpha \in \Delta} \sum_{h \in H} b_m(s_1, \ldots, s_k; h\alpha)$$
$$= \frac{1}{|A|} \sum_{\bar{\alpha} \in \bar{A}} b_m(s_1, \ldots, s_k; \bar{\alpha}).$$

Now it follows from equality (6.2.6) that

$$b_m(s_1, \ldots, b_k) = \frac{1}{|A|} \sum_{\alpha \in A} b_m(s_1, \ldots, s_k; \alpha). \tag{6.2.7}$$

Consider the generating function

$$\Phi_m(y_1, \ldots, y_k; \alpha) = \sum_{(s_1, \ldots, s_k)} b_m(s_1, \ldots, s_k; \alpha) y_1^{s_1} \cdots y_k^{s_k},$$

where the summation is over all configurations which are invariant with respect to the substitution α. We see from (6.2.3) and (6.2.7) that

$$\Phi_m(y_1, \ldots, y_k) = \frac{1}{|A|} \sum_{\alpha \in A} \Phi_m(y_1, \ldots, y_k; \alpha). \tag{6.2.8}$$

Let $f(\alpha x) = f(x)$, $\alpha \in A$, and let X_j be the orbit where a cycle α_j of length l of the substitution α acts. Then by virtue of the equalities

$$f(\alpha_j x) = f(\alpha_j^2 x) = \cdots = f(\alpha_j^l x) = f(x),$$

the configuration f is constant on X_j. Consequently, the set of configurations invariant under the action of the substitution $\alpha \in A$ can be obtained by independent choice of values of f on the orbits of α. If $\alpha = \alpha_1 \alpha_2 \ldots \alpha_d$ is a decomposition of α into a product of independent cycles and $b_m(s_1^{(j)}, \ldots, s_k^{(j)}; \alpha)$ is the number of configurations with characteristic $(s_1^{(j)}, \ldots, s_k^{(j)})$ which are the restrictions to X_j of the set of configurations invariant under the action of α_j, then

$$b_m(s_1, \ldots, s_k; \alpha) = \sum_{S_{kd}} b(s_1^{(1)}, \ldots, s_k^{(1)}; \alpha_1) \cdots b(s_1^{(d)}, \ldots, s_k^{(d)}; \alpha_d), \tag{6.2.9}$$

where the summation is over the set S_{kd} determined by the equations

$$s_1^{(1)} + s_1^{(2)} + \cdots + s_k^{(d)} = s_1,$$
$$\cdots$$
$$s_k^{(1)} + s_k^{(2)} + \cdots + s_k^{(d)} = s_k.$$

Consider generating functions of the form

$$\varphi(y_1, \ldots, y_k; \alpha_i) = \sum_{(s_1,\ldots,s_k)} b(s_1, \ldots, s_k; \alpha_i) y_1^{s_1} \cdots y_k^{s_k}.$$

It follows from equality (6.2.9) that

$$\Phi_m(y_1, \ldots, y_k; \alpha) = \varphi(y_1, \ldots, y_k; \alpha_1) \cdots \varphi(y_1, \ldots, y_k; \alpha_d). \quad (6.2.10)$$

Let $f(\alpha x) = f(x)$, and let $\bar{y}$ be the constant value which the configuration f takes on the orbit X_j corresponding to the cycle α_j of length l of the substitution α. If $[\bar{y}] = (s_1, \ldots, s_k)$, then the characteristic of the restriction of the configuration f to the orbit X_j is equal to $(s_1 l, \ldots s_k l)$, where $s_j l$ is the composition of l identical elements of the Abelian group. Consequently,

$$\varphi(y_1, \ldots, y_k; \alpha_j) = \sum_{(s_1,\ldots,s_k)} a(s_1, \ldots, s_k) y_1^{s_1 l} \cdots y_k^{s_k l}.$$

Thus, the generating function $\varphi(y_1, \ldots, y_k; \alpha_j)$ depends on the length l of the cycle α_j only, and, moreover,

$$\varphi(y_1, \ldots, y_k; \alpha_j) = F(y_1^l, \ldots, y_k^l). \quad (6.2.11)$$

From (6.2.8), (6.2.10) and (6.2.11) it follows that

$$\Phi_m(y_1, \ldots, y_k) = \frac{1}{|A|} \sum_{j_1+2j_2+\cdots+mj_m=m} C(j_1, \ldots, j_m; A) \times F^{j_1}(y_1, \ldots, y_k) \cdots F^{j_m}(y_1, \ldots, y_k).$$

With regard to (6.2.4) this equality is equivalent to (6.2.5). The theorem is proved. □

6.2.2 The necklace problem

There is a set of beads consisting of c_i of color i, $i = 1, \ldots, k$. Let m beads chosen from the set be placed on m equidistant points of a circle provided that $n = c_1 + \cdots + c_k \geqslant m$. If $X = \{1, \ldots, m\}$ and Y is the set of n beads, then such an allocation of beads on the circle is determined by a configuration $f : X \to Y$. A class of allocations of beads on the circle differing by a shift is called a necklace. It is clear that the necklaces are put in a one-to-one correspondence with the equivalence classes on the set Y^X induced by the cyclic group C_m acting on X.

If $y_i \in Y$ corresponds to a bead of the ith color, then we take as the characteristic of y_j the k-dimensional vector $[y_j] = (0,\ldots,0,1,0,\ldots,0)$, where the unit is on the ith place, $i = 1,\ldots,k$. The generating function of the characteristics of the set Y is equal to

$$F(y_1,\ldots,y_k) = c_1y_1 + \cdots + c_ky_k.$$

Using formula (6.1.5) for the cycle index of the group C_m, and Pólya's theorem, we find the generating function of the necklaces:

$$\Phi_m(y_1,\ldots,y_k) = \frac{1}{m}\sum_{d|m} \varphi(d)(c_1y_1^d + \cdots + c_ky_k^d)^{m/d}. \tag{6.2.12}$$

Hence it follows that if $N(s_1,\ldots,s_k)$ is the number of necklaces with primary specification $[y_1^{s_1}\ldots y_k^{s_k}]$, where y_i means the ith color, $i = 1,\ldots,k$, then

$$N(s_1,\ldots,s_k) = \frac{1}{m}\sum \varphi(d)\frac{(m/d)!\,c_1^{s_1/d}\cdots c_k^{s_k/d}}{(s_1/d)!\cdots(s_k/d)!}, \tag{6.2.13}$$

where the summation is over all divisors d of the greatest common divisor of the numbers $s_1,\ldots,s_k$. In particular, putting $y_1 = \cdots = y_k = 1$ in formula (6.2.12), we obtain the solution of the problem on the number of cyclic sequences:

$$N_{nm} = \frac{1}{m}\sum_{d|m} \varphi(d)n^{m/d}.$$

6.2.3 *m*-samples

Let $X = \{1,\ldots,m\}$, $Y = \{y_1,\ldots,y_n\}$ and let a group A of substitutions acting on the set X determine equivalence classes on the set Y^X which will be referred to as m-samples. Define the characteristics of the elements of Y in the same way as in the preceding subsection and consider the corresponding generating function

$$F(t_1,\ldots,t_n) = t_1 + \cdots + t_n.$$

The common value of the characteristics of the elements of an equivalence class is assumed to be the characteristic of the class. Denote by $C_m(s_1,\dots,s_n;a)$ the number of equivalence classes with characteristic $(s_1,\dots,s_n)$ and introduce the generating function

$$\Phi_m(t_1,\dots,t_n;A) = \sum_{s_1+\cdots+s_n=m} C_m(s_1,\dots,s_n;A)t_1^{s_1}\cdots t_n^{s_n}.$$

Using Pólya's theorem, we obtain

$$\Phi_m(t_1,\dots,t_n;A) = Z(t_1+\cdots+t_n, t_1^2+\cdots+t_n^2,\dots t_1^m+\dots,+t_n^m;A). \tag{6.2.14}$$

If $C_{nm}(A)$ is the number of orbits generated by the group A on the set Y^X, then

$$C_{nm}(A) = Z(n,\dots,n;A). \tag{6.2.15}$$

Put $A = S_m$, where S_m is the symmetric group of degree m. Then formula (6.2.15) gives the expression of the number of m-samples in the commutative asymmetric n-basis:

$$C_{nm} = Z(n,\dots,n;S_m) = \frac{1}{m!}\sum_{k=1}^{m} C(m,k)n^k, \tag{6.2.16}$$

where $C(m,k)$ is the number of substitutions of degree m with exactly k cycles. Using the identity

$$\sum_{k=1}^{m} C(m,k)n^k = n(n+1)\cdots(n+m-1),$$

and equality (6.2.16), we obtain the well-known formula

$$C_{nm} = \binom{n+m-1}{m}.$$

6.3 Trees and chemical trees

6.3.1 Cayley's relation

We denote by r_n the number of rooted trees with $n-1$ identical vertices and a root, and consider the generating function

$$r(x) = \sum_{n=1}^{\infty} r_n x^n. \tag{6.3.1}$$

Every tree with m edges at the root can be associated with m rooted trees. Thus, if $X = \{1,2,\dots,m\}$ and Y is the set of rooted trees whose one-dimensional characteristics are equal to the numbers of their vertices,

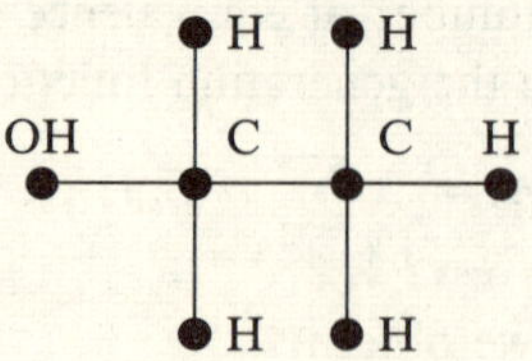

Fig. 6.3.1. The tree of ethyl alcohol

then every tree with m edges at the root can be put in one-to-one correspondence with an equivalence class on the set Y^X determined by the symmetric group S_m. If r_{nm} is the number of trees with n vertices and m edges incident to the root, and

$$r_m(x) = \sum_{n=m+1}^{\infty} r_{nm} x^n, \tag{6.3.2}$$

then, by virtue of Pólya's theorem,

$$r_m(x) = xZ(r(x), r(x^2), \ldots, r(x^m); S_m). \tag{6.3.3}$$

In addition, it is obvious that

$$r(x) = \sum_{m=0}^{\infty} r_m(x), \qquad r_0(x) = x. \tag{6.3.4}$$

Formulae (6.3.3) and (6.3.4) yield the well-known Cayley relation

$$r(x) = x \exp\left\{\sum_{m=1}^{\infty} \frac{r(x^m)}{m}\right\}.$$

6.3.2 Chemical trees

Let us represent the molecules of carbon chemical compounds called alcohols as rooted trees. The vertices of the trees labeled by C represent the atoms of carbon. Each such vertex is joined to four other vertices. All end vertices, excluding the root, are labeled by the letter H and represent the atoms of hydrogen; the root is labeled by the hydroxyl group OH. For example, the tree of ethyl alcohol is presented in Figure 6.3.1. Two alcohols are considered to be equivalent if identical trees correspond to them. Let Y be the set of all alcohol molecules. For values $n_1, n_2, \ldots$ of

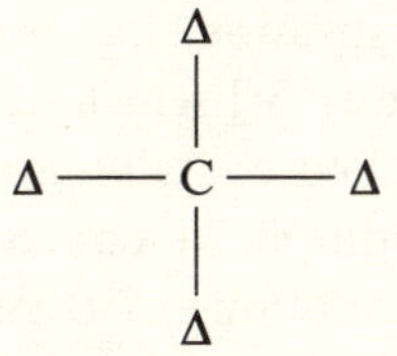

Fig. 6.3.2. Molecule of an organic compound

the characteristics of alcohol molecules, which coincide with the number of carbon atoms in the molecule, we define the generating function

$$F(x) = 1 + \sum_{r=1}^{\infty} F_{n_j} x^{n_j},$$

where F_{n_j} is the number of molecules with n_j atoms of carbon. The unity corresponds to the case where the molecule has no carbon, i.e. it is the water molecule.

If we delete the vertex closest to the root labeled by C in the tree representing an alcohol molecule, then this tree is decomposed into three others which, after addition of the roots labeled by OH, become rooted. Therefore, the number of such trees is equal to the number of equivalence classes on Y^X, $X = \{1,2,3\}$, induced by the symmetric group S_3 acting on X. Since

$$Z(t_1, t_2, t_3; S_3) = \frac{1}{6}(t_1^3 + 3t_1t_2 + 2t_3),$$

by Pólya's theorem (Cayley, 1874),

$$F(x) = 1 + \frac{x}{6}(F^3(x) + 3F(x)F(x^2) + 2F(x^3)).$$

Let us consider the class of molecules of organic compounds of the structure presented in Figure 6.3.2. Here C means the atom of carbon, and Δ denotes one of the following four components: methyl CH_3, ethyl C_2H_5, hydrogen H, chlorine Cl. Every molecule of the structure under consideration can be represented as a regular tetrahedron whose center is the atom of carbon and the vertices are labeled by Δ. Calculating the number of distinct molecules means calculating the number of distinct tetrahedra with labeled vertices. Tetrahedra are considered to be distinct if they do not match while rotating about the symmetry axes. If $X = \{1,2,3,4\}$ and $Y = \{CH_3, C_2H_5, H, Cl\}$, then the number of distinct molecules coincides with the number of equivalence classes induced on Y^X by the tetrahedron rotation group A.

Along with the unity substitution, the group A contains eight substitutions of the cycle structure $[1^1 3^1]$ which correspond to the rotations of the tetrahedron around the line containing its center and perpendicular to one of its faces. In addition, A contains two two-cyclic substitutions corresponding to the rotations of the tetrahedron around the line containing the midpoints of the opposite edges. Thus (Read, 1972),

$$Z(t_1, t_2, t_3; A) = \frac{1}{12}(t_1^4 + 8t_1t_3 + t_2^2).$$

Formula (6.2.15) implies that the number of distinct molecules is equal to

$$C_{4,4}(A) = \frac{1}{12}(4^4 + 8 \cdot 4 \cdot 4 + 3 \cdot 4^2) = 36.$$

6.4 Classes of functions and automata

6.4.1 Operations on groups of substitutions

(a) Product of groups Let A and B be groups of substitutions of degrees m and n acting on disjoint sets X and Y respectively. The product $A \otimes B$ of these groups is a group of substitutions consisting of ordered pairs (α, β), $\alpha \in A$, $\beta \in B$, and acting on the Cartesian product $X \times Y$. The action of $(\alpha, \beta) \in A \otimes b$, $\alpha \in A$, $\beta \in B$, is defined by the relation

$$(\alpha, \beta)(x, y) = (\alpha x, \beta y).$$

It is clear that the degree of $A \otimes B$ is equal to mn and $|A \otimes B| = |A||B|$. Denote by $j_k(\alpha, \beta)$ the number of cycles of length k in the substitution (α, β). Harary proved that

$$j_k(\alpha, \beta) = \sum_{[p,q]} j_p(\alpha) j_q(\beta) \langle p, q \rangle, \tag{6.4.1}$$

where $[p, q]$ is the least common multiple and $\langle p, q \rangle$ is the greatest common divisor of the numbers p and q, $j_p(\alpha)$ and $j_q(\beta)$ are the numbers of cycles of lengths p and q in the substitutions α and β, respectively, and the summation is carried out over all p and q such that $[p, q] = k$.

To prove formula (6.4.1), suppose that $(\alpha, \beta)^k(x, y) = (x, y)$ where k is the least number with this property. This means that $\alpha^k(x) = x$, $\beta^k(y) = y$. If x belongs to a cycle of length p and y belongs to one of length q, then $[p, q]$ divides k, and, consequently, $k = [p, q]$. The number of different cycles of length k which can be constructed from fixed cycles of lengths p and q is equal to $\langle p, q \rangle$. Formula (6.4.1) is proved.

Formula (6.4.1) lets us write the cycle index of the group $A \otimes B$ in the form

$$Z(t_1, \ldots, t_{mn}; A \otimes B) = \frac{1}{|A||B|} \sum_{(\alpha,\beta)\in A\otimes B} \prod_{k=1}^{mn} t_k^{j_k(\alpha,\beta)}.$$

In particular (see (Harary and Palmer, 1973)),

$$Z(S_m \otimes S_n) = \frac{1}{m!\,n!} \sum_{(\alpha,\beta)} \prod_{k,s=1}^{m,n} t_{[k,s]}^{j_k(\alpha) j_s(\beta) \langle k,s \rangle}. \tag{6.4.2}$$

Formula (6.4.2) follows from the remark that exactly $\langle k, s\rangle$ cycles of length $[k, s]$ correspond to a substitution $(\alpha, \beta) \in A \otimes B$ on the pairs of elements (x, y), where $x \in X_k$, $y \in Y_s$ and X_k and Y_s are orbits of the substitutions $\alpha \in A$ and $\beta \in B$ of sizes k and s.

(b) Direct product of groups Let A and B be groups of substitutions acting on sets X and Y, respectively, and $X \cap Y = \varnothing$. The direct product of the groups A and B, denoted by $A \times B$, is the group of degree $|X| + |Y|$ and order $|A||B|$ acting on the set $U = X \cup Y$ in the following way: if $\alpha \times \beta \in A \times B$ and $u \in U$, then

$$\alpha \times \beta(u) = \begin{cases} \alpha(u), & u \in X, \\ \beta(u), & u \in Y. \end{cases}$$

Pólya showed (Pólya, 1937) that the cycle index $Z(A \times B)$ of the group $A \times B$ is equal to

$$Z(A \times B) = Z(A)Z(B),$$

where $Z(A)$ and $Z(B)$ are the cycle indices of the groups A and B. This formula follows from the obvious fact that if $\alpha \in A$ and $\beta \in B$ belong to the cycle classes $[1^{j_1(\alpha)} 2^{j_2(\alpha)} \ldots]$ and $[1^{j_1(\beta)} 2^{j_2(\beta)} \ldots]$, respectively, then $\alpha \times \beta$ belongs to the cycle class $[1^{j_1(\alpha)+j_1(\beta)} 2^{j_2(\alpha)+j_2(\beta)} \ldots]$.

(c) Power groups Recall that the power group of substitutions B^A, defined for groups of substitutions A and B of degree m and n, respectively, acts on the set of function Y^X if A acts on X and B acts on Y, $|Y| > 1$. If $\alpha \in A$, $\beta \in B$, then the action of the element $(\alpha, \beta) \in B^A$ on the element $f \in Y^X$ is determined by the equality

$$(\alpha, \beta) f(x) = \beta f(\alpha x)$$

for all $x \in X$. The degree of B^A is equal to n^m, and the order is $|A||B|$. It is not difficult to establish an isomorphism between the groups B^A and $A \times B$.

The number $j_l(\alpha;\beta)$ of cycles of length l of the element $(\alpha,\beta) \in B^A$ can be represented (Harary and Palmer, 1973) as

$$j_1(\alpha;\beta) = \prod_{k=1}^{m} \left(\sum_{s|k} s j_s(\beta) \right)^{j_k(\alpha)}, \tag{6.4.3}$$

$$j_l(\alpha,\beta) = \frac{1}{l} \sum_{s|l} \mu(l/s) j_1(\alpha^s;\beta^s), \qquad l > 1, \tag{6.4.4}$$

where μ is the Möbius function.

Let us prove formulae (6.4.3) and (6.4.4). Let α_l be a cycle of length l of the substitution α, and let $X_l \subseteq X$ be the corresponding orbit. If $r_l(\beta)$ is the number of functions in Y^{X_l} which are invariant under the action of the substitution $(\alpha_l;\beta)$, then

$$j_1(\alpha;\beta) = \prod_{l=1}^{m} (r_l(\beta))^{j_l(\alpha)}. \tag{6.4.5}$$

Let us calculate $r_l(\beta)$. It is clear that the configuration $f \in Y^{X_l}$, which is invariant under the action of $(\alpha_l;\beta)$, takes values from an orbit Y_s of some cycle β_s of length s of the substitution β. The equality

$$(\alpha_l;\beta)^l f(x) = \beta^l f(x) = f(x)$$

implies that s is a divisor of l. The element $y \in Y_s$ can be chosen in s ways; therefore exactly s functions correspond to the orbit Y_s. Consequently,

$$r_l(\beta) = \sum_{s|l} s j_s(\beta). \tag{6.4.6}$$

Formula (6.4.3) is proved.

Note that for any integer k a function $f \in Y^X$ satisfies the equality

$$(\alpha^k;\beta^k) f(x) = f(x)$$

for all $x \in X$ if and only if f is an element of an orbit of a cycle of length s of the substitution $(\alpha;\beta)$ and $s \mid k$. Thus,

$$j_1(\alpha^k;\beta^k) = \sum_{s|k} s j_s(\alpha;\beta).$$

Applying the Möbius inversion formula to this equality, we obtain (6.4.4).

The cycle index of the group B^A is of the form

$$Z(t_1,\ldots,t_m;B^A) = \frac{1}{|A||B|}\sum_{(\alpha;\beta)\in B^A}\prod_{k=1}^{n^m} t_k^{j_k(\alpha;\beta)}.$$

6.4.2 Equivalence classes of functions

Let $f \sim f_1$, where $f, f_1 \in Y^X$, if there exists a substitution $(\alpha;\beta) \in B^A$ such that $(\alpha;\beta)f(x) = f_1(x)$ for all $x \in X$. The equivalence relation $\sim$ is called the equivalence relation determined by the power group B^A on the set of function Y^X. Denote by $N_{mn}(B^A)$ the number of corresponding equivalence classes assuming that $|X| = m$, $|Y| = n$. It follows from Burnside's lemma that

$$N_{mn}(B^A) = \frac{1}{|A||B|}\sum_{(\alpha;\beta)\in B^A} j_1(\alpha;\beta). \tag{6.4.7}$$

Using formulae (6.4.5) and (6.4.6), we obtain the following theorem due to de Bruijn.

Theorem 6.4.1 *The number of equivalence classes generated by the power group B^A on the set of functions Y^X with $|X| = m$ and $|Y| = n$ is equal to*

$$N_{mn}(B^A) = \frac{1}{|B|}\sum_{\beta\in B} Z(r_1(\beta),\ldots,r_m(\beta);A), \tag{6.4.8}$$

where

$$r_k(\beta) = \sum_{s|k} s j_s(\beta).$$

6.4.3 Weights of classes of functions

Let X and Y be finite sets. Consider the equivalence relation determined on the set Y^X by the power group B^A, where the group A acts on X and the group B acts on Y. Assign a weight $W(f)$ to any $f \in Y^X$, where $W(f)$ takes values from a commutative ring K and $W(f) = W(f_1)$ if $f \sim f_1$. The common value of weights of the elements of an orbit F of the group B^A defined on Y^X is called the weight of the orbit and is denoted by $W(F)$. The following lemma was proved by de Bruijn.

Lemma 6.4.1 *The total weight of the orbits of the power group B^A on the set Y^X is equal to*

$$\sum_F W(F) = \frac{1}{|A||B|} \sum_{\alpha \in A} \sum_{\beta \in B} \sum_{f\,:\,\beta f \alpha = f} W(f), \qquad (6.4.9)$$

where $\beta f \alpha = f$ means $\beta f(\alpha x) = f(x)$ for all $x \in X$.

This lemma follows from the Burnside–de Bruijn lemma if the set Y^X is taken as X and the group B^A is taken as A.

If the mapping $f : X \to Y$ determining a function is injective, then the corresponding function is called injective. For finite sets X and Y denote by $J(A, B)$ the number of equivalence classes determined by the power group B^A on the set of injective functions. It was shown by de Bruijn that

$$J(A, B) = Z\left(\frac{\partial}{\partial z_1}, \frac{\partial}{\partial z_2}, \ldots ; A\right) Z(1 + z_1, 1 + 2z_2, \ldots ; B)\Big|_{z_1 = z_2 = \cdots = 0}, \qquad (6.4.10)$$

where $Z(A)$ is a differential operator acting on the operand $Z(B)$.

To prove formula (6.4.10), first define the weight of a function $f \in Y^X$ as

$$W(f) = \begin{cases} 1, & f \text{ is injective,} \\ 0, & f \text{ is not injective.} \end{cases}$$

The following property is valid for the equivalence relation determined on Y^X by the group B^A: if $f \sim f_1$, then $W(f) = W(f_1)$. Therefore, if F is an equivalence class, then

$$J(A, B) = \sum_F W(F),$$

and we can use de Bruijn's lemma given above. To this end we need to calculate the sum

$$\sum_{f\,:\,\beta f \alpha = f} W(f),$$

that is, the number of injective mappings f satisfying the equation $\beta f \alpha = f$. Let $j_k(\alpha)$ and $j_s(\beta)$ be the number of cycles of length k in the substitution α and the number of cycles of length s in the substitution β respectively. Choose elements $x \in X$ and $y \in Y$ such that $y = f(x)$ and denote by X_k the orbit containing the element x of a cycle α_k of length k in the substitution $\alpha \in A$, and by Y_s, the orbit containing the element y of a cycle β_s of length s in the substitution $\beta \in B$. It follows from the results of Subsection 6.4.1 that $s \mid k$, and since f is injective, we obtain $s = k$. Thus, an injective function f maps any orbit of the group A into one and only one orbit of the same size of the group B. The number of such mappings is equal to the size k of the orbit. The total number of mappings of $j_k(\alpha)$ orbits of size k in the substitution α into $j_k(\beta)$ orbits of the same size in β is equal to

$$k^{j_k(\alpha)}(j_k(\beta))_{j_k(\alpha)} = k^{j_k(\alpha)} j_k(\beta)(j_k(\beta)-1)\cdots(j_k(\beta)-j_k(\alpha)-1),$$

where $(v)_\mu = 0$ for $\mu > v$.

Thus,

$$\sum_{f\,:\,\beta f\alpha} W(f) = \prod_k k^{j_k(\alpha)}(j_k(\beta))_{j_k(\alpha)}.$$

Note that

$$k^{j_k(\alpha)}(j_k(\beta))_{j_k(\alpha)} = \left.\frac{d^{j_k(\alpha)}}{dz^{j_k(\alpha)}}(1+kz)^{j_k(\beta)}\right|_{z=0}.$$

Therefore, for fixed $\alpha \in A$ and $\beta \in B$

$$\sum_{f\,:\,\beta f\alpha = f} W(f)$$

$$= \left.\left(\frac{\partial}{\partial z_1}\right)^{j_1(\alpha)}\left(\frac{\partial}{\partial z_2}\right)^{j_2(\alpha)}\cdots(1+z_1)^{j_1(\beta)}(1+2z_2)^{j_2(\beta)}\cdots\right|_{z_1=z_2=\cdots=0}.$$

Formula (6.4.10) easily follows from this equality.

The following formula for the total number of equivalence classes generated by the power group B^A on the set Y^X was obtained by de Bruijn:

$$I(A,B) = Z\left(\frac{\partial}{\partial z_1}, \frac{\partial}{\partial z_2}, \ldots ; A\right)$$
$$\times Z\left.\left(\exp\left(\sum_{j=1}^{\infty} z_j\right), \exp\left(2\sum_{j=1}^{\infty} z_{2l}\right), \ldots ; B\right)\right|_{z_1=z_2=\cdots=0}, \quad (6.4.11)$$

where $Z(A)$ is a differential operator acting on the operand $Z(b)$.

Put $W(f) = 1$ for all $f \in Y^X$ and note that

$$I(A,B) = \sum_F W(F), \quad (6.4.12)$$

where F are the equivalence classes. For fixed $\alpha \in A$ and $\beta \in B$, find the number of functions such that $\beta f \alpha = f$. As for the injective functions, if $y = f(x)$, then the orbit X_k containing x and corresponding to a cycle

α_k of the substitution α is mapped into the orbit Y_s containing y and corresponding to a cycle β_s of the substitution β, and in this case $s \mid k$. There exist s mappings X_k into Y_s determined by the choice of the initial element in Y_s. Thus,

$$\sum_{f\,:\,\beta f\alpha=f} W(f) = \prod_k \left(\sum_{s|k} s j_s(\beta)\right)^{j_k(\alpha)}. \tag{6.4.13}$$

It is not difficult to check that

$$\prod_s \left(\sum_{s|k} s j_s(\beta)\right)^{j_k(\alpha)}$$
$$= \left(\frac{\partial}{\partial z_1}\right)^{j_1(\alpha)} \left(\frac{\partial}{\partial z_2}\right)^{j_2(\alpha)} \cdots \exp\left(\sum_{k=1}^{\infty} z_k \sum_{s|k} s j_s(\beta)\right)\Bigg|_{z_1=z_2=\cdots=0}, \tag{6.4.14}$$

since

$$\left(\frac{\partial}{\partial z}\right)^{j_k(\alpha)} \exp\left(z \sum_{s|k} s j_s(\beta)\right)\Bigg|_{z=0} = \left(\sum_{s|k} s j_s(\beta)\right)^{j_k(\alpha)}.$$

Note also that

$$\sum_{k=1}^{\infty} z_k \sum_{s|k} s j_s(\beta) = \sum_s s j_s(\beta) \sum_{k=1}^{\infty} z_{ks}. \tag{6.4.15}$$

Now from (6.4.12)–(6.4.15) we obtain the equality

$$I(A,B) = \frac{1}{|A||B|} \sum_{\alpha\in A} \sum_{\beta\in B} \left(\frac{\partial}{\partial z_1}\right)^{j_1(\alpha)} \left(\frac{\partial}{\partial z_2}\right)^{j_2(\alpha)} \cdots$$
$$\times \exp\left(j_1(\beta) \sum_{k=1}^{\infty} z_k\right) \exp\left(2 j_2 \sum_{k=1}^{\infty} z_{2k}\right) \cdots \Bigg|_{z_1=z_2=\cdots=0},$$

which implies formula (6.4.11).

6.4.4 Equivalence classes of pairs of functions

Consider pairs of functions (f_1, f_2), where $f_1 \in Y_1^{X_1}$ and $f_2 \in Y_2^{X_2}$. Let a group A_i act on the set X_i and a group B_i act on the set Y_i, $i = 1, 2$. Consider the product of the power groups $C = B_1^{A_1} \otimes B_2^{A_2}$. The group C acts on the set of pairs of functions $Z = Y_1^{X_1} \times Y_2^{X_2}$ and determines the equivalence relation on the set Z under which $(f_1, f_2) \sim (f'_1, f'_2)$ if there exists an element

$$\gamma = ((\alpha_1; \beta_1), (\alpha_2; \beta_2)) \in C$$

such that for all $x_1 \in X_1$, $x_2 \in X_2$,

$$\gamma(f_1(x_1), f_2(x_2) = (f'_1(x_1), f'_2(x_2))).$$

The last equality means that

$$\beta_i f_i(\alpha_i x_i) = f'_i(x_i), \quad \alpha_i \in A_i, \quad \beta_i \in B_i, \quad i = 1, 2.$$

Harary and Palmer proved the following theorem.

Theorem 6.4.2 *The number of equivalence classes determined by the group* $C = B_1^{A_1} \otimes B_2^{A_2}$ *on the set of functions* $Y_1^{X_1} \times Y_2^{X_2}$ *is equal to*

$$N(C) = \frac{1}{|C|} \sum_{\gamma \in C} \prod_{i=1}^{2} \prod_{k=1}^{m_i} \left(\sum_{s|k} s j_s(\beta_i) \right)^{j_k(\alpha_i)}, \qquad (6.4.16)$$

where $\gamma = ((\alpha_1; \beta_1), (\alpha_2; \beta_2))$, $j_v(\delta)$ *is the number of cycles of length* v *in the substitution* δ, $|X_i| = m_i$, $i = 1, 2$.

Proof According to Burnside's lemma

$$N(C) = \frac{1}{|C|} \sum_{\gamma \in C} j_1(\gamma). \qquad (6.4.17)$$

Put $C_i = B_i^{A_i}$, $i = 1, 2$, and note that by virtue of the condition $C = C_1 \otimes C_2$ formula (6.4.1) yields the equality

$$j_1(\gamma) = j_1(\delta_1) j_1(\delta_2), \quad \delta_i \in C_i, \quad i = 1, 2.$$

Using (6.4.3) for the calculation of $j_1(\delta_1)$ and $j_2(\delta_2)$, we obtain equality (6.4.16) from (6.4.17). □

6.4.5 Isomorphism of abstract automata

Let X, Y and S be finite sets usually referred to as the input and output alphabets and the set of states. An abstract automaton $\langle X, Y, S, f_1, f_2 \rangle$ is defined by an ordered pair of functions (f_1, f_2) of the form

$$f_1 : X \times S \to S, \qquad f_2 : X \times S \to Y.$$

The function f_1 is the transition function and the function f_2 is the output function of the automaton.

Let $|X| = m$, $|Y| = n$ and $|S| = k$ and let the symmetric groups S_m, S_n and S_k act on the sets X, Y and S respectively.

Automata $\langle X, Y, S, f_1, f_2\rangle$ and $\langle X, Y, S, g_1, g_2\rangle$ are called isomorphic if there exist $\alpha \in S_m$, $\beta \in S_n$ and $\gamma \in S_k$ such that for all $x \in X$ and $s \in S$

$$f_1(x,s) = \gamma^{-1} g_1(\alpha x, \gamma s),$$
$$f_2(x,s) = \beta^{-1} g_2(\alpha x, \gamma s).$$

It is clear that the isomorphism of two automata means that the corresponding pairs of functions belong to the same equivalence class if the equivalence relation on the set $S^{X\times S} \times Y^{X\times S}$ is determined by the group

$$F = S_k^{S_m \otimes S_k} \otimes S_n^{S_m \otimes S_k}.$$

From (6.4.2) and (6.4.16) we obtain the following formula for the number $N(m,n,k)$ of non-isomorphic automata with input and output alphabets consisting of m and n symbols, respectively, and with k states (Harrison, 1965):

$$N(m,n,k) = \frac{1}{m!\,n!\,k!} \sum I(\alpha,\gamma,\gamma^{-1}) I(\alpha,\gamma,\beta^{-1});$$

here the summation is over all $\alpha \in S_m$, $\beta \in S_n$, $\gamma \in S_k$ and

$$I(\alpha,\gamma,\beta) = \prod_{p=1}^{m} \prod_{q=1}^{k} \left(\sum_{s|[p,q]} s j_s(\beta) \right)^{j_p(\alpha) j_q(\gamma) \langle p,q\rangle},$$

where $[p,q]$ and $\langle p,q\rangle$ are the least common multiple and the greatest common divisor of p and q.

Bibliography

Aigner, M. (1979). *Combinatorial Theory*, Springer, Berlin.

Aleksandrov, A. D. (1938). The theory of geometric volumes of convex bodies IV, *Math. USSR Sbornik* **3**: 227–251. In Russian.

Anderson, I. (1967). On primitive sequences, *J. London Math. Soc.* **42**: 137–148.

Andre, D. (1881). Sur les permutations alternées, *J. Math. Pures Appl.* **7**: 167–184.

Andrews, G. E. (1976). *The Theory of Partitions*, Addison-Wesley, London.

Austin, T. L. (1960). The enumeration of point-labelled chromatic graphs and trees, *Canadian J. Math.* **12**: 535–545.

Austin, T. L., Fagen, R. E., Penney, W. F. and Riordan, J. (1959). The number of components in random linear graphs, *Ann. Math. Statist.* **30**: 747–754.

Bammel, S. E. and Rothstein, J. (1975). The number of 9×9 Latin squares, *Discrete Math.* **11**: 93–95.

Beckenbach, E. E. (1964). *Applied Combinatorial Mathematics*, Wiley, New York.

Bell, E. T. (1934). Exponential polynomials, *Ann. Math.* **35**: 258–277.

Bell, E. T. (1940). Postulational bases for the umbral calculus, *American J. Math.* **62**: 717–724.

Berge, C. (1958). *Théorie des Graphes et ses Applications*, Dunod, Paris.

Berge, C. (1968). *Principes de Combinatoire*, Dunod, Paris.

Blakley, G. R., Coppage, W. E. and Dixon, R. D. (1967). A set of linearly independent permutation matrices, *American Math. Monthly* **74**: 1084–1085.

Bleick, W. E. and Wang, P. C. C. (1974). Asymptotics of Stirling numbers of the second kind, *Proc. American Math. Soc.* **42**: 575–580.

Bollobás, B. (1978). *Extremal Graph Theory*, Academic Press, London.

Bollobás, B. (1985). *Random Graphs*, Academic Press, London.

Bolotnikov, Y. V., Sachkov, V. N. and Tarakanov, V. E. (1980). Asymptotic normality of some variables connected with the cyclic structure of random permutations, *Math. USSR Sbornik* **36**: 87–99.

Bose, R. C. (1938). On the application of the properties of Galois fields to the problem of construction of hyper-Graeco-Latin squares, *Sankhya* **3**: 323–338.

Bregman, L. M. (1973). Some properties of non-negative matrices and their permanents, *Soviet Math. Dokl.* **211**: 27–30. In Russian.

Brualdi, R. A. and Csima, J. (1992). Butterfly embedding proof of a theorem of König, *American Math. Monthly* **99**: 228–230.

Brualdi, R. A. and Newman, M. (1965). Inequalities to permanents and permanental minors, *Proc. Cambridge Phil. Soc.* **61**: 3.

Bruck, R. H. (1955). Difference sets in a finite group, *Trans. American Math. Soc.* **78**: 464–481.

Bruck, R. H. and Ryser, H. J. (1949). The nonexistence of certain finite projective planes, *Canadian J. Math.* **1**: 88–93.

Burnside, W. (1911). *Theory of Groups of Finite Order*, Cambridge Univ. Press, Cambridge.

Carlitz, L. (1969). *Generating Functions*, Duke Univ. Press, Durham.

Carlitz, L., Roselle, D. P. and Scoville, R. A. (1966). Permutations and sequences with repetitions by number of increases, *J. Combinatorial Theory* **1**: 350–374.

Cashwell, E. D. and Everett, C. J. (1963). Formal power series, *Pacific J. Math.* **13**: 45–64.

Cayley, A. (1874). On the mathematical theory of isomers, *Philos. Mag.* **67**: 444–446.

Cayley, A. (1889). A theorem on trees, *J. Pure and Appl. Math. Quarterly* **23**: 376–378.

Clarke, R. J. (1990). Covering a set by subsets, *Discrete Math.* **81**: 147–152.

de Bruijn, N. G. (1958). *Asymptotic Methods in Analysis*, North-Holland, Amsterdam.

Dénes, J. (1959). The representation of permutation as the product of minimal number of transpositions and its connection with the theory of graphs, *Publ. Math. Inst. Hungar. Acad. Sci.* **4**: 63–70.

Deshouillers, J.-M. (1974). Valeur maximale des coefficients p-nomiaux d'orde n, *C. R. Acad. Sci. Paris, Ser. A* **278**: 401–404.

Dickson, L. E. (1966). *History of the Theory of Numbers*, Chelsea, New York.

Dickson, T. I. (1969). On problem concerning separating systems of a finite set, *J. Combinatorial Theory* **7**: 191–196.

Doubilet, P., Rota, G.-C. and Stanley, R. (1972). On the foundations of combinatorial theory (VI): The idea of generating function, *Proc. 6th Berkeley Symp. on Math. Statist. and Probab.*, Univ. California Press, Berkeley, pp. 267–318.

Egorychev, G. P. (1980). A solution of the van der Waerden problem for permanents, *Preprint IFSO-13M*, The Kirenskii Institute of Physics of the Siberian Branch of the Academy of Sciences of the USSR, Krasnoyarsk. In Russian.

Erdős, P. and Kaplansky, I. (1946). The asymptotic number of Latin rectangles, *American J. Math.* **68**: 230–236.

Erdős, P. and Spencer, J. (1974). *Probabilistic Methods in Combinatorics*, Akadémiai Kiadó, Budapest.

Euler, L. (1753). De partitione numerorum, *Novi Commentari Acad. Sci. Petropolianae* **3**: 125–169.

Evgrafov, M. A. (1957). *Asymptotic Estimates and Entire Functions*, Gostekhizdat, Moscow. In Russian.

Falikman, Y. I. (1981). A proof of the van der Waerden conjecture on the permanent of a doubly stochastic matrix, *Math. Notes* **29**: 931–938.

Faure, R., Kaufmann, A. and Denis-Papin, M. (1964). *Mathematiques Nouvelles*, Vol. 1, Dunod, Paris.

Fikhtengoltz, G. M. (1969). *Differential and Integral Calculus*, Vol. 1, 2, 3, Nauka, Moscow. In Russian.

Flajolet, P. and Odlyzko, A. (1990). Random mapping statistics, *Lecture Notes in Computer Sci.* **434**: 329–354.

Frobenius, G. (1912). Über Matrizen mit nicht negativen Elementen, *S. B. Berliner Akad.* **23**: 456–477.

Gantmakher, F. R. (1959). *The Theory of Matrices*, Chelsea, New York.

Gavrilov, G. P. (1979). *Enumerative Problems of Combinatorial Analysis*, Mir, Moscow. In Russian.

Gavrilov, G. P. and Sapozhenko, A. A. (1989). *Selected Problems in Discrete Mathematics*, Mir, Moscow.

Gelfond, A. O. (1967). *Calculus of Finite Differences*, Nauka, Moscow. In Russian.

Gilbert, E. N. (1956). Enumeration of labelled graphs, *Canadian J. Math.* **8**: 405–411.

Gilbert, E. N. (1959). Random graphs, *Ann. Math. Statist.* **30**: 1141–1144.

Gill, A. (1962). *Introduction to the Theory of Machines*, McGraw–Hill, New York.

Godsil, G. D. and McKay, B. D. (1984). Asymptotic enumeration of Latin rectangles, *Bull. American Math. Soc.* **10**: 91–92.

Goldman, J. R. and Rota, G.-C. (1970). On the foundations of combinatorial theory IV: Finite vector spaces and Eulerian generating functions, *Studies in Appl. Math* **49**: 239–258.

Goncharov, V. L. (1944). On the field of combinatorics, *Soviet Math. Izv., Ser. Math.* **8**: 3–48. In Russian.

Gould, H. W. (1973). An expansion of the operator $\left(x \underset{x,h}{\Delta}\right)^n$, *Glasnik Math.* **8**: 259–272.

Goulden, J. P. and Jackson, D. M. (1983). *Combinatorial Enumeration*, Wiley, New York.

Hall, M. (1945). An existence theorem for Latin squares, *Bull. American Math. Soc.* **51**: 387–388.

Hall, M. (1958). *A Survey of Combinatorial Analysis*, Wiley, New York.

Hall, M. (1959). *The Theory of Groups*, MacMillan, New York.

Hall, M. (1967). *Combinatorial Theory*, Blaisdell, Waltham.

Hall, M. and Ryser, H. J. (1956). A survey of difference sets, *Proc. American Math. Soc.* **7**: 975–986.

Hall, P. (1935). On representatives of subsets, *J. London Math. Soc.* **10**: 26–30.

Harary, F. (1969). *Graph Theory*, Addison–Wesley, Reading, Mass.

Harary, F., Mowshowitz, A. and Riordan, J. (1970). Labeled trees with unlabeled end-points, *J. Combinatorial Theory* **8**: 99–103.

Harary, F. and Palmer, E. (1966). The power group enumeration theorem, *J. Combinatorial Theory* **1**: 157–173.

Harary, F. and Palmer, E. (1967). Enumeration of finite automata, *Information and Control* **10**: 499–508.

Harary, F. and Palmer, E. M. (1973). *Graphical Enumeration*, Academic Press, New York.

Harris, B. and Schoenfeld, L. (1967). The number of idempotent elements in symmetric semigroups, *J. Combinatorial Theory* **3**: 122–135.

Harrison, M. (1965). A census of finite automata, *Canadian J. Math.* **17**: 100–113.

Hsu, L. C. (1948). Note on an asymptotic expansion of the *n*-th difference of zero, *Ann. Math. Statist.* **19**: 273–277.

Jordan, C. (1965). *Calculus of Finite Differences*, Chelsea, New York.

Kaluzhnin, L. A. (1973). *Introduction to Algebra*, Nauka, Moscow. In Russian.

Kargapolov, M. I. and Merzlyakov, Y. I. (1972). *Foundations of the Group Theory*, Nauka, Moscow. In Russian.

Katz, L. (1955). Probability of indecomposability of random mapping function, *Ann. Math. Statist.* **26**: 512–517.

Kaufmann, A. (1968). *Introduction à la Combinatorique en vue des Applications*, Dunod, Paris.

Klass, M. J. (1976). A generalization of Burnside's combinatorial lemma, *J. Combinatorial Theory* **20**: 273–278.

Kolchin, V. F. (1971). A problem of the allocation of particles in cells and cycles of random permutations, *Theory Probab. Appl.* **16**: 74–90.

Kolchin, V. F. (1973). A new proof of asymptotic normality of the logarithm of the order of a random permutation, *Combinatorial and Statistical Analysis*, Krasnoyarsk, pp. 82–93. In Russian.

Kolchin, V. F. (1986). *Random Mappings*, Springer, New York.

Kolchin, V. F. (1988). *Systems of Random Equations*, MIEM. In Russian.

Kolchin, V. F. and Chistyakov, V. P. (1975a). Combinatorial problems of probability theory, *J. Soviet Math.* **4**: 217–243.

Kolchin, V. F. and Chistyakov, V. P. (1975b). The cyclic structure of random permutations, *Math. Notes* **18**: 1139–1144.

Kolchin, V. F., Sevastyanov, B. A. and Chistyakov, V. P. (1978). *Random Allocations*, Wiley, New York.

Kolesova, G., Lam, G. W. H. and Thiel, L. (1990). On the number of 8×8 Latin squares, *J. Combinatorial Theory (A)* **54**: 143–148.

König, D. (1950). *Theorie der endlichen und unendlichen Graphen*, Chelsea, New York.

Kurosh, A. G. (1973). *Lectures on Algebra*, Nauka, Moscow. In Russian.

Kurosh, A. G. (1975). *Higher Algebra*, Mir, Moscow.

Kuznetsov, D. S. (1965). *Special Functions*, Vysshaya Shkola, Moscow. In Russian.

Leibniz, G. W. (1880). Dissertatio de arte combinatoria, *Die Philosophische Schriften*, Berlin, pp. 27–102.

London, D. (1971). Some notes on the van der Waerden conjecture, *Linear Algebra Appl.* **4**: 155–160.

Lubell, D. (1966). A short proof of Sperner's theorem, *J. Combinatorial Theory* **1**: 299.

MacMahon, P. A. (1915). *Combinatorial Analysis*, Vol. 1, 2, Cambridge Univ. Press, Cambridge.

Mann, H. B. (1942). The construction of orthogonal Latin squares, *Ann. Math. Statist.* **13**: 418–423.

Mann, H. B. (1943). On the construction of sets of orthogonal Latin squares, *Ann. Math. Statist.* **14**: 401–414.

Mann, H. B. (1949). *Analysis and Design of Experiment*, Dover, New York.

Marcus, M. and Minc, H. (1964). *A Survey of Matrix Theory and Matrix Inequalities*, Allyn & Bacon, Boston.

Marcus, M. and Minc, H. (1965). Permanents, *American Math. Monthly* **72**: 577–591.

Marcus, M. and Newman, M. (1959). On the minimum of the permanent of a doubly stochastic matrix, *Duke Math. J.* **26**: 61–72.

Marcus, M. and Newman, M. (1962). Inequalities for permanent function, *Ann. Math.* **75**: 47–62.

Markushevich, A. I. (1966). *A Brief Course of Theory of Analytic Functions*, Nauka, Moscow. In Russian.

Medvedev, Y. I. and Ivchenko, G. I. (1965). Asymptotic representations of finite differences of the power function at a given point, *Theory Probab. Appl.* **10**: 151–156.

Menon, V. V. (1973). On maximum of Stirling numbers of the second kind, *J. Combinatorial Theory* **15A**: 11–24.

Meshalkin, L. D. (1963). A generalization of Sperner's theorem on the number of subset of a finite set, *Theory Probab. Appl.* **8**: 219–220.

Minc, H. (1963). Upper bounds for permanents of (0, 1)-matrices, *Bull. American Math. Soc.* **69**: 789–791.

Minc, H. (1967). An inequality for permanents of (0, 1)-matrices, *J. Combinatorial Theory* **2**: 321–326.

Minc, H. (1978). *Permanents*, Addison–Wesley, Reading, Mass.

Mirsky, L. and Perfect, H. (1966). Systems of representatives, *J. Math. Anal. Appl.* **15**: 520–568.

Moon, J. W. (1969). Connected graphs with unlabeled end-points, *J. Combinatorial Theory* **6**: 65–66.

Moser, L. and Wyman, M. (1955a). An asymptotic formula for the Bell numbers, *Trans. Royal Soc. Canada* **49**(3): 49–54.

Moser, L. and Wyman, M. (1955b). On solutions of $x^d = I$ in symmetric groups, *Canadian J. Math.* **7**: 159–168.

Moser, L. and Wyman, M. (1958). Asymptotic development of the Stirling numbers of the first kind, *J. London Math. Soc.* **33**: 133–146.

Netto, E. (1901). *Lehrbuch der Kombinatorik*, Teubner, Leipzig.

Nörlund, N. (1954). *Vorlesungen über Differenzenrechnung*, Neudruck, Berlin.

Norton, H. W. (1939). The 7×7 squares, *Ann. Eugenics* **9**: 269–307.

Ore, O. (1962). *Theory of Graphs*, American Math. Soc., Providence.

Percus, J. (1969). One more technique for dimer problem, *J. Math. Phys.* **10**: 1881–1884.

Perfect, H. (1964). An inequality for the permanent function, *Proc. Cambridge Phylos. Soc.* **60**: 1030–1031.

Pólya, G. (1937). Kombinatorische Anzahlbestimmungen für Gruppen, Graphen und chemische Verbindungen, *Acta Math.* **68**: 145–254.

Postnikov, A. G. (1971). *An Introduction to the Analytic Number Theory*, Nauka, Moscow. In Russian.

Pro (1973). Problems of cybernetics, Proc. All-Union Seminar on Combinatorial Math. Publ. Acad. Sci. USSR, Sci. Council on the Complex Problem 'Cybernetics', Moscow. In Russian.

Pro (1976). Problems of cybernetics., Proc. All-Union Seminar on Combinatorial Math. Publ. Acad. Sci. USSR, Sci. Council on the Complex Problem 'Cybernetics', Moscow. In Russian.

Prüfer, H. (1918). Neuer beweis eines satzes über permutationen, *Arch. Math. Phys.* **27**: 742–744.

Pupkov, K. A. (1974). *Foundations of Cybernetics*, Vysshaya Shkola, Moscow. In Russian.

Rabin, H. and Sitgreaves, R. (1954). Probability distributions related to random transformations of a finite set, *Technical Report 19A*, Stanford Univ.

Rademacher, H. (1937). A convergent series for the partition function $p(n)$, *Proc. Nat. Acad. Sci.* **23**: 78–84.

Rado, R. (1942). A theorem on independence relations, *J. Pure and Appl. Math. Quarterly* **13**(50–51): 83–89.

Rado, R. (1967). On the number of systems of distinct representatives of sets, *J. London Math. Soc.* **42**: 107–109.

Read, R. C. (1972). Some recent results in chemical enumeration, *Lect. Notes Math.* **303**: 243–259.

Redfield, J. H. (1927). The theory of group-reduced distributions, *American J. Math.* **49**: 433–455.

Rényi, A. (1959). Some remarks on the theory of trees, *Publ. Math. Inst. Hungar. Acad. Sci.* **4**: 73–83.

Riordan, J. (1958). *An Introduction to Combinatorial Analysis*, Wiley, New York.

Riordan, J. (1960). The enumeration of trees by height and diameter, *IBM J. Research and Develop.* **4**: 473–478.

Riordan, J. (1962). Enumeration of linear graphs for mappings of finite sets, *Ann. Math. Statist.* **33**: 178–185.

Riordan, J. (1968). *Combinatorial Identities*, Wiley, New York.

Riordan, J. (1969). Ballots and trees, *J. Combinatorial Theory* **6**: 408–411.

Rota, G.-C. (1964a). The number of partitions of a set, *American Math. Monthly* **71**: 498–504.

Rota, G.-C. (1964b). On the foundations of combinatorial theory I: Theory of Möbius functions, *Zeitschrift Wahrscheinlichkeitstheorie und Verw. Gebiete* **2**: 340–368.

Rybnikov, K. A. (1972). *Introduction to Combinatorial Analysis*, Moscow Univ. Press, Moscow. In Russian.

Ryser, H. J. (1960). Matrices of zeros and ones, *Bull. American Math. Soc.* **66**: 442–464.

Ryser, H. J. (1963). *Combinatorial Mathematics*, Wiley, New York.

Ryser, H. J. and Jurkat, W. B. (1966). Matrix factorizations of determinants and permanents, *J. Algebra* **3**: 1–27.

Sachkov, V. N. (1971a). The distribution of the number of fixed points among the elements of the symmetric semigroup under the restriction $\sigma^{h+1} = \sigma^h$, and the number of trees of heights not exceeding h, *Theory Probab. Appl.* **16**: 676–687.

Sachkov, V. N. (1971b). Distributions of the number of distinct elements of a symmetric basis in a random mA-sample, *Theory Probab. Appl.* **16**: 504–513.

Sachkov, V. N. (1972). Mappings of a finite set with restraints on contours and height, *Theory Probab. Appl.* **17**: 640–656.

Sachkov, V. N. (1982). *An Introduction to Combinatorial Methods of Discrete Mathematics*, Nauka, Moscow. In Russian.

Sachkov, V. N. (1993a). The asymptotic behaviour of the number of t-minimal coverings, *Discrete Math. Appl.* **3**: 265–273.

Sachkov, V. N. (1993b). Random minimal coverings of sets, *Discrete Math. Appl.* **3**: 201–212.

Sachkov, V. N. and Oshkin, I. B. (1993). Exponents of classes of non-negative matrices, *Discrete Math. Appl.* **3**: 365–375.

Sachkov, V. N. and Tarakanov, V. E. (1994). *Combinatorics on Non-Negative Matrices*, TVP, Moscow.

Sade, A. (1951). An omission in Norton's list of 7×7 squares, *Ann. Math. Statist.* **22**: 306–307.

Schrijver, A. (1978). A short proof of Minc's conjecture, *J. Combinatorial Theory (A)* **25**: 80–81.

Spee, J. W. (1990). Finding all minimal covers of a set using implicit enumeration, *Technical report*, Centre for Math. and Computer Science, Amsterdam.

Spencer, J. (1970). Minimal completely separating systems, *J. Combinatorial Theory* **8**: 446–447.

Sperner, E. (1928). Ein Satz über Untermengen einer endlichen Menge, *Math. Zeitschrift* **27**: 544–548.

Stam, A. J. (1985). Regeneration points in random permutations, *Fibonacci Quarterly* **23**: 49–56.

Stanley, R. P. (1986). *Enumerative Combinatorics*, Vol. 1, Wadsworth, Monterey, California.

Steiner, J. (1853). Kombinatorische Aufgabe, *Zeitschrift Reine Angew. Math.* **45**: 181–182.

Stepanov, V. E. (1969). Limit distributions of certain characteristics of random mappings, *Theory Probab. Appl.* **14**: 612–626.

Stevens, W. (1939). The completely orthogonalized Latin square, *Ann. Eugenics* **9**: 82–93.

Sveshnikov, A. G. and Tikhonov, A. N. (1967). *Theory of Functions of Complex Variables*, Nauka, Moscow. In Russian.

Szekeres, G. and Binet, F. E. (1957). On Borel fields over finite sets, *Ann. Math. Statist.* **28**: 494–498.

Tarakanov, V. E. (1985). *Combinatorial Problems and* $(0,1)$*-matrices*, Nauka, Moscow. In Russian.

Tutte, W. T. (1975). On elementary calculus and the Good formula, *J. Combinatorial Theory (B)* **18**: 97–137.

van der Waerden, B. L. (1926). Aufgabe 45, *Jber. Deutsch. Math. Verein.* **35**: 117.

van Lint, J. H. (1981). Notes on Egoritjev's proof of van der Waerden conjecture, *Linear Algebra Appl.* **39**: 1–8.

Vatutin, V. A. and Mikhailov, V. G. (1982). Limit theorems for the number of empty cells in the equiprobable scheme of allocating particles by groups, *Theory Probab. Appl.* **37**: 684–692.

Vinogradov, I. M. (1949). *Foundations of Number Theory*, Gostekhizdat, Moscow. In Russian.

Wells, M. B. (1967). The number of Latin squares of order eight, *J. Combinatorial Theory* **3**: 98–99.

Yablonskii, S. V. and Lupanov, O. B. (1974). *Discrete Mathematics and Mathematical Problems of Cybernetics*, Nauka, Moscow. In Russian.

Yamamoto, K. (1951). On asymptotic number of Latin rectangles, *Japan J. Math.* **21**: 113–119.

Zariski, O. and Samuel, P. (1958). *Commutative Algebra*, Vol. 2, Van Nostrand, Amsterdam.

Index